Alice

in

Quantumland

AN ALLEGORY OF QUANTUM PHYSICS

Alice
in
Quantumland

Robert Gilmore

COPERNICUS
An Imprint of Springer-Verlag

Published in the United States by Copernicus, an imprint of Springer-Verlag New York, Inc. by arrangement with Birkhäuser Boston.

Copernicus
Springer-Verlag New York, Inc.
175 Fifth Avenue
New York, NY 10010

Library of Congress Cataloging-in-Publication Data
Gilmore, Robert, 1941-
    Alice in Quantumland : an allegory of quantum physics / Robert
Gilmore.
      p.   cm.
    Includes bibliographical references and index.
    ISBN 0-387-91495-1 (acid-free paper)
    1. Quantum theory—Fiction.   I. Title
PS3557.I4595A45   1995
813'.54—dc20   95-10163

Manufactured in the United States of America.
Printed on acid-free paper.

9 8 7 6 5 4 3 2 1

ISBN 0-387-91495-1

Preface

In the first half of the twentieth century, our understanding of the Universe was turned upside down. The old classical theories of physics were replaced by a new way of looking at the world—quantum mechanics. This is in many ways at variance with the ideas of the older Newtonian mechanics; indeed, in many ways it is at variance with our common sense. Nevertheless, the strangest thing about these theories is their extraordinary success in predicting the observed behavior of physical systems. However nonsensical quantum mechanics may at times appear to us, that seems to be the way that Nature wants it—and so we have to play along.

This book is an allegory of quantum physics, in the dictionary sense of "a narrative describing one subject under the guise of another." The way that things behave in quantum mechanics seems very odd to our normal way of thinking and is made more acceptable when we consider analogies to situations with which we are familiar, even though the analogies may be inexact. Such analogies can never be very true to reality as quantum processes are really quite different from our normal experience.

An allegory is an extended analogy, or series of analogies. As such, this book follows more in the footsteps of *Pilgrim's Progress* or *Gulliver's Travels* than of *Alice in Wonderland*. "Alice" appears the more suitable model, however, when we examine the world that we inhabit.

The Quantumland in which Alice travels is rather like a theme park in which Alice is sometimes an observer, while sometimes she behaves as a sort of particle with varying electric charge. This Quantumland shows the essential features of the *quantum world:* the world that we all inhabit.

Much of the story is pure fiction and the characters are imaginary, although the "real-world" notes described below are true. Throughout the narrative you will find many statements that are obviously nonsensical and quite at variance with common sense. For the most part these are true. Neils Bohr, the father figure of quantum mechanics in its early days, is said to have remarked that anyone who did not feel dizzy when thinking about quantum theory had not understood it.

Seriously, Though, . . .

The description of the world that is given by quantum mechanics is undoubtedly interesting and remarkable, but are we seriously expected to believe that it is true? Amazingly, we find that we must. To underline this assertion, throughout this book you will find brief notes which emphasize the importance of quantum mechanics in the real world. The notes look like this:

These notes summarize the relevance to our world of the quantum topics Alice encounters in each chapter. They should be sufficiently unobtrusive that you may ignore them as you read the story of Alice's adventures, but when you wish to discover the real significance of these adventures, the notes are conveniently nearby.

There are also some longer, end-of-chapter, notes. These amplify some of the trickier points in the text and are denoted thus:

See end-of-chapter note 1

Much of the way that quantum theory describes the world may seem at first sight to be nonsense—and possibly it may seem so at the second, third, and twenty-fifth sight as well. It is, however, the only game in town.

The old classical mechanics of Newton and his followers is unable to give any sort of explanation for atoms and other small systems. Quantum mechanics agrees very well with observation. The calculations are often difficult and tedious, but where they have been made, they have agreed perfectly with what has really been seen.

It is impossible to stress too strongly the remarkable practical success of quantum mechanics. Although the outcome of one measurement may be random and unpredictable, the predictions of quantum theory agree consistently with the average results obtained from many measurements. Any large-scale observation will involve very many atoms and thus very many observations on the atomic scale. We again find that quantum mechanics is successful, in that it automatically agrees with the results of classical mechanics for large objects. The converse is not true.

Quantum theory was developed to explain observations made on atoms. Since its conception, it has successfully been applied to atomic nuclei, to the strongly interacting particles which derive from the nucleus, and to the behavior of the quarks of which these are composed. The application of the theory has been extended over a factor of some hundred thousand million. The systems considered have both decreased in size and increased in energy by this factor. This is a long way to extrapolate a theory from its original conception, but so far quantum mechanics appears to be quite able to deal with these extreme systems.

Insofar as it has been investigated, quantum mechanics appears to be of universal applicability. On a large scale, the predictions of quantum theory lose their random aspect and agree with those of classical mechanics, which works very well for large objects. On a small scale, however, the predictions of quantum theory are consistently borne out by experiment. Even those predictions, which seem to imply a nonsensical picture of the world, are supported by experimental evidence. Intriguingly, as is discussed in Chapter 4, quantum mechanics would appear to be in the strange position of agreeing with all observations made, while disputing that any observations can actually *be* made at all. It seems that the world is stranger than we imagine and perhaps stranger than we *can* imagine.

For the present, however, let us accompany Alice as she begins her journey into Quantumland.

Robert Gilmore

Contents

1 Into Quantumland

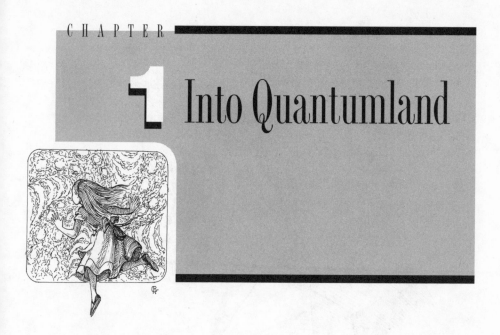

Alice was bored. All her friends were on holiday or visiting relations and it was raining, so that she was marooned indoors watching television. So far that afternoon she had watched part five of a series on introductory Esperanto, a program on gardening, and a paid political broadcast. Alice was really bored.

She looked down at the book lying on the floor beside her chair. It was a copy of *Alice in Wonderland,* which she had been reading earlier and had dropped there when she finished it. "I do not know why there cannot be more cartoons and interesting programs on the television," she wondered idly to herself. "I wish I could be like that other Alice. She was feeling bored and then she found her way to a land full of interesting creatures and strange happenings. If I could shrink down somehow and float through the television screen perhaps I might find all sorts of fascinating things."

She stared in frustration at the screen, which at that moment carried a picture of the Prime Minister telling her how, all things considered, everything was really far better than it had been three years ago, even if it didn't always seem that way. As she watched she was mildly surprised to see the picture of the Prime Minister's face slowly break apart into a mist of bright dancing speckles which all seemed to be rushing inward, as if they were beckoning her. "Why," said Alice, "I do believe that they want me to follow them in!" She leapt to her feet and

started toward the television, but tripped on the book which she had discarded so untidily on the floor and fell headlong.

As she fell forward she was amazed to see the screen grow enormously, and she found herself in among the swirling speckles, rushing with them down into the picture. "I cannot see anything with these dots swirling all around me," thought Alice. "It is just like being lost in a snowstorm; why I cannot even see my feet. I wish I could see just a little. I could be anywhere."

At that moment Alice felt her feet strike something solid and she found herself standing on a hard, flat surface. All around her the swirling dots were fading away and she found that she was surrounded by a number of vague shapes.

She looked more closely at the one nearest to her and observed a small figure, coming roughly up to her waist. It was exceedingly difficult to make out, as all the time it kept hopping rapidly to-and-fro, moving so fast that it was very difficult to see at all clearly. The figure seemed to be carrying some sort of stick, or possibly a rolled umbrella, which was pointing straight up in the air. "Hello," Alice introduced herself politely. "I am Alice. May I ask who you are?"

"I am an electron," said the figure. "I am a spin-up electron. You can readily tell me apart from my friend there who is a spin-down electron,

so, of course, she is quite different." Under his breath he added some-thing which sounded rather like "*Vive la différence!*" As far as Alice could see, the other electron looked very much the same, except that her umbrella, or whatever it was, was pointing down toward the ground. It was very difficult to tell for sure, as this figure also was jigging to-and-fro as rapidly as the first.

Particles at the atomic level differ from large-scale objects. Electrons are very small and show no distinguishing fea-tures, being all completely identical to one another. They do have some form of rotation, although what it is that is rotating you cannot say. A peculiar feature is that every electron is rotating at exactly the same speed, no matter in what direction you choose to measure the rotation. They only difference is that some rotate in one direction and some in the other. Depending on their direction of rotation, the electrons are known as *spin-up* or *spin-down*.

"Oh please," said Alice to her first acquaintance. "Would you be good enough to stand still for a moment, as I really cannot see you at all clearly?"

"I am good enough," said the electron, "but I am afraid there is not *room* enough. However I will try." So saying he slowed his rate of jiggling. But as he moved more slowly, he began to expand sideways and become more and more diffuse. Now, although he was no longer moving at all quickly, he looked so fuzzy and quite out of focus that Alice could no more see what he looked like than she had been able to before. "That is the best I can do," he panted. "I am afraid that the more slowly I move, the more spread out I become. That is the way things are here in Quantumland: The smaller the space you occupy, the faster you have to move. It is one of the rules, and there is nothing I can do about it."

"There isn't really room to slow down here," continued Alice's companion as he began once more to leap rapidly around. "The platform is becoming so crowded that I have to be more compact." Sure enough, the space in which Alice stood had now become very crowded indeed, being closely packed with the small figures, each dancing feverishly to-and-fro.

"What strange beings," thought Alice. "I do not think I shall ever be able to see quite what they look like if they will not stand still for a minute, and there does not seem to be much chance of that." Since it did not look as if she could get them to slow down she tried another topic. "Would you tell me please what sort of platform we are on?" she asked.

"Why a railway platform, of course," replied one of the electrons cheerfully (it was very hard for Alice to say which had spoken; they really did all look very much the same). "We are going to take the wave train to the screen you see. You will change there to the photon express I expect, if you want to go any farther."

"Do you mean the television screen?" asked Alice.

"Why of course I do," cried one of the electrons. Alice could have sworn that it was not the same one which had just spoken, but it was very difficult to be certain. "Come on! The train is here and we have to get on."

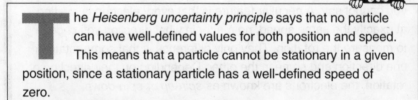

The *Heisenberg uncertainty principle* says that no particle can have well-defined values for both position and speed. This means that a particle cannot be stationary in a given position, since a stationary particle has a well-defined speed of zero.

Sure enough, Alice could see a line of small compartments drawn up at the platform. They were very small. Some were empty, some had one electron in, and some two. All of the empty compartments were filling rapidly—in fact there did not seem to be any left—but Alice noticed that not one of the compartments held more than two electrons. As they passed by any of these compartments, the two occupants would cry out "No room! No room!"

"Surely you could squeeze more than two into a compartment, seeing as the train is so crowded?" Alice asked her companion.

"Oh no! Never more than two electrons together, that is the rule."

"I suppose we shall have to get into different compartments then," declared Alice regretfully, but the electron reassured her.

"There's no problem there for you, no problem at all! You can get into any compartment that you want, of course."

"I am sure that I do not see why that should be," Alice replied. "If a compartment is too full to hold you, then it must surely be too full for me as well."

"Not at all! The compartments are only allowed to hold two electrons, so almost all the places for *electrons* may be taken up, but you are not an electron! There is not a single other *Alice* on the train, so there is plenty of room for an Alice in any of the compartments."

This did not seem to follow so far as Alice could see, but she was afraid that the train would start to move off before they got seats, so she began looking for an empty space that could take another electron. "How about this one?" she asked her associate. "Here is a compartment with only one other electron already in it. Can you get in here?"

"Certainly not!" he snapped, sounding quite horrified. "That is another spin-up electron. I cannot share a compartment with another spin-up electron. What a suggestion! It is quite against my principle."

"Don't you mean against your principles?" Alice asked him.

"I mean what I say, against my principle, or rather Pauli's principle. It forbids any two of us electrons from doing exactly the same thing, which

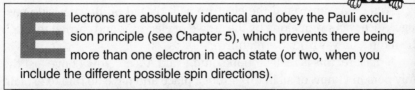

E lectrons are absolutely identical and obey the Pauli exclusion principle (see Chapter 5), which prevents there being more than one electron in each state (or two, when you include the different possible spin directions).

includes being in the same space *and* having the same spin," he responded crossly.

Alice did not really know why she had upset him, but she looked around hastily to find another compartment which might suit him better. She managed to find one that held a single electron who was of the *spin-down* variety, and Alice's companion leaped into this one readily enough. Alice was surprised to find that although the tiny compartment now seemed full there was somehow enough room for her to fit in quite easily.

No sooner were they settled in than the train moved off. The journey was uneventful and the scenery not very interesting, so Alice was rather glad when the train began to slow down. "This must be the screen, I suppose," thought Alice. "I wonder what will happen here."

As they alighted at the screen there was an enormous bustle everywhere. "Whatever is going on?" Alice wondered aloud. "Why does everyone seem to be so excited?" Her questions were answered by an announcement which appeared to come from the air all around her.

"The screen phosphor is presently being excited by the incoming electrons, and we shall be having photon emission soon. Stand by for the departure of the photon express." Alice looked around to see if she could see the express arriving, when there was a rush of bright shining shapes across the platform. Alice was caught up in the middle of the crowd and carried along with them as they all crowded into one compartment. "Well, they do not seem to be worried by any principle, Pauli or otherwise," thought Alice as they crowded in around her. "These ones are certainly not worried about all being in the same place. I suppose the express is going to start soon. I wonder where . . .

". . . we shall end up," she concluded as she stepped out onto the platform. "My, that was certainly a fast journey. Why, it seemed to take no time at all." (Alice was quite right about this. The journey did indeed take no time at all, as time is effectively frozen for anything which is traveling at the velocity of light.) Once again she found herself surrounded by a crowd of electrons, all rushing away from the platform.

"Come along!" one of them cried to her as it rushed off. "We must get out of the station now if we are to get anywhere."

"Excuse me," Alice asked it tentatively, "are you the same electron that I was talking to before?"

"Yes I am," answered the electron as it darted off down a side passage. Alice was swept along by the crowd of electrons and carried through the main gateway from the platform.

"I declare, this is really too irksome," said Alice. "Now I have lost the only person I know at all in this strange place and have no one to explain what is happening."

"Don't worry Alice," said a voice from about knee level. "I will show you where to go." It was one of the electrons.

"How do you know my name?" asked Alice in surprise.

"That's simple. I am the same electron that spoke to you before."

"You cannot be!" exclaimed Alice. "I saw that electron go off in a different direction. Perhaps he was not the same one I was talking to before?"

"Certainly he was."

"Then you cannot be the same one," said Alice reasonably. "You cannot both be the same one you know."

"Oh yes we can!" replied the electron. "He is the same. I am the same. We are all the same, you know, exactly the same!"

"That is ridiculous," argued Alice. "You are here beside me, while he has run off somewhere over there, so you cannot both be the same person. One of you must be different."

"Not at all," cried the electron, jumping up and down even faster in its excitement. "We are all identical; there is no way whatsoever that you can tell us apart, so you see that he must be the same and I am the same too."

At that point the crowd of electrons which surrounded Alice all began to cry out, "I am the same," "I am the same too," "I am just the same as you are," "I am too, just the same as you." The tumult was dreadful, and Alice closed her eyes and put her hands over her ears until the noise died down again.

When it was quiet again Alice opened her eyes and lowered her hands. She found there was no sign of the crowd of electrons which had been clustering around her and that she was walking out of the station entrance all alone. Looking around she found herself in a street which at first sight seemed quite normal. She turned left and began to walk along the sidewalk.

Before she had gone very far she came across a figure standing dejectedly in front of a doorway and searching though his pockets. The figure was short and very pale. His face was difficult to make out distinctly, as was the case for everyone Alice had met recently, but he did look, Alice thought, rather like a rabbit. "Oh dear, oh dear, I am late and I cannot find my keys anywhere. I *must* get inside straightaway!" So saying he stepped back a few paces and then ran quickly toward the door.

He ran so very fast that Alice was not able to see him in any one position, but saw instead a string of afterimages which showed him at all the different positions he passed through along his path. These extended from his starting point to the door, but there, instead of stopping as Alice would have expected, they continued on *into* the door, getting smaller and smaller until they were too small to be seen. Alice had scarcely had time to register this strange series of images when he rebounded backward just as rapidly, once again leaving a series of images. This time they ended abruptly with the unfortunate person sprawled on his back in the gutter. Apparently in no way discouraged, he picked himself up and raced toward the door again. Again there was the series of afterimages, shrinking away into the door, and again he bounced off and ended up on his back.

As Alice hurried toward him he repeated this action several more times, throwing himself at the door and then falling back again. "Stop, stop," cried Alice. "You must not do that; you will surely hurt yourself."

The person stopped his running and looked at Alice. "Why, hello my dear. I must do this I'm afraid. I am locked out and I must get inside quickly, so I have no choice but to try and *tunnel* through the barrier."

Alice looked at the door, which was very large and solid. "I do not think you have much chance of getting through that by running at it," she said. "Are you trying to break it down?"

"Oh no, certainly not! I do not want to destroy my beautiful door. I just want to tunnel through it. I am afraid that what you say is true, though. The probability of my managing to get through is indeed not very high at all, but I have to try." As he said this he charged at the door again.

Alice gave him up as a bad job and walked off, just as he came staggering back once more.

After she had walked a few paces, Alice could not resist looking back to see if by any chance he had abandoned his efforts, and she saw again the series of images rushing toward the door and shrinking down when they got to it. She waited for the rebound. Previously this had followed immediately after, but this time it did not happen. The door stood there looking solid and rather deserted, but there was no sign of her acquaintance. After a few seconds had passed with nothing happening, Alice heard a rattling of bolts and chains from behind the door and then it swung open. Her vanished companion looked out and waved to her. "I was really in luck!" he called. "The probability of penetrating a barrier this thick is very small indeed, and I was amazingly fortunate to get through so quickly." He closed the door with a solid thump and that seemed to end the encounter, so Alice walked on up the street.

A little farther along she came to an empty plot by the side of the road, where a group of builders was clustered around a pile of bricks. Alice assumed they were builders, as they were unloading more bricks from a small cart. "Well at least these people seem to be behaving in a sensible manner," she thought to herself. Just then another group came running around a corner carrying what looked like a very large rolled-up carpet and proceeded to spread it out on the site. When it was unrolled Alice could see that it was some sort of building plan. It did seem to be rather a large plan since it covered most of the available space. "Why, I do believe it must be exactly the same size as the building they are going to put up," said Alice, "but how will they manage to build anything if the plan is already taking up all the room?"

The builders had finished easing the plan into position and had retreated to the pile of bricks. They all picked up bricks and began throwing them at the plan, apparently quite at random. All was confusion— some fell in one place, some in another—and Alice could see no purpose

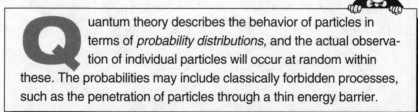

Quantum theory describes the behavior of particles in terms of *probability distributions,* and the actual observation of individual particles will occur at random within these. The probabilities may include classically forbidden processes, such as the penetration of particles through a thin energy barrier.

in it at all. "What are you doing?" she asked a person who was standing to one side. He appeared to be doing nothing, and she assumed him to be the foreman. "You are just making untidy piles of bricks. Aren't you supposed to be putting up a building?"

"Ah, sure, and we are, me darling," answered the foremen. "It's true so it is that the random fluctuations are still large enough to hide the pattern, but since we have laid down the probability distribution for the result we are after needing, we'll be getting there, never fear."

Alice felt that this display of optimism was not very convincing, but she kept her peace and watched as the shower of bricks continued to descend onto the site. Gradually, to her amazement, she noted that more bricks were falling in some regions than in others, and she could begin to make out the patterns of walls and doorways. She watched in fascination as the recognizable shape of rooms began to appear out of the initial chaos. "Why, that is amazing," she cried. "How have you managed to do that?"

"Well now, haven't I already told you," smiled the foreman. "You watched us lay down the probability distribution before we began. This specifies where there should be bricks and where there should be none. We must do this before we start bricklaying as we cannot tell where each brick will go when we throw it, you know," he continued.

"I do not see why!" Alice interrupted him. "I am used to seeing bricks being laid in place one after another in neat lines."

"Well now, that is not the Quantum way. Here we cannot control where each individual brick goes, only the probability that it will go one place or another. This means that when you have only a few bricks, they can go almost anywhere and seem to have no sort of pattern at all. As the number becomes large, however, you find that there are bricks only where there is some probability that they should be there, and where the probability is higher, there you get more bricks. When you have large numbers of bricks involved it all works out very nicely in the end, so it does."

Alice found this all very peculiar, although the foreman spoke so definitely that it sounded as if it might make some sort of strange sense. She did not ask any more questions at this time, as his answers only made her feel more confused than ever, so she thanked him for his information and went on down the road.

Before long she came to a window in which was displayed a large notice:

> Are you dissatisfied with your State?
> Would you like to move up to a higher level?
> We will help you to make the Transition for only 10 eV.
> (Offer subject to normal Pauli exclusion limitation)

"That all sounds very enticing I am sure, but I have no idea what it is talking about, and if I were to ask someone I am sure the answer would leave me even worse off than I am now," exclaimed Alice in desperation. "I have not really understood anything that I have seen so far. I wish I could find someone to give me a good explanation of what is going on around me."

She had not realized that she had spoken aloud, until she was answered by a passerby. "If you want to understand Quantumland you will need to find someone to explain to you about quantum mechanics. For that you ought to go to the Mechanics Institute," she was advised.

"Oh, will they be able to help me understand what is happening here?" cried Alice in delight. "Will they be able to explain all the things I have seen, such as that notice in the window there, and to tell me what these 'eV' are?"

"I should think the Mechanics will be able to give you an explanation for most of it," answered her informant, "but as 'eV' are units of energy you had probably best start by asking about them at the Heisenberg Bank, particularly as it is just across the road there."

Alice looked across where he was pointing and saw a large building with a very formal frontage, obviously designed to impress. It had a tall portico with stone pillars and over the top, in large letters, was carved the name THE HEISENBERG BANK. Alice crossed the road, climbed the long flight of stone steps which led up to the lofty doorway, and passed through.

CHAPTER 2

The Heisenberg Bank

When Alice stepped through the doorway she found herself in a large pillared hall with marble walls. It looked much like other banks that she had seen only more intensely so, as it were. There was a line of cashiers' windows along the far wall, and the vast floor area was divided up by portable tape barriers so that the customers would be guided into regular lines as they waited to be served. At the moment however the place appeared to be quite empty of any customers at all. Apart from the cashiers behind their counters and a bank guard standing by the door, Alice did not see anyone.

As she had been advised to ask for information at the Bank, she began to walk purposefully toward the distant line of windows. "Now just you wait a minute!" called the guard by the door. "Where do you think you are going, young miss? Can't you see that there is a line?"

"I am sorry," replied Alice, "but actually I *can't* see a line. There are no people here."

"There certainly are, and a lot of them!" answered the guard emphatically. "We seem to have quite a rush of 'no people' today. Usually though we refer to them as *virtual*. I have seldom seen quite so many virtual particles waiting to get their energy loans."

Alice had a by-now-familiar feeling that things were not going to become all that much clearer very quickly. She looked over at the windows and saw that, although the room still appeared to be quite empty, the cashiers were all very busy. As she watched, she saw bright figures appear, one at a time, in front of one till or another and then rush quickly from the Bank. At one till she saw a pair of figures materialize together in front of a grill. One she recognized as an electron; the other was very similar, but was a sort of photographic negative of the first, opposite in every way to the electrons she had seen previously.

"That is a positron, an *antielectron*," murmured a voice in her ear. Alice looked around and saw a severe-looking, smartly dressed young woman.

"Who are you?" she asked.

"I am the Bank Manager," replied her companion. "I am in charge of the distribution of energy loans to all the virtual particles here. Most of them are photons, as you can see, but sometimes we get pairs of particles and antiparticles who come along together to ask for a loan, like the electron and positron pair that you were looking at just now."

"Why do they need an energy loan?" asked Alice. "And why can't I see them before they get it?"

"Well you see," replied the Manager, "in order for a particle to exist properly, so that it can be a *free* particle and able to move around and be observed normally and so on, it has to have, at the very least, a certain minimum energy which we call its *rest mass energy*. These poor virtual particles do not have even that energy. Most of them have no energy at all, so they do not really exist. Fortunately for them, they can get a *loan* of energy here at the Bank and this allows them to exist for a little while." She pointed to a notice on the wall, which read:

CONDITIONS OF LOAN
$$\Delta E\, \Delta t = \hbar/2$$
Prompt repayment
would be appreciated.

"That is called the Heisenberg relation. It governs all our transactions. The value $\hbar$ is called *Planck's constant,* the correctly reduced value, of course. The relation gives the rate of exchange for our energy loans. The quantity ΔE is the amount of energy which is borrowed, and Δt is the period for which the loan is made you see."

"You mean," said Alice, trying to follow what the Manager was saying, "that it is like an exchange rate between different types of money, so that the more time there is, the more energy they can have?"

"Oh no! Quite the reverse! It is the energy and time *multiplied together* which are constant, so the greater the amount of energy, the *shorter*

the amount of time they are allowed to keep it. If you want to see what I mean just look at that exotic particle and antiparticle which have just taken out a loan at window #7."

Alice looked where she was directed and observed a striking sight. In front of the window was a pair of figures; one was the opposite of the other, in much the same way as for the electron and positron that she had seen earlier. This pair, however, were bright, flamboyant figures, taking up so much space with their presence that they quite obscured the cashier behind them. Alice could not but be impressed by the extravagance of the two, but as she opened her mouth to make a comment they grew hazy and then vanished completely.

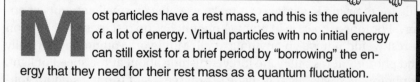

Most particles have a rest mass, and this is the equivalent of a lot of energy. Virtual particles with no initial energy can still exist for a brief period by "borrowing" the energy that they need for their rest mass as a quantum fluctuation.

"That is an illustration of what I was saying," continued the Manager calmly. "That pair took out an enormous energy loan to support the huge rest mass that they needed for their lifestyle. Because the loan was so large, the repayment time was very short indeed, so short that they did not even manage to leave the counter before it had to be repaid. Because such heavy particles cannot get very far before they have to repay their energy loan, they are known in the trade as *short-range* particles," she added.

"Is the relation between time and energy the same for everyone then?" asked Alice, who felt that she might have discovered something definite at last.

"Yes indeed! The Planck constant is always the same whenever and wherever it applies. It is what is called a *universal constant,* which simply means it is always the same everywhere.

"We deal with energy at the Bank here," continued the Manager, "because energy acts as the currency here in Quantumland. As you would express your currency in pounds or dollars, the unit of energy that we use most of the time is called the *eV.* How much energy a particle has determines what it is able to do; how fast it can go, what state it can get into, how much it will be able to affect other systems, these all depend on the energy it has.

"Not all particles are completely destitute like the ones that are lining up. Many of them do have sufficient energy of their own, and in that case they can keep it for as long as they like. Those are the ones which you may see moving around outside. Any particle which needs to have a mass has to have energy just to exist at all."

She pointed at another framed notice on the wall, which read:

Mass Is Energy.
Energy Is Mass.

"If a particle wants to have mass then it must find the energy to support it somehow. If it has any energy left over then it can use it to do other things. Not that all particles bother with mass," she added. "There are some free-and-easy, bohemian particles which do not have any rest mass at all. They are not tied down like most particles who have to provide for their mass, so they can make use of even small amounts of energy. Photons are a case in point. A photon has no rest mass, so a photon at rest would not weigh anything at all. Mind you, you do not normally find photons at rest; they are forever rushing about at the speed of light, as photons *are* what light is made of you see. Light is not a smooth continuous stream. It is made up of a lot of *quanta,* little packets of energy, so that the flow of light is lumpy. These quanta, or particles, of light are called photons. Practically everything comes in quanta of some size. This gives quantum physics its name, you know. Look at all those photons leaving the Bank now. Basically photons are all the same, exactly like one another in the way that electrons are all the same, but you may notice that many of these photons seem quite different. That is because they have different amounts of energy. Some of them have very little energy, like those radio frequency photons going out now."

Alice looked down at a crowd of photons which were rushing past her, flowing around her feet and on out through the door. As they went, she heard snatches of music, dramatic voices, and something about "doing lunch on Thursday." "I didn't know that radio waves were made up from photons," admitted Alice. "Oh yes indeed. They are very long wavelength photons of course, with low frequency and very little energy. They are very gregarious because if they are to have any noticeable effect you need a lot of them at once. Friendly little fellows aren't they?" smiled Alice's companion. "Visible photons now, the ones which make up the light that people use to see by, they have higher frequency and more energy. One of those can have quite a noticeable effect. The really afflu-ent ones though, the big spenders, are the X-ray and gamma photons. Each one of those carries a lot of energy around with it and they can really make their presence felt on their surroundings if they choose to interact."

"That is certainly very interesting," said Alice, not entirely untruth-fully, "but I still feel confused about the whole idea of energy. Can you tell me what energy actually *is*?"

"Well now," replied the Manager with satisfaction, "that is a very sen-sible question to ask. Unfortunately it is not a very easy one to answer. Come into my office and I will try to give you an explanation."

The Manager led Alice briskly across the tiled floor of the main hall and through an unobtrusive but rather forbidding door in one corner. Within was a large modern office. Motioning Alice to sit on a deep com-fortable chair placed in front of the wide desk, the Manager went round and sat in the chair behind it.

"Well," she began, "energy is a little bit like money in your world and it is not too easy to say exactly what that is either."

"I should think that was quite easy," responded Alice. "Money is coins, like my pocket money, or it can be bank notes."

"That is *cash*, which is certainly one form of money, but money does not have to be in notes and coins. It can be in a savings account, for example, or in stocks and shares, or even invested in a building. In much the same way energy can take many forms, which seem quite different from one another.

"The most obvious form is *kinetic energy*," said the Manager, as she settled more comfortably into her chair and her voice took on the complacent tone of someone who is about to give a long lecture to a captive audience.

"A particle, or any other object for that matter, will have kinetic energy if it is moving about. Kinetic just means moving, you know. There are other forms of energy as well. There is *potential energy*, such as the gravitational energy which a stone has if it is up a hill and so is in a position to roll down it. You can also have electrical energy, or chemical energy, which is just potential energy which the electrons have when they are inside atoms. Then, as I have already mentioned, there is the *rest mass energy* which many particles must have just to exist, so that they

Energy comes in many forms. It may appear as the rest mass energy of a particle, as the kinetic energy which is involved in the movement of any object and as various types of potential energy. One form of potential energy is the gravitational potential energy of an object which decreases when the object falls.

can have some mass. One form of energy can *convert* into another, just as you can pay cash into your deposit account. I can illustrate that for you if you will just look through the round window." She leaned over and pressed a button on her desk, and a round window on the wall in front of Alice opened up. Through it Alice could see a fairground roller coaster. As she watched, a carriage climbed to the top of one hump and paused there momentarily before it rushed down the far side.

"That carriage, as you can see, is not moving at the moment, so it has *no* kinetic energy, but it is high up so it has potential energy because of its position. Now as it starts to fall down into the dip it is losing height, so it loses some of that potential energy. This is converted to kinetic energy, so as it falls it goes faster and faster." Alice could vaguely hear the happy excited shrieks of the distant passengers in the carriage as it thundered down the track.

"If the track were very smooth and the wheels ran without friction," the lecturer continued dispassionately, "then the carriage would come to rest again at exactly the same height." She leaned over and fiddled with something on her desk again. The distant figures on the roller coaster cried out in surprise as the next hump in the track suddenly surged up before them to a much greater height. Their carriage slowed and came to a complete stop before it had reached the top. "How did you do that?" exclaimed Alice in amazement. "Never underestimate the influence of a bank," muttered her companion. "Now see what happens."

The carriage began to roll backward down the track, accompanied by more shrieks, still excited but not quite as happy as last time. It gathered speed until it shot through the lowest point and then climbed the oppo-

site slope, slowing as it went. It came to rest just at the peak where Alice had first seen it and then began to slip back down once again.

"This will go on indefinitely now, with the energy of the carriage changing from potential energy to kinetic energy and back again, but you get the idea." The Manager pressed another button on her desk and the window closed on the scene.

"That is the sort of obvious way in which you see energy in Classic-World. It will change from one form to another in a smooth continuous manner. You saw how the carriage got steadily faster as it rolled steadily down the slope, with no big jumps, and there are no obvious restrictions on the exact amount of energy which any object *might have*. Here in Quantumland it is often not like this. In many situations a particle is only allowed to have one of a restricted set of values and it can only accept or give up energy in large lumps, which we call *quanta*. In ClassicWorld all energy payments are made on the installment plan, with very frequent and very very tiny payments, but here they often have to be made as a lump sum.

"As you saw, kinetic energy is a dramatic, showoff sort of energy—something which a body has just because it is moving. The more massive it is the more kinetic energy it has, and the faster it moves the more kinetic energy it has, but the amount does not depend at all on the *direction* in which it is moving, only on the speed. In this respect it is different from another important quantity which tells us how a particle moves. This is something we call momentum. *Momentum* is a sort of measure of

I n quantum theory it is as important to consider energy and momentum as it is to consider position and time. Rather more important in fact, as it is easier to measure the energy of an atom than to determine where it is. Energy is in a sense the physical world's equivalent of money. Energy is defined classically as the "capacity for doing work," and it is necessary for particles to have energy in order to do something: to make transitions from one state to another. Momentum is a quantity more like velocity. It is going in a specific direction while energy has only a size. When you have said how *much* energy there is, then there is nothing left to say about it. Electrons moving from right to left and from left to right at the same speed have the same kinetic energies, but opposite momenta.

a particle's obstinacy. Every particle is determined that it is going to keep on moving in exactly the same way as it was before, without any change at all. If something is moving fast it takes a lot of force to slow it down. It also takes a lot of force to make it move in a different *direction,* even if its speed does not change. Now a change in direction does not cause a particle to lose any of its precious kinetic energy, as this depends only on how fast it is traveling, but it still does not want to change because its momentum would have to be different. Particles are rather *conservative* that way.

"It is all a question of what we call *parameters,*" continued the Manager enthusiastically. "When you want to describe a particle, you have to use the right parameters. If you want to say where it is you must talk about its position and time, for example."

"I would have thought that you would just need to say what its position was," objected Alice. "That will tell you where it is, surely?"

"No, certainly not. You must give the time as well as the position. If you want to know where something is now, or where it will be tomorrow, it is no good my only telling you a position if that is where it was last week. You must know the position *and* the time, as things tend to move around you know. Just as if you want to know what a particle is *doing* you must describe that in terms of its momentum and energy, in general you need to give both position and time if you want to know where a particle is."

There are many varieties of energy. Kinetic energy is directly due to movement: A moving cannonball has energy which a stationary one does not have. Rest mass energy is another form. The rest mass energy of any object is large. In Newtonian mechanics there was no need to consider rest mass energy because it never changed, so it did not affect any energy transfer. In quantum processes the masses of particles often do change, and the change in rest mass energy may be released to other forms. A release of much less than 1 percent of the rest mass for a small fraction of the material occurs in a nuclear weapon, for example. This is not a large energy change per particle when compared to many processes investigated in particle physics, but it is devastating when it is released by a significant number of particles into our everyday world.

"Here in Quantumland the parameters tend to be related. If you try to see *where* something is then that has an effect on its momentum, how fast it is moving. It is another form of the Heisenberg relation which I pointed out to you in the Bank."

"Oh!" cried Alice, remembering a previous encounter. "Was that the reason that the electron I saw earlier could not stand still to let me see him without becoming all fuzzy?"

"Yes, undoubtedly. The uncertainty relations affect all particles that way. They always seem a bit indefinite, and you can never pin them down too precisely.

"I know what I shall do! I shall get the Uncertain Accountant to explain it to you," exclaimed the Manager. "His job is to try and balance the accounts, so he has to worry all the time about quantum fluctuations." She reached out an elegant finger and pressed yet another of the buttons with which her desk was so well supplied.

There was a short pause, and then one of the doors which were spaced around the room opened and a figure entered. He looked rather like a picture of Ebenezer Scrooge from an illustrated copy of *A Christmas Carol*, except that he had a rather bemused expression on his face and an uncontrollable nervous twitch. He was carrying an enormous ledger whose covers bulged, not to say wriggled as if the contents were in continuous motion.

"I believe I have done it," he cried triumphantly, twitching so violently that he almost dropped the book. "I have gotten the accounts to balance! Apart from the residual quantum fluctuations, of course," he added, less enthusiastically.

"Very good," answered the Manager absently. "Now I should like you to take this little girl, Alice, here and explain to her about quantum uncertainty and fluctuations in the energy of a system and all that sort of

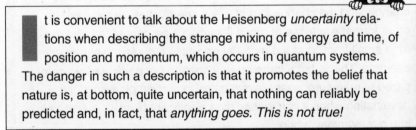

I t is convenient to talk about the Heisenberg *uncertainty* relations when describing the strange mixing of energy and time, of position and momentum, which occurs in quantum systems. The danger in such a description is that it promotes the belief that nature is, at bottom, quite uncertain, that nothing can reliably be predicted and, in fact, that *anything goes. This is not true!*

thing." With a wave of farewell to Alice, the Manager turned back to her desk and began doing something particularly complicated with all the buttons on it. The Accountant led Alice out quickly before anything further could happen.

They came to a much smaller, more cluttered office which contained a tall, old-fashioned desk covered in ledgers and with scraps of paper piled all over the floor. Alice looked at one of the open ledgers. The page was covered with columns of figures, much like other accounts ledgers she had seen, except that in this one the figures were continually changing slightly as she looked at them.

"Right!" said the rather Victorian figure in front of Alice. "You want to know about Uncertainty do you, young lady?"

"Yes please, if it is not too much trouble," replied Alice politely.

"Well now," he began, seating himself at his desk. He steepled his fingers together in the traditional magisterial manner to increase the dignity of his appearance, but this was not a good idea as just then he gave such a particularly violent jerk that he got his fingers all tangled up, and he had to stop to unravel them.

"Well now," he repeated, thrusting his hands deep into his pockets for safety. "The thing you must remember about energy is that it is *conserved*, which is to say that there is always the same amount of it. It may convert from one form to another but the *total* amount is always the same. At least it is if you take the long view," he added wistfully and sighed, staring mournfully into the distance.

"Isn't it true in the short term then?" asked Alice, who felt that she should say something to keep the conversation going.

"Well no, not entirely. In fact, not at all, if the term is short enough. You saw the Heisenberg relation on the notice outside in the Bank didn't you?"

"Oh yes. I was told it gave the terms for the energy loans."

"Well, so it does, in a way, but where do you think the energy for the loans comes from?"

"Why, from the Bank of course."

"Dear me, no!" said the Accountant, looking slightly horrified. "Most certainly not! It would be a fine thing if the Bank started lending out energy from its own stock!

"No," he went on conspiratorially, looking around him carefully, "It is not widely known, but the energy does not come from the Bank. In fact it does not really *come from* anywhere. It is a quantum fluctuation. The amount of energy that any given system has is not absolutely definite, but will vary up and down, and the shorter the time over which you measure it the more it is likely to vary.

"In this respect energy is not really at all like money. Money is well conserved in the short term. If you want to have money for some purpose, you have to get it from somewhere, don't you? You may take it out of a bank account, or borrow it from someone, or you might even steal it!"

"I wouldn't do that!" cried Alice indignantly, but the Accountant continued, ignoring her.

"No matter where you get it, it has to come from somewhere. If you get more, then someone else has less. That is what happens in the immediate short term at any rate.

"In the long term it is different; you may get inflation and find there is more and more money about. Everyone has more, but it does not seem to buy as much as it did. Energy is quite the reverse in a way. In the long term it is conserved, the total amount stays the same, and you get nothing like economic inflation. Every year you will need the same amount of energy on average to transfer from one state in an atom to another. In the short term, though, energy is not well conserved. A particle can pick up the energy it needs for some purpose without it having to come from *anywhere* else; it just appears as a quantum fluctuation. These fluctuations are a consequence of the uncertainly relation: The amount of energy you have is *uncertain,* and the shorter the time you have it the more uncertain the amount you have."

"That sounds terribly confusing," said Alice.

"You do not have to tell me!" answered her companion emphatically. "It is! How would you like to be an accountant when the figures you are trying to balance are fluctuating all the time?"

"That sounds terrible," cried Alice sympathetically. "How do you manage?"

"Well, I usually try to take as long as I possibly can to do the accounts. That helps a bit. The longer the period of time that I spend the smaller the residual fluctuations, you see. Unfortunately people will get impatient and come to me asking if I am planning to take forever to balance the accounts. That would be the only way to do it, you see," he went on earnestly. "The longer I take, the smaller the energy fluctuations, so if I did take *forever,* why then there would be *no* fluctuations and my accounts would balance perfectly," he cried triumphantly. "Unfortunately they just won't let me alone. Everyone is much too impatient and anxious to be off making transitions from one state to another all the time."

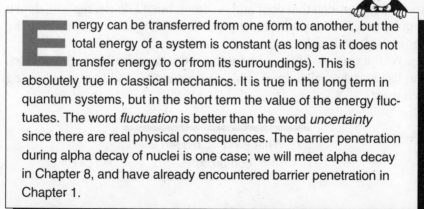

Energy can be transferred from one form to another, but the total energy of a system is constant (as long as it does not transfer energy to or from its surroundings). This is absolutely true in classical mechanics. It is true in the long term in quantum systems, but in the short term the value of the energy fluctuates. The word *fluctuation* is better than the word *uncertainty* since there are real physical consequences. The barrier penetration during alpha decay of nuclei is one case; we will meet alpha decay in Chapter 8, and have already encountered barrier penetration in Chapter 1.

"That is another thing that I wanted to ask about," remembered Alice. "What are all these states that I keep hearing of? Would you explain them to me please?"

"I am not really the best person to do that. It is all part of Quantum Mechanics, so you really ought to go to the Mechanics Institute and ask them there."

"That is what I was told before," said Alice. "If that is the best place to ask, would you please tell me how I might get there?"

"I am afraid that I cannot actually *tell* you how to get there. That is not the way we do things here. But I can arrange that it is very *probable* that you will get there.

He turned to the far wall of his office, which was covered with a dusty curtain. When he drew this aside with a sudden jerk, Alice could see a row of doors spaced along the wall. "Where does each of those lead?" she asked. "Does one of them lead to this Institute you were talking about?"

"*Each* of them could lead you almost anywhere, including, of course, to the Institute. But the point is that *all* of them will be *very* likely to lead you to the door of the Institute."

"I do not understand," complained Alice, with an all-too-familiar feeling of increasing confusion. "What is the difference? If each of them can lead almost anywhere, it is the same as saying that they all could lead almost anywhere."

"Not at all! It is a different thing entirely. If you were to go through any *one* door, why then you might end up almost anywhere, but if you go through them *all* at once then you will most probably end up where you want to be, at the peak of the interference pattern."

"What nonsense!" cried Alice. "I cannot possibly go through all the doors at once. You can only go through one door at a time you see."

"Ah, that is different! Of course, if I *see* you going through a door, then you *will* go through that door and no other, but if I do *not* see you, then it is quite possible for you to have gone through any door. In that case the general rule will apply."

With a wave of his hand he indicated a large, striking notice which was fixed on the wall in front of his desk, where he could not avoid seeing it. It read:

> What is not forbidden is
> compulsory!

"That is one of the most basic rules that we have here. If it is possible to do several things, you do not just do one of them, you have to do them all. That way it saves having to make your mind up very often. So off you go, just go out through all the doors and when you have, then set off in every direction at once. You will find it is quite easy and very soon you will have got to the right place."

"This is ridiculous!" protested Alice. " There is no way that I can go through several doors at once!"

"How can you say that until you have tried? Have you never done two things at the same time?"

"Well, of course I have" answered Alice. "I have watched television while I was doing my homework, but that is not the same thing at all. I have never gone in two directions at the same time."

"I suggest then that you try it," replied the Accountant, rather huffily. "You never know whether you can do something until you try. That is the sort of negative thinking which is always holding back progress. If you want to get anywhere here you have to do *everything* that you possibly can and do it all at the same time. You do not have to worry about where it will take you, the interference will take care of that!"

"How do you mean? What is interference?" cried Alice.

"No time to explain. The Mechanics will tell you all about that. Now off you go and they will explain when you arrive."

"This is really too bad!" thought Alice to herself. "Everyone I speak to rushes me on somewhere else and promises me that I will get an explanation as soon as I get there. I wish that someone would just explain things properly, once and for all! I am sure that I do not know how I can possible go several ways at the same time. It seems to me to be quite impossible, but he is so certain that I shall be able to manage it here that I had better try, I suppose."

Alice opened a door and stepped through.

Alice's Many Paths

Alice stepped through the left-hand door and found herself in a small cobbled square with three narrow alleys leading out of it. She walked down the left-hand alley. Before she had gone very far, she found herself on the edge of a broad paved area. In the center rose a tall dark building with no windows on the lower levels. It looked very forbidding.

Alice stepped through the left-hand door and found herself in a small cobbled square with three narrow alleys leading out of it. She walked down the right-hand alley. Before she had gone very far she came to a park, with weed-choked gravel paths winding between dismal drooping trees. Tall iron railings surrounded the park and a dank mist obscured the scenery within.

Alice stepped through the left-hand door and found herself in a small cobbled square with three narrow alleys leading out of it. She walked down the middle alley. Before she had gone very far she came to another small square, in front of a rather shabby-looking building.

Alice stepped through the right-hand door and found herself in a narrow alleyway with two others branching off it. She walked down the left-hand alley. Before she had gone very far she found herself on the edge of a broad paved area. In the center rose a tall, dark building with no windows on the lower levels. It looked very forbidding, and she had a distinct feeling that she ought not to be there.

Alice stepped through the right-hand door and found herself in a narrow alleyway with two others branching off it. She walked down the right-hand alley. Before she had gone very far she came to a park, with weed-choked gravel paths winding between dismal drooping trees. Tall iron railings surrounded the park and a dank mist obscured the scenery within. She had a very strong feeling that she ought not to be there.

Alice stepped through the right-hand door and found herself in a narrow alleyway with two others branching off it. She walked on down the central alley. Before she had gone very far she came to another small square, in front of a rather shabby-looking building. Somehow it seemed to her that this was the right place to be.

Alice stepped through the center door and found herself facing a wall with three arched gateways which led to alleys beyond. She walked down the left-hand alley. Before she had gone very far she found herself on the edge of a broad paved area. In the center rose a tall, dark building with no windows on the lower levels. It looked very forbidding. She now felt very strongly that she ought not to be there.

Alice stepped through the center door and found herself facing a wall with three arched gateways which led to alleys beyond. She did not walk down the right-hand alley at all, as that route somehow seemed to be completely wrong.

Alice stepped through the center door and found herself facing a wall with three arched gateways which led to alleys beyond. She walked through the gateway to the central alley. Before she had gone very far she came to another small square, in front of a rather shabby-looking building. She now felt quite sure that this was the place where she ought to be.

Alice looked more closely at the building. On a faded board by the door she could make out the words "Mechanics Institute." This was indeed where she had intended to come!

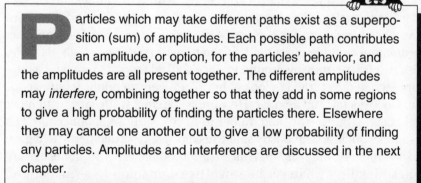

Particles which may take different paths exist as a superposition (sum) of amplitudes. Each possible path contributes an amplitude, or option, for the particles' behavior, and the amplitudes are all present together. The different amplitudes may *interfere,* combining together so that they add in some regions to give a high probability of finding the particles there. Elsewhere they may cancel one another out to give a low probability of finding any particles. Amplitudes and interference are discussed in the next chapter.

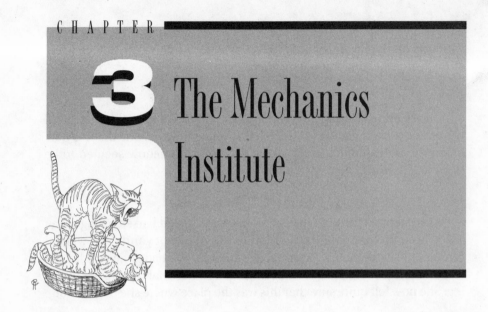

CHAPTER 3
The Mechanics Institute

Alice examined the building in front of her. It was unremarkable, a plain brick structure now rather the worse for wear. In front of her was the board which stated that this was "The Mechanics Institute." Beside this was a wooden door on which someone had pinned a note: "Don't knock. Just come in." Alice tried the door and found it was not locked, so she opened it and walked through.

Inside she found herself in a large, dark room. In the middle of the room there was an area of light and clarity. Within this limited area it was possible to make out a reasonable amount of detail. Beyond this there was a seemingly limitless expanse of darkness within which nothing meaningful could be discerned. In the pool of light was a billiards table, with two figures moving around it. Alice walked toward them, and as she approached they turned to look at her. They were an oddly assorted couple. One was tall and angular. He wore a starched white shirt with a tall stiff collar, a narrow tie, and, rather to Alice's surprise, a boiler suit. His face was aquiline, with bushy side whiskers. He regarded her with a gaze of such piercing intensity that Alice felt he could clearly distinguish every tiniest detail in whatever he saw. His companion was smaller and younger. He had a round face decorated with large, round metal-rimmed glasses. Behind the glasses his eyes were strangely hard to see; it was difficult to say where he was looking, or even exactly where his eyes were. He was wearing a white laboratory coat, which was open to display beneath it a T-shirt with a picture of something vaguely atomic on the front. It was not easy to say exactly what it was meant to be as the colors appeared to have run in the wash.

"Excuse me, is this the Mechanics Institute please?" asked Alice, mostly for the sake of making conversation. She knew from the notice outside that it must be.

"Yes, my dear girl," said the taller and more impressive looking of the two. "I myself am a Classical Mechanic from ClassicWorld, and I am visiting my colleague here, who is a Quantum Mechanic. Whatever your problem is, I am sure that between us we will be able to assist you, if you would just wait a moment while we finish our shots."

Both men turned back to the billiards table. The Classical Mechanic took careful aim, clearly judging all the angles involved to within a tiny fraction of a degree. At last, he very deliberately played his shot. The ball

bounced to and fro in a remarkable series of ricochets, ending in a collision with the red ball and knocking it squarely into the center of a hole. "There you are," he exclaimed with satisfaction as he retrieved the ball from the pocket. "That is the way to do it, you know; careful and exact observation followed by precise action. If you do things that way you can produce any result you choose."

His companion did not respond, but took his place at the table and made a vague stab with his cue. After her previous recent experiences Alice was not really surprised to discover that the ball shot off in every direction at once, so that there was no part of the table where she could say definitely that the ball had not gone, though equally she could in no way say where it actually was. After a moment the player went over and peered into one of the pockets, then reached in and drew out a red ball.

"If you do not mind my saying so," said Alice, "you do seem to play the game very differently."

"Quite so," replied the Classical Mechanic. "I hate the way he plays his shots like that. I like everything to be done very carefully and precisely and to be planned in every detail in advance. However," he added, "I imagine that you did not come here to watch us play billiards, so tell us what you wanted to know."

Alice recounted all her experiences since she came into Quantumland and explained how confusing she found it and how everything seemed so strange and somehow indefinite. "And I do not even know how I came to find this building," she finished. "I was told that the interference would probably bring me to the right place, but I do not understand what happened at all."

"Well now," began the Classical Mechanic, who seemed to have appointed himself as the spokesman for the two. "I cannot say that I really understand all of it either. As I have said, I like things to be clear-cut, with cause following effect in a sensible fashion and everything clear and predictable. If truth be told, not a lot that goes on here makes much sense to me," he whispered confidentially to her. "I am just visiting from ClassicWorld. That is a splendid place where everything happens with mechanical precision. Cause follows effect in a wonderfully predictable fashion, so it all makes sense and you know what is going to happen. What is more, the trains all run on time," he added as an afterthought.
See end-of-chapter note 1

"That sounds very impressive," said Alice politely. "If it is so well organized, is everything run by computers?"

"Well, no," answered the Classical Mechanic. " We do not use computers at all. In fact electronics will not work in ClassicWorld. We are better with steam engines. I do not really feel at home in Quantumland. My friend here is much more familiar with quantum conditions.

"However," he went on more confidently, "I can tell you what interference is. That happens in classical mechanics as well. Just follow me and I shall demonstrate how it works."

He led Alice out through a door, down a short corridor, and into another room. This one was well illuminated, with a clear light which was equally bright everywhere and did not seem to come from any particular source. They stood on a narrow wooden walkway which ran around the edges of the room. The floor in the center was covered with some sort of shimmering grayish material, which did not look solid. It was shot through with random flashes of light, rather like a television set when there is no picture being received.

Her guide explained, "This is the *gedanken* room, which means a 'thinking room.' You know that many gentlemen's clubs have a writing room and a reading room. Well, we have a thinking room. In here one's thoughts can take on substance, so that anyone can look at them. It allows us to do *thought experiments*. These allow us to work out what would happen in various physical cases, and they are much cheaper than real experiments of course."

"How does it work?" asked Alice. "Do you just think of something and it appears?"

"That is correct; in essence that is all you have to do."

"Oh please, may I try?" asked Alice.

"Yes certainly, if you wish."

Alice thought very intensely at the shifting, flickering surface. To her surprise and delight, where there had before been a featureless area there was now a group of small furry rabbits hopping about.

"Yes, very pretty," said the Mechanic rather impatiently. "But this is not helping to explain interference." He made a gesture and all the rabbits vanished, all but one little one who remained unnoticed in a corner of the area.

"Interference," he began authoritatively, "is something which happens with waves. You can have all kinds of waves in physical systems, but it will be simplest to consider water waves." He stared hard at the floor, which turned before Alice's eyes into a sheet of water, with gentle ripples running over the surface. In the corner the rabbit vanished below the surface with a "plop" as the floor turned to water beneath it. It quickly struggled out again and glared at them. Then it shook itself, looked mournfully at its damp fur, and vanished.

"Now we want some waves," continued the Classical Mechanic, paying no attention to the unhappy rabbit. Alice obligingly thought at the floor and a long curling wave came sweeping across the surface and broke dramatically upon a beach at one end.

"No, that is not the sort of wave that we want. Those large breaking waves are too complicated. We want the sort of gentler ripple which spreads out when you throw a stone into water." As he spoke a series of circular ripples spread out from the middle of the water.

"But we need to think about what are called *plane waves* where they all move in the same direction." The circular ripples changed to a series of long, parallel furrows, like a wet plowed field, all moving across the floor from one side to the other.

"Now we put a barrier in the middle." A low fence sprang up across the center, dividing the floor in two. The waves flowed up to the barrier and lapped up and down against it, but there was no way for them to get through and the water beyond was now still and calm.

"Now we make a hole in the barrier, so that the waves can get through there." A neat little gap appeared just to the left of the fence's center point. Where the ripples struck this narrow gap they could pass through and spread out in circular ripples into the calm region beyond.

"And now, see what happens when we have two holes in the barrier," cried the Mechanic. Abruptly there were holes both to the right and to the left of the center. Circular ripples spread out from both of these. Where they crossed, Alice could see that in some places the water was surging up and down much more than it had when there was only one hole open, whereas in other places it hardly moved at all and was locally quite still.

"You can see what is happening if we freeze the motion. We can do that of course in a thought experiment." All motion on the water stopped, and the patterns of ripples were frozen into position, as if the whole area had turned abruptly to ice.

"Now we shall mark regions of maximum and minimum amplitude," continued the Classical Mechanic determinedly. "The amplitude is the amount by which the water moves from the surface level it had when calm." Two fluorescent arrows appeared, hanging in space above the surface. One was an apple green color and was pointing down at a point where the disturbance was greatest. The other was a pale red and pointed to a spot where the surface was almost undisturbed.

"You will be able to see what is happening if we now look at the effect of only one hole at a time," he said, with steadily increasing enthusiasm. One of the gaps in the fence vanished, and there were left only the circular ripples spreading out from the other one, though still frozen in position as if they were made from glass. "Now we will switch to the other hole." Alice could see very little difference when this happened. The position of the gap had moved and the pattern of circular ripples coming from it had moved very slightly, but overall it looked much the same. "I am afraid that I cannot understand what you are trying to show me," she said. "The two cases look just the same to me."

"It will help you to see the difference if we cut quickly from one case to the other." Now the gap in the fence leapt to and fro, first to the right, then to the left. As it moved, the pattern of ripples on the surface shifted slightly back and forth.

"Look at the wave patterns under the green arrow," cried the Mechanic, who seemed to Alice to have become quite unnecessarily excited about the subject. However, she did as requested and saw that at the point indicated there was a hump in the water in each case. "Each gap in the fence has produced a wave which is high at this particular point, so when both gaps are open the wave is twice as high here and the overall rise and fall of the water is much greater than it is for one gap alone. This is called *constructive* interference.

"Now look at the wave patterns under the red arrow." Here Alice saw that, while one gap gave rise to a hump at that point, the other produced a trough in the surface. "You can see that in this position the wave from one gap goes up and that from the other goes down, so when you have the two present together, they cancel one another out and you get no overall effect. This is called *destructive* interference.

"That is all there is to wave interference really. When two waves overlap and combine with one another, their amplitudes, the amounts by which they go up or down, combine with one another. In some places the contributing waves are all going in the same direction, so the disturbances add up and you get a large effect. At other positions they go in different directions and cancel one another out."

"Yes, I think that I follow that," said Alice. "So you are saying that the doors in the Bank acted rather like the gaps in the fence here and gave rise to some sort of large effect in the place where I needed to go and canceled one another out in other positions. I do not see how that can apply to my case though. With your water wave you say that there is more of the wave in one place and less in another because of this interference, but the wave is spread out over the whole area, while I am always in just one place at any time."

"Exactly!" cried the Classical Mechanic triumphantly. "That is the problem. As you say, you are in one place. You are more like a particle than a wave, and particles behave quite differently in a sensible classical world. A wave is spread over a wide area and you look at only a small portion of it at any position. Because of interference you may get more or less of it at different positions, but it is only a small part of the whole wave wherever you look. A particle, on the other hand, is located at some point.

nterference is classically a property of waves. It happens when amplitudes, or disturbances, from different sources come together, since they may add in some places and subtract or cancel in others. This will result in regions of intense or of low activity respectively. You can see such an effect in the pattern produced when the wakes left by two boats cross one another. Interference effects can also result in poor television reception when reflections from a building interfere with the direct signal. Interference requires extended, overlapping distributions. Classically particles are in one single position and do not interfere.

If you look in various positions you will either find the whole particle or it is simply not there. In classical mechanics there is no question of particles showing interference effects, as we can show."

He turned to the floor of the *gedanken* room and stared firmly at it. The surface turned from water to a smooth area of steel armor, with armored barriers around the edges, high enough for them to hide behind. Across the middle of the floor, where the low fence had stretched across the water, was now a tall armored wall, with a narrow slit slightly to the left of center. "Now we can look at the same setup, but I have changed it so that we can look at fast particles. These are something like bullets from a gun, so that is what we will use."

He gestured toward one end of the room where there appeared an unpleasant-looking machine gun with many boxes of ammunition stacked beside it. "This gun has an unsteady mounting, so that it will not always shoot in the same direction. Some of the bullets will strike the gap in the wall and pass through, as part of the wave did in our last thought experiment. Most of them, of course, will hit the steel wall and bounce off. Oh that reminds me," he added abruptly. "We had better wear these in case we are struck by ricocheting bullets." He produced a pair of steel helmets and handed one to Alice.

"Do we really need these?" asked Alice. "If this is only a thought experiment, surely these are thought bullets, and can't do us any harm."

"Well, perhaps so. But you might still *think* that you had been hit by a bullet, and that would not be very nice you know."

Alice put the helmet on. She could not feel it on her head and did not

think that it would be the least bit of use, but there did not seem to be much point in arguing any further. The Mechanic stood upright and gave an imperious wave of his hand, and the gun began firing very noisily. The bullets shot out in an unsteady stream; most hit the armored screen and whined off in all directions, but a few got through the slits in the barrier and hit the wall opposite. Alice was intrigued to note that when a bullet hit this wall, it immediately came to a stop and then rose slowly into the air to hang suspended in space, directly above the point where it had struck the wall.

"As you can see, whereas the water wave was spread out all over the far wall, a bullet will hit it in one position only. However, in this experiment there is a greater *probability* that the bullet will strike the far wall opposite the slit in the screen than there is that it will bounce off the slit edge and end up a long way off to the side. If we wait for a little we will see how the probability varies for different points along the wall." As time passed and the air became full of flying bullets, the number which were suspended above the wall grew steadily. As she watched, Alice could make out a distinct trend developing.

"There, you see how the bullets which have passed through the slit are distributed along the wall," remarked the Mechanic as the gun fell silent. "Most have ended up directly opposite the hole, and the number falls off steadily on either side. Now see what happens when the slit is off-set to the right." With another wave of his hand the hovering bullets dropped to the ground, and the gun began to fire again. Though the demonstration was noisy and rather unsettling, as far as Alice could see the end result was just the same as last time. Frankly, it was disappointing.

"As you can see," said the Mechanic with misplaced confidence, "the distribution is similar to the previous one, but displaced slightly to the right because the center is now opposite the new position of the slit." Alice could not see any difference at all, but she was prepared to take his word for it.

"Now," continued the Mechanic dramatically, "see what happens when *both* slits are open." As far as Alice could see it did not make the slightest difference, except that, since two slits were now open, more bullets got through to hit the far wall. This time she decided to comment. "I am afraid that it looks just the same to me each time," she said rather apologetically.

"Exactly!" replied the Mechanic with satisfaction. "Except that, as you will of course have observed, the center of the distribution is now centered between the two slits. We had one distribution for the probability that bullets will pass through the left-hand slit and another distribution for the probability that bullets will pass through the right-hand slit. When we have both slits open, then bullets may pass through either slit, so the overall distribution is given by the sum of the probabilities that we got for the two slits on their own, since the bullets must have passed through one *or* the other. They cannot have passed through *both* you know," he added, addressing the Quantum Mechanic, who had just come into the room.

"You say that," replied his colleague, "but how can you be so sure? Just look what happens when we repeat your *gedanken* experiment with electrons."

In his turn, the Quantum Mechanic waved his hand at the floor of the room. His gestures were not so decisive as his companion's, but they seemed to work just as well. The gun and the armored walls all disappeared. The floor returned to the shimmering material which Alice had first seen, but the now-familiar wall with two slits near its middle was still there, stretching across the center of the floor. At the far end of the floor was a wide screen with a greenish glow. "That is a fluorescent screen," muttered the Mechanic in her ear. "It gives a flash of light when an electron hits it, so it can be used to detect where they are."

At the opposite end of the floor, where the machine gun had been placed before, was another gun. This was a small stubby affair, like a very small version of the cannons from which people are sometimes shot during circus performances. "What is that?" asked Alice.

"Why, it's an *electron gun,* of course." As Alice looked more carefully, she could see a short flight of steps leading up to the mouth of the cannon and a line of electrons waiting to be fired from it. They seemed to be a great deal smaller than when she had last seen them. "But of course," she told herself, "these are only *thought* electrons."

As she looked at them, she was surprised to see the electrons all turn and wave to her. "I wonder how they know me?" she asked herself. "But then I suppose that they are *all* the same electron that I met before!"

"Commence firing!" commanded the Quantum Mechanic, and the electrons hurried up the steps into the gun and shot out in a steady stream. Alice could not make them out at all when they were in flight, but she saw a bright flash where each one hit the screen. As each flash died, it left a small glowing star which rose up the screen and remained behind to provide a marker for the position where the electron had landed.

As had been the case for the machine gun before it, the electron gun continued to fire out its stream of electrons and the stacks of little glowing stars began to build up a recognizable distribution. At first Alice could not be too sure what she was seeing, but as the number of little stars displayed became larger it was clear that their distribution was quite different from that represented by the previous stacks of bullets.

Instead of a slow, steady decrease from a maximum number in the center, the stars were now arranged in bands, with dark gaps between where there were few if any of the glowing markers. Alice realized that this was in a way like the case she had seen for the water waves, where there had been regions of high activity with calm areas in between. Now there were regions where many electrons had been detected, with very few in the areas between. It consequently came as no great surprise to her when Quantum Mechanic said, "There you see a clear interference effect. With the water waves you had regions of greater and lesser motion at the surface. Now each electron will be detected at one position only, but the *probability* of detecting an electron varies from one position to another. The distribution of different wave intensities which you saw before is replaced by a probability distribution. With one or two electrons such a distribution is not obvious, but when you use a lot of electrons you will find more of them in the regions of high probability. With one slit alone we would have seen that the distribution would decrease smoothly to either side, much as the bullets or the water waves did when there is only one slit. In this case we see that, when there are two slits open, the amplitudes from the two slits are interfering and are producing obvious peaks and troughs in the probability distribution. The behavior of the electrons is quite different from that of my friend's bullets."

"I do not understand," said Alice. This seemed to her to be the only thing she ever said. "Do you mean that there are so many electrons going

The strongest experimental evidence for quantum behavior is given by the phenomenon of interference. When an observed result may come about in several ways, then an amplitude for each possible way will in fact be present. Further, if these amplitudes are in some way brought together then they can add or subtract and the overall probability distribution shows distinctive maxima and minima: alternating intense and empty bands. This effect is seen in practice wherever it might be expected. A form of interference produces the sets of distinct energy states which occur in atoms. Only those states which "fit neatly" within the potential will interfere positively to give a strong maximum in the probability. Any other states would cancel themselves out and hence do not exist.

through that somehow the electrons which go through one hole are interfering with the ones which go through the other?"

"No, that is not what I mean. Not at all. You shall now see what happens when there is only one electron in flight at any time." He clapped his hands and cried "OK! Let's do it again, but slowly this time." The electrons sprang into action or rather, to be strictly accurate, one climbed up into the cannon and shot off. The others continued to sit around where they were. A little later another electron climbed in and was fired on its way. This continued for some time, and Alice could see the same pattern of clumps and gaps appearing. These clumps and gaps were not so clear this time as they had been before because the slow rate at which the electrons were arriving meant that there were not very many in the clumps, but the pattern was clear enough. "There, you see that the interference effect works just as well even when there is only one electron present at any time. One electron on its own can show interference. It can go through both slits and interfere with itself, so to speak."

"But that is silly!" cried Alice. "One electron cannot go through both slits. As the Classical Mechanic said, it just isn't sensible." She went up to the barrier and peered more closely, to try and see where the electrons went as they passed through the slits. Unfortunately the light was poor and the electrons moved by so quickly that she could never quite make out which slit any one had passed through. "This is ridiculous," thought Alice. "I need more light." She had forgotten that she was in the "thinking

room" and was startled when an intense spotlight mounted on a stand appeared by her elbow. Quickly she directed the light toward the two slits and was pleased to find that now there was a visible flash near one hole or the other when the electron passed through. "I have done it!" she cried. "I can see the electrons as they go through the slits, and it is just as I said it must be. Each one does go through just one slit."

"Aha!" replied the Quantum Mechanic meaningfully. "But have you looked to see what is happening to the interference pattern?" Alice looked back toward the far screen and was amazed to see that now the distribution of little stars fell smoothly from a central maximum, just like the distribution that she had seen for the classical bullets. It didn't seem fair somehow.

"That is how it always happens; there is nothing that you can do about it," said the Quantum Mechanic soothingly. "If you don't have any observation to show which hole the electrons go through, then you get interference between the effects of the two holes. If you do observe the electrons, then you find that indeed they *are* in one place or the other, not both, *but* in that case they also act as you would expect if they had come through one hole only and you do not get any interference. The problem is that there is no way in which you can look at the electrons without disturbing them, as when you shone that light on them, and the very act of making the observation forces the electrons to choose one course of action. It doesn't matter whether or not you make a note of which hole the electron came through. It does not matter whether you are aware which hole it came through. Any observation which *could* tell you this will disturb the electron and stop the interference. The interference effects only happen when there is no way that you could know which slit the electron went through. Whether or not you do know does not matter.

"So you see, when there is interference it seems as if each electron is going through both slits. If you try and check on this, you will find that the electrons go through only one slit, but then the interference vanishes. You can't win!"

Alice thought about this for a bit. "That is utterly ridiculous!" she decided.

"Certainly it is," replied the Mechanic with a rather smug smile. "Quite ridiculous I agree, but as it also happens to be how Nature works we have to go along with it. Complementarity, that's what I say!"

"Would you please tell me what you mean by *complementarity*?" asked Alice.

"Why of course. By complementarity I mean that there are certain things you cannot know, not all at the same time anyhow."

"Complementarity doesn't mean that," protested Alice.

"It does when I use it," replied the Mechanic. "Words mean what I choose. It is a question of who is to be master, that is all. Complementarity, that's what I say."

"You said that before," pointed out Alice, who was not entirely convinced by his last assertion.

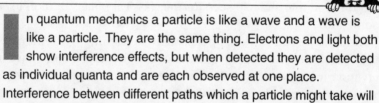

I n quantum mechanics a particle is like a wave and a wave is like a particle. They are the same thing. Electrons and light both show interference effects, but when detected they are detected as individual quanta and are each observed at one place.

Interference between different paths which a particle might take will give a probability distribution with pronounced maxima and minima, where it is more *likely* that a particle will be detected in one position than another.

"No, I didn't," said the Mechanic. "This time it means that there are questions you cannot ask of a particle, such as where it is and, at the same time, how fast it is going. In fact it may not be really meaningful to talk about an electron *having* an exact position."

"That is a great deal for one word to mean!" said Alice tartly.

"Why, to be sure," answered the Mechanic, "but when I make a word do extra work like that I always pay it more. I am afraid that I cannot really explain what is happening to the electrons. An explanation is usually required to make sense in terms of things you already know about and quantum physics doesn't do that. It seems to make nonsense but it works. It is probably safe to say that no one really understands quantum mechanics, so I cannot *explain,* but I can tell you how we *describe* what goes on. Come into the back room and I will do my best."

See end-of-chapter note 2

They left the *gedanken* room, whose floor had returned to its original shimmering aspect, and walked down the corridor to another room furnished with scattered armchairs. When they had seated themselves, the Quantum Mechanic continued. "When we talk about a situation like the electrons passing through the slits, we describe it with an *amplitude*. This is something like the waves that you looked at, and indeed it is often called a *wave function* instead. The amplitude can pass through both the slits, and it is not always positive, like a probability. The lowest probability that you can have is zero, but the amplitude may be negative or positive, so the parts from different paths can cancel or add and give interference, again just like the water wave."

"So where are the particles?" asked Alice. "Which slit do they actually go through?"

"The amplitude doesn't really tell you about that. However if you *square* the amplitude, that is multiply it by itself so that it gives something that is always positive, then it gives you a *probability distribution*. If you choose any position this will tell you the probability that, when you observe a particle, you will find it at that position."

"Is that all it can tell you ?" exclaimed Alice. "I must say that it sounds very unsatisfactory. You would never know where anything is going to be."

"Yes, that is true enough. For *one* particle you cannot tell where it will be found, except that it will not be at a position where there is zero probability of course. If you have a large number of particles, though, then you can be fairly sure that you will find more where the probability is high and far fewer where it is low. If you have a *very* large number of particles, then you can say quite accurately how many will end up where. That was the case with those builders you were telling us about. They knew what

they would get because they used a large number of bricks. For really large numbers the overall reliability is very good."
See end-of-chapter note 3

"And there is no way you can say what each particle is doing until it is observed?" repeated Alice, just to get this clear.

"No, no way at all. When the thing that you actually observe could have come about in several different ways, then you have an amplitude for each possible way, and the overall amplitude is given by adding all of these together. You have a *superposition of states*. In some sense the particle is doing all the things which it could possibly be doing. It is not just that you do not *know* what the particle is doing. The interference shows that the different possibilities are all present and affect one another. In some way they are all equally real. Everything that is not forbidden is compulsory."

"Oh, I saw that on a notice in the Bank. It looked very stern."

"You had better believe it! It is one of the main rules here. Where there are several things which might happen, they all do. Look at the Cat, for example."

"What cat?," asked Alice, looking around her in confusion.

"Why Schrödinger's Cat over there. He left it with us to look after."
Alice looked over in the corner where the Mechanic was pointing and saw a large tabby cat sleeping in a basket in the corner. As if awakened by

hearing its name the cat stood up and stretched. Or rather, it did and it didn't. Alice could see that, as well as the slightly hazy figure of the cat standing with back arched in the basket, there appeared to be another identical cat which was still lying on the bottom. It was very stiff and motionless and lay in a rather unnatural position. From the look of it, Alice would have sworn that it was dead.

"Schrödinger devised a *gedanken* experiment in which an unfortunate cat was enclosed in a box, together with a flask of poison gas and a mechanism which would break the flask should a sample of radioactive material happen to decay. Now such a decay is definitely a quantum process. The material might or might not decay, so according to the rules of quantum physics you would have a superposition of states, in some of which the decay would have happened and in others it would not. Of course, for those states where a decay had happened the cat would have been killed, so you would have a superposition of cat-states, some dead and some alive. When the box was opened someone would observe the cat, and from that time on it would be either alive or dead. The question which Schrödinger posed was, 'What was the state of the cat before the box was opened?'"

"And what did happen when the box was opened?" asked Alice.

"Well actually, everyone was so engrossed in discussing the question that no one ever did open the box, which is why the Cat was left like that."

Alice peered closely into the basket, where one aspect of the Cat was busily licking itself. "He looks pretty lively to me," she observed. No

sooner were the words out of her mouth than the Cat became fully solid and the dead version vanished. With a satisfied purr the Cat leapt out of the box and began to stalk a mouse which had just popped out of the wall. Alice noted that there was no mouse hole visible—the mouse had simply come out of the solid wall. The Quantum Mechanic followed the direction of her gaze. "Ah, yes. That is an example of barrier penetration; we get it happening all the time. Where you have a region that a particle could not enter at all according to classical mechanics, the amplitude does not necessarily stop abruptly at the boundary, though it does die away rapidly inside the region. If the region is very narrow, then there is still some small amplitude left at the other side, and this gives a slight probability that the particle may appear there, having apparently tunneled through an impassible barrier. It happens quite often."

Alice had been thinking through what she had seen and had noted a difficulty. "How is it that I was able to make an observation and fix the condition of the Cat if it was not able to do it for itself? What is it that decides when an observation is actually made and who is able to make one?"

"There you have a good question," replied the Quantum Mechanic, "but we are only mechanics after all, so we do not worry too much about such things. We just get on with the job and use ways that we know will work in practice. If you want someone to discuss the *measurement problem* with you, you will need to go somewhere more academic. I suggest that you go to a class at the Copenhagen School."

"And how do I get there?" asked Alice, resigned to being passed on somewhere else once again. In answer the Mechanic led her out into the corridor and opened yet another door. This did not lead into the alleyway from which she had entered, but into a wood.

Notes

1. Quantum mechanics is usually contrasted with classical or Newtonian mechanics. The latter covers the detailed description of moving objects which was developed before the early years of the twentieth century and was based on the original work of Galileo, Newton, and others both before and since. Newtonian mechanics works very well on a large scale. The motion of the planets can be predicted over long times and with great accuracy. It works almost as well for artificial planets and the various exploratory space missions: Their positions may be predicted years ahead. It also works pretty well for falling apples.

In the case of a falling apple there will be significant resistance from the air that surrounds it. Classical mechanics describes this as the collision of vast numbers of air molecules bouncing off the apple. When you ask about air molecules you are told that they are small groups of atoms. When you ask about atoms there is an embarrassing silence.

Classical mechanics had virtually no success in describing the nature of the world on the scale of atoms. Things must somehow be different for small objects from how they seem to be for large ones. If you argue in this way, then you must ask: large or small relative to what? There must be some dimension, some fundamental constant which fixes the size at which this new behavior becomes obvious. It is a definite change in the way things are observed to behave, and it is universal. Atoms in the sun and in distant stars emit light with a spectrum which is like that from a lamp on a table beside us. The onset of quantum behavior is not something that just happens to take place locally; there is some fundamental property of Nature involved. This is given by the universal constant $\hbar$, which features in most equations of quantum mechanics. The world is *grainy* on the scale defined by this constant, $\hbar$. On this scale energy and time, position and momentum are blurred together. It need hardly be pointed out that, on the human scale of perception, $\hbar$ is very small indeed and most quantum effects are not at all obvious.

2. What the Heisenberg uncertainty relations are telling us is that we are looking at things in the wrong way. We have a preconception that we *ought* to be able to measure the position and momentum of a particle at the same time, but we find that we cannot. It is not in the nature of particles for us to be able to make such a measurement on them, and the theory tells us that we are asking the wrong questions, questions for which there is no viable answer. Neils Bohr used the word *complementarity* to express the fact that there may be concepts which cannot be precisely defined at the same time: such pairs of concepts as justice and legality, emotion and rationality.

    There is, apparently, something fundamentally wrong with our belief that we *should* be able to talk about the position and momentum of a particle, or of its exact energy at a given time. It is not clear why it should be meaningful to talk simultaneously of two such different qualities, but it appears that it is not.

3. Quantum mechanics is not really about definite particles in the traditional classical sense; instead you talk about *states* and *amplitudes*. If you *square* an amplitude (i.e., multiply it by itself), then you get a

probability distribution which gives the *probability* of obtaining various results when you make an observation or measurement. The actual value that you get for any one measurement appears to be quite random and unpredictable. So it does look as if the suggestion made earlier that nature is uncertain and "anything goes" must, after all, be true, does it not?

Well, no—if you make many measurements the *average* result is accurately predictable. Bookmakers do not know which horse will win each race, but they confidently expect to make a profit at the end of the day. They do not anticipate large surprise losses even though they have to work with rather small numbers, so that the averaging is not too reliable. The number of gamblers will be a mere few thousand people rather than the 1,000,000,000,000,000,000,000,000 or more atoms you will get in even a tiny speck of matter. This looks less like a number than a repetitive wallpaper pattern, but it is undeniably large. The overall statistical fluctuations to be expected for measurements made on such a large number of atoms are negligible, even though the result for each individual atom may be quite random.

Quantum-mechanical amplitudes may be calculated very accurately and compared with experiments. An often quoted result is for the magnetic moment of the electron. Electrons spin like little tops and they also have electrical properties: They behave rather like tiny bar magnets. The magnetic strength and the electron spin are related, and their ratio may be calculated using suitable units.

A classical calculation gives the result 1 (with rather arbitrary assumptions about the distribution of the electric charge in an electron).

The quantum calculation gives the result 2.0023193048 (± 8) (the error is in the last decimal place).

A measurement has given the result 2.0023193048 (± 4).

This is good agreement! The probability of getting by chance a value which is in such good agreement is similar to the probability of throwing a dart at random and hitting the bull's-eye on a dartboard—when the dartboard is as far away as the Moon. This particular result is often given as an example of the success of quantum theory. It is possible to calculate accurately the amplitudes for other processes just as accurately, but there are very few quantities which you can *measure* to this precision.

4 The Copenhagen School

A lice entered the wood and made her way along a path which wound between the trees, until she came to a place where it forked. There was a signpost at the junction, but it did not appear very helpful. The arm pointing to the right bore the letter "A," that to the left the letter "B," nothing more. "Well, I declare," exclaimed Alice in exasperation. "That is really the most unhelpful signpost I have ever seen." She looked around to see if there were any clues as to where the paths might lead, when she was a little startled to see that Schrödinger's Cat was sitting on the bough of a tree a few yards off.

"Oh Cat," she began rather timidly. "Would you tell me please which way I ought to go from here?"

"That depends a good deal on where you want to get to," said the Cat.

"I am not really sure where. . . ," began Alice.

"Then it doesn't matter which way you go," interrupted the Cat.

"But I have to decide between these two paths," said Alice.

"Now that is where you are wrong," mused the Cat. "You do not have to decide, you can take all the paths. Surely you have learned that by now. Speaking for myself, I often do about nine different things at the same time. Cats can prowl around all over the place when they are not observed. Talking of observations," he said hurriedly, "I think that I am about to be obs. . ." At that point the Cat vanished abruptly.

"What a strange cat," thought Alice, "and what a strange suggestion.

He must have been referring to that superposition of states that the Mechanic was talking about. I think that it must be something like the time that I left the Bank. Somehow I managed to go in many different directions that time, so I suppose I shall just have to try and do it again."

◆ ❖ ◆ ❖ ◆ ❖ ◆

State: Alice (A1)

Alice turned right at the sign and walked farther along the winding path, looking around her at the trees as she passed. She had not gone very far when she came to another fork in the path; this time the signpost had two arms, labeled "1" and "2." Alice turned to the right and continued on her way.

As she walked along, the trees thinned out and she found herself trudging up a steep, rocky track. It became steeper and steeper as she went on, until she found herself climbing the side of a lonely mountain. The track brought her along a narrow ledge running along the side of a precipitous cliff. This finally ended at a little grassy-floored area with vertical sides. Before her eyes was a yawning mouth in the cliff face, from which a passage led in and down.

The passage was very dark, but to her surprise Alice found herself creeping on down it. It had a smooth floor and sides and ran straight ahead, sloping gently downward toward a dimly visible distant glow. As she went, the light steadily became brighter and also redder, and the tunnel got hotter. Wisps of vapor floated past her, and she heard a sound like some vast animal snoring in its sleep.

At the end of the tunnel Alice peeped out into a great cellar. Its dark vastness could only be guessed dimly, but rising from close below her feet was a great glow. There lay a vast reddish-gold dragon fast asleep with his huge tail coiled around him. Beneath him, forming his bed, was an enormous pile of gold and silver, jewels, and marvelously carved objects, all red-stained in the ruddy light.

State: Alice (A2)

Alice turned right at the sign and walked farther along the winding path, looking around her at the trees as she passed. She had not gone very far when she came to another fork in the path; this time the signpost had two arms, labeled "1" and "2." Alice turned to the left and continued on her way.

As she was walking along, she looked down and found that the path she was on had changed from a forest track to a narrow road paved with yellow bricks. She followed this through the trees until the wood opened out into a wide meadow. It was very wide, extending as far as Alice could see, and the whole field was covered with bright poppies. The yellow brick road ran through the middle of the meadow up to the gates of a distant city. From where she stood, Alice could see that the high walls of the city were a brilliant green and the gates were studded with emeralds.

State: Alice (B1)

Alice turned left at the sign and walked farther along the winding path. There was nothing very remarkable to see as yet. She turned a corner and came to another fork in the path; this time the signpost had two arms, labeled 1 and 2. Alice turned to the right and continued on her way.

The undergrowth between the trees became thicker, and it was difficult to see anything that was at all far from the path, though the path itself was still quite clear as it wound between the closely packed trees. Alice turned a corner and came suddenly upon an open space. In the center of the clearing stood a little building with a steeply pitched roof and a small belfry at one end. The words "Copenhagen School" were carved deeply into the stone lintel over the door.

"This must be the place I was told to go to," Alice remarked to herself. "I am not sure that I want to go to a school though! I spend quite enough time at school as it is. But maybe a school here will be quite different from the one that I am used to. I will go in and see!" Without knocking, she opened the door and went in.

State: Alice (B2)

Alice turned left at the sign and walked farther along the winding path. There was nothing very remarkable to see as yet. She turned a corner and came to another fork in the path; this time the signpost had two arms, labeled 1 and 2. Alice turned to the left and continued on her way.

A little further along, the path began to rise, and Alice climbed up the side of a little hill. At the top of the hill she stood for some minutes looking out in all directions over the country—and a most curious country it was. There were a number of little brooks running across it from side to side, and the ground between was divided up into squares by a number of hedges, which reached from brook to brook.

"I declare. It's marked out just like a large chessboard," Alice said at last.

♦ ❖ ♦ ♦ ❖ ♦

"Ah, come in, my dear," a voice called softly, and Alice realized that she had been observed. She stepped through the door and looked around the schoolroom. It was quite a large room with high windows all round. There were rows of desks down the middle of the room. At one end there was a blackboard and a large table behind which stood the Master.

"It does look very much like an ordinary school," Alice admitted to herself as she turned to look at the children in the class. She found that the desks were not occupied by children, however, but by a most remarkable selection of beings who clustered around the front of the room. There was a mermaid, with long flowing hair and a scaly fish's tail. There was a uniformed soldier who, on closer examination, appeared to be made out of tin and a ragged little girl with a tray full of matches. There was a dramatically ugly duckling and a haughty looking man of regal bearing who for some reason was clad only in his underclothes.

"Or is he?" Alice wondered to herself. As she looked again she fancied she could see him wearing rich embroidered garments and a thick flowing velvet robe. When she looked once more, however, all she could see was a rather portly man dressed in his undergarments.

"Hello, my dear," said the Master, who was a kindly looking avuncular figure with bushy eyebrows. "Have you come to join our discussion?"

"I am afraid that I do not know how I have come to be here," said Alice. "I seemed to be in several other places just a moment ago, and I am not at all sure why I have ended up here, rather than in one of the others."

"That is because we have observed you to be here, of course. You were in a superposition of quantum states, but once you had been observed to be here, why you were here, naturally. Obviously you were not observed in any of the other places."

"What would have happened if I had been?" asked Alice curiously.

"Why then your set of states would have collapsed to that other one. "You would not be here, but would instead be at the position where you had been observed to be, of course."

"I really do not see how that can be," replied Alice, who was once again feeling terribly confused. "What difference does it make whether I was observed or not? Surely I must be in one place or the other no matter who sees me."

"Not at all! After all, you cannot say what is happening in any system if you do not observe it. There may be a whole range of things that it might be doing and you could give a probability that it is or is not doing any one of them as long as you do not look. In fact the system will be in a mixture of states corresponding to all the things that it might be doing. That will be the situation up to the point when you look to see what it is doing. At that point of course one possibility is selected and the system will then be doing only that."

"Then what happens to all the other things it was doing?" asked Alice. "Do they just vanish?"

"Well, there are more things that it *may* be doing than things it *was*

T he "orthodox" picture of quantum mechanics is the *Copenhagen interpretation* (named after the Danish physicist Neils Bohr, not Hans Christian Andersen). Where different things might happen in a physical system, there will be present an *amplitude* for each, and the total state of the system is given by a sum, or superposition, of all these amplitudes.

When an observation is made it will find a value which corresponds to one of these amplitudes, and the excluded amplitudes will vanish, a process called *reduction of the amplitudes.*

doing, but yes," answered the Master, beaming at her. "You have got it exactly. All the others states just vanish. The *land of maybe* becomes the *land that never was*. At that point all the other states cease to be in any way real. They become, if you like, just dreams or fantasies, and the observed state is the real one. This is called *reduction of the quantum states*. You will soon get used to it."

"Does that mean that when you look at something you can choose what you will see?" asked Alice in some disbelief.

"Oh no, you do not get any choice in the matter. What you are *likely* to see is determined by the probabilities for the various quantum states. What you actually *do* see is a matter of random chance. You do not get to choose what will happen; the quantum amplitudes only give the probability of different results, but they do not fix what *will* happen. That is pure chance and only becomes fixed when an observation is made." The Master said this very earnestly, though so quietly that Alice had to strain to catch everything that he said.

"Making an observation seems to be very important then," Alice mused, half to herself. "But then who can make an observation? Obviously the electrons are not able to observe themselves as they go through the slits in an interference experiment, as they seem to go through both slits. Or should I say that amplitudes for both slits are present?" she corrected herself, copying the way of speaking which she had heard so much recently. "Apparently I did not observe myself properly when I was in a superposition of states just now.

"In fact," Alice said abruptly, struck by a sudden thought, "if quantum mechanics says that you must do everything you can do, then surely you must observe *all* the possible results of any measurement you make. If your quantum superposition principle is to work everywhere, then it is not possible to make measurements at all! Any measurement you tried to make could have several possible results. You might observe *any* of these results and, according to your rules, if you could observe any of them you would have to observe *all* of them. The results of your measurement would all be present in a new version of this superposition of states you talk about. You could never actually observe anything, or rather there would never be anything you could fail to observe."

Alice paused for breath, quite carried away with this new thought, and noticed that everyone in the room was staring at her intently. When she stopped they all stirred a little uneasily.

"Of course you have a very important point there," said the Master kindly. "It is known as the *measurement problem* and is the very subject that we have been considering in the class here."

See end-of-chapter note 1

The Master continued: "It is important to remember that it is a real problem. There must exist a mixture of amplitudes such as we describe for systems of one or two electrons, as in the two slit interference experiment you saw, because there is interference between the amplitudes. This is not just a way of saying that the electron may be in one state, but that you do not happen to *know* what state that is. That situation could not give any interference, so we are forced to accept that, in some sense, each electron is in *all* the states. I believe that it is not a proper question when you ask what the electron is really doing because there is no way you can ever find out. If you try to check you will alter the system, so that you are examining something different.

"As you point out, there seems to be a problem here. Atoms, and systems containing a small number of particles, always do everything they possibly can, and they never make any decisions. We, on the other hand, always do one thing or the other and do not observe more than one outcome from any given situation. The students have each prepared a short talk about the measurement problem. They consider at what point, if any, the quantum behavior which allows all states to be present at the same time ceases to operate, so that unique observations may be made. You might like to sit and listen to their presentations." This seemed to Alice to be a good opportunity, so she sat at one of the desks and leaned forward expectantly.

"The first talk," announced the Master, his quiet voice managing to quell the expectant buzz of comment from the students, "is from the Emperor." The portly gentleman in the tasteful purple underwear, whom Alice had noticed when she first entered the classroom, got up and walked to the front of the class.
See end-of-chapter note 2

The Emperor's Theory (Mind over Matter)

"Our hypothesis," he began, with a haughty glance around the room, "is that it is all in the mind.

"The laws that are obeyed by quantum systems," he continued, "the description of physical states by amplitudes, and the superposition of these amplitudes when there is more than one possible condition—these laws apply to every material thing in the world. We say 'every material thing,'" he repeated, "as Our contention is that such a superposition is not experienced by the conscious mind. The physical world is governed at every stage by quantum behavior, and any purely material system, large or small, will always be in a combination of states, with an amplitude present for everything which might be or might have been. It is only

when the situation comes to the attention of the sovereign will of a conscious mind that a choice is made.

"For the mind is a thing outside, or in Our case above, the laws of the quantum world. We are not tied by the need to do everything that might possibly be done; instead We are free to make selections. When We observe something, then that thing is observed; it knows that We have observed it, the Universe knows that We have observed it, and it remains thereafter in the condition in which We have observed it. It is Our very act of observation which imposes a unique and definite form upon the world. We may not have the choice to select what We will see, but whatever We do observe has become uniquely real at that point."

He paused and looked commandingly around the room once more. Alice found herself strangely impressed by his authoritative delivery, despite his purple underwear. "For example, when We look at Our magnificent new imperial clothing We observe that We are of course exquisitely attired." He looked down at himself and abruptly he was clad from head to foot in rich garments. His coat and waistcoat were smothered in fine embroidery and he wore a flowing velvet robe trimmed with ermine. "Now it is conceivable that, when Our attention was diverted from Our garments they might have been less tangibly real than they are now seen to be, but if that had been so, now that We have observed them they are seen by all to be of the finest cut and that is in reality what they are."

The Emperor raised his head again and looked out at the class. Alice was intrigued to note that, although his observation of the clothes had fully established their rich aspect, as soon as he had looked away they gradually became hazy-looking once more and his tastefully mono-grammed underclothes began to show through.

"That then is Our thesis. The whole material world is indeed gov-erned by the laws of quantum mechanics, but the human mind is outside the material world and not so restricted. We have the ability to see things uniquely. We cannot choose what We will see, but what We do see becomes reality in the world, at least for the time We observe it. When We have finished Our observation, then of course the world can once again begin to enter its customary set of mixed states."

He stopped and looked around with a satisfied air. "Thank you for an interesting talk," said the Master, "that was very, very interesting. Does anyone have any questions?"

Alice discovered that she did. Perhaps the school atmosphere was affecting her after all. She put her hand up. "Yes," said the Master, point-ing to her, "what is the question that you would like to ask?"

"There is one thing that I do not understand," said Alice. (This was not strictly true, as there were many things that she did not understand, and the number was becoming larger at a most alarming rate, but there was one particular thing about which she wished to ask a question.) "You say the world is customarily in this strange mixture of different states, but it reduces to one unique condition when you, as a conscious mind, happen to look at it. I suppose that any person is able to make something become real in this way, so what happens about *other* peo-ple's minds?"

"We do not believe that We understand what you mean," replied the Emperor crushingly, but the Master cut in at this point.

"Perhaps I might enlarge on the young lady's question. We were talk-ing earlier about electrons passing through two slits. Suppose I were to take a photograph which would show an electron in the act of passing through either one slit or the other. If I follow what you say, you would maintain that, as the photograph might show that the electron was in either slit, it would have to show that it was in both. The photographic plate has no conscious mind and would be unable to reduce the wave function, so a superposition of two different images would be present on the film. Now suppose that I were to make a number of copies of this photograph, without of course looking at any of them. Would you say that each print would now also have a mixture of different images on it, each corresponding to the different slits which the electron might have passed through?"

"Yes," replied the Emperor cautiously. "We believe that that would be the case."

"If that is so and if the prints were all posted to different people, then the first one to open his envelope and look at the picture would cause one image from the mixture to become the real one and all the others would vanish?" Again the Emperor agreed cautiously. "But in that case, the photographs which the other people received would then each have to reduce to the same image, even though they might be in different cities miles apart. We know from experience that copies of a photograph do indeed show the same thing as the original, and if it was the occasion when the first person looked at a copy which caused one possibility to become uniquely real, presumably this act affected all the other copies, as they must subsequently agree with the first one. So one person who looked at a copy in one city would make all the other copies in other cities all over the world suddenly change to show the same thing. It would turn into a peculiar sort of race, with the first person to open the envelope fixing the images on all the other people's prints before they opened them. I think that was what the young lady meant," he finished.

"Naturally such a consideration would not present any problem in Our case," responded the Emperor, "since no one would presume to look at such a photograph before We had examined it first. However, We see that such a situation might arise among people of the lower orders, and in that case the situation would indeed be as you describe."

Alice was so startled at having this apparently ridiculous argument accepted that she did not notice the Emperor return to his seat and the little mermaid come up. The mermaid was unable to stand in front of the class, as she did not have any feet, so she sat on the Master's table, swinging her tail in front of her. Alice's attention returned to the proceedings as the mermaid began to speak.

The Little Mermaid's Theory (Many Worlds)

"As you know," she began in a liquid, musical voice, "I am a creature of two worlds. I live in the sea and am equally at home upon the land. But this is as nothing compared with the number of worlds which we all inhabit, for we are all citizens of many worlds—many, many worlds.

"The previous speaker told us that the quantum rules apply to the whole world, apart from the minds of the people who live in it. I tell you that they apply to the whole world, to everything. There is no limit to the

idea of the superposition of states. When an observer looks at a superposition of quantum states you would expect him or her to see all of the effects that are appropriate to the selection of states present. This is what does happen; one observer does see all the results, or rather the observer also is in a superposition of different states, and each state of the observer has seen the result that goes with one of the states, in the original mixture. Each state is simply extended to include the observer in the act of seeing that particular state.

"This is not the way that it *seems* to us, but that is because the different states of the observer are not aware of one another. When an electron passes through a screen with two slits in it, then it might pass through to the left or to the right. What you observe to happen is pure chance. You might see that the electron has gone to the left, but there will be another *you* that will have seen the electron go to the right. At the point at which you observe the electron, you split into two versions of yourself, one to see each possible result. If these two versions never get together again, then each remains totally unaware of the other's existence. The world has split into two worlds with slightly different versions of you in them. Of course, as these different versions of you will then talk to other people, you need different versions of them also, so what you have is a splitting of the entire universe. In this case it would split into two, but for a more complex observation it would split into a larger number of versions."

"But surely this would happen rather often," Alice could not help herself from saying, interrupting the flow of the mermaid's talk.

"It always happens," replied the mermaid calmly. "Whenever you have a situation where a measurement could give different results, then all the possible results *will* be observed, and the world will split into the appropriate number of versions.

"Mostly the split worlds would remain separate and would diverge without ever being aware of one another, but sometimes they will come together again at some point and give interference effects. It is the presence of these interference effects between the different states which shows that they can and do all exist together."

The mermaid stopped speaking and sat there combing the myriad strands of her long hair as they fell, side by side but separate, down over her shoulders.

"It must mean there are an awful lot of universes. There would have to be as many as there are grains of sand on all the beaches on Earth," Alice protested.

"Oh, there would be far more than that. Far more!" replied the mermaid dismissively. "Far, far more," she went on dreamily. "Far, far, far. . . ."

"That theory," cut in the Master, "has the advantage of being rather economical with assumptions, but it is very extravagant with universes!" He went on to ask for the next speaker. This was the Ugly Duckling, who had to stand on top of the Master's table so that he could be more clearly seen.

The Ugly Duckling's Theory (It Is All Too Complicated)

The Duckling began his speech, and Alice observed that, as well as being very ugly, he appeared to be very cross as well. His speech was so full of quacks and spluttering that she was hard put to make out what he said. As far as she could tell he was saying that the superposition of different states only worked for rather small systems, with just a few electrons

or atoms. He said that you need only argue that systems were often in mixtures of states because interference happened, since a single, unique, state would have nothing with which it could interfere.

He further argued that you do not actually know that interference does happen for objects which contain many particles. People know that interference and hence the superposition of states can occur for groups of a few particles, so they think the same must still be true for compli-cated things, like ducklings. He would be quacked if he believed that.

A duckling contains a lot of quacking atoms, he went on, and before any superimposed states can interfere, all the atoms in each separate state must combine exactly with the appropriate atom in the other states. There are so many atoms that this is not quacking likely. Any effects would average out, and you could not see any net result. So how, he asked, can you be so quacking sure that ducklings are ever in a superposition of states? Answer me that if you are so quacked clever. All this superposition of states is fine and quacking for a few particles at a time, but it stops well short of ducklings.

He went on to say that he quacking well knew when he saw something and when he quacking didn't. He knew that he was not in any quacking superposition of states, he was in only one, worse luck. So when he changed, he continued forcefully, he really changed from one definite state to another. The change was irreversible and there was no question of going back to combine with other states. Nothing was going to quacking interfere with him he concluded. At this point his quacking became so extreme that Alice could not follow him at all and was not really surprised when he became so angry that he fell off the table, out of her sight.

There was a pause and a moment of silence. This ended as a long graceful neck rose from behind the desk, followed by a snowy white feathered body. It was a swan.

"How beautiful!" exclaimed Alice. "May I stroke you?"

The swan hissed at her furiously and clapped his wings in a threatening manner. Alice decided that, though his change was certainly irreversible, it did not appear to have changed his temper very much.

At this point there was a disturbance at the back of the classroom, and Alice heard a voice shouting "Stop this charade, you are all wrong!" She looked across and saw a tall figure striding angrily down the space between the desks. It was the Classical Mechanic. His progress was considerable hampered by the fact that he was carrying a pinball machine, much as Alice had previously seen in cafes. (They might more often be found in bars, but of course Alice was too young to have seen them there.)

The Classical Mechanic's Theory (Wheels Within Wheels)

The Classsical Mechanic marched to the front of the room and set his machine down by the Master's table. It was labeled "Electron Interceptor" and had the form of a sloping table, with two slits at the top through which the particles would be fired and a row of pockets along the bottom which were alternately marked "Win" and "No Win." The surface of the table, though brightly painted, appeared strangely free of the usual selection of obstacles and flippers which Alice had previously seen on pinball machines.

"You are all deceiving yourselves," the Classical Mechanic announced firmly. "I have looked carefully at this device, which is basically a normal two-slit electron interference setup, and I believe I see what is really going on."

Alice could see that, apart from its garish decoration, it was indeed a smaller version of the experiment which she had been shown in the Mechanics' *gedanken* room. The Classical Mechanic quickly demonstrated its operation by firing a stream of electrons from the two slits. At least Alice presumed that they must have come through those slits as they were the only ones present, although she was not able to see clearly where the electrons actually were until their arrival registered along the bottom of the table. As she had by now come to expect, the electrons clustered in a series of heaps, with gaps between the heaps where very few were detected. Alice was intrigued to see that these gaps in the interference pattern corresponded closely with the pockets marked Win.

"You see that interference is produced and you would argue that this

shows the electrons have somehow each come through both slits, so that the combination of the amplitudes for the two slits is producing the interference pattern we see. I tell you now that the electrons are in fact each going through just one slit, in a perfectly sensible way. The interference is due to *hidden variables!*"

Alice found it very hard to follow exactly what happened at that point. The best she could say afterward was that the Classical Mechanic seemed to pull from the pinball table a dust cover which had not apparently been there before. However it had happened, she now saw that the surface of the table was covered with a pattern of deep ridges and grooves, leading away from the two slits. "Behold, hidden variables!" cried the Mechanic.

"They are not very well hidden," remarked Alice, looking critically at the complicated surface now revealed.

"My contention," began the Classical Mechanic, pointedly ignoring Alice's remark, "is that electrons and other particles behave in a perfectly rational and indeed classical fashion, very much like the particles to which I am accustomed in ClassicWorld. The only difference is that here, as well as the normal forces which act upon particles, they are also affected by a special *quantum force,* or *pilot wave.* This causes the strange effects which you interpret as due to interference. In my demonstration with the electron pinball here each electron really does enter by one slit or the other. It then moves over the table in a respectable and predictable fashion. Any randomness in the setup comes from the different directions and speeds which the electrons happen to have initially. When the electrons cross over the dips that you see here in the quantum potential, then the quantum force will deflect them, like a bicycle wheel caught in a trolley rail, so that most of the electrons end up in clumps. This gives your so-called interference effects."

"Well now," said the Master, "that is certainly a very interesting theory—very, very interesting indeed. However, if you do not mind my saying so, you seem to have removed those difficulties you had with the electron's behavior at the expense of some very peculiar behavior for your *quantum potential.*

"Because your quantum force has to produce the effects which we say are due to interference, it must be affected by things that happen in quite different places. If a third slit were to open in your table, then the quantum forces on the particles would change, even if none of the particles had gone through that hole. It must do this because the interference for three holes is different from that for two, and your force has to reproduce all those interference effects that we know to occur. Further your quantum potential, or network of quantum forces, must be very complicated indeed. In this theory you have nothing like the reduction of the wave functions which occurs in the normal quantum theory, so your potential must be affected by all the possibilities of everything which might have happened—ever. It is like the Many Worlds theory in that way. In your theory you say that what is observed will depend on how the particles happened to be traveling when they were affected by your pilot wave, but the pilot wave itself will retain information from all the possible things which might have happened and there is no way of removing it. Your wave would have to be incredibly complicated, like the sum of *all* the worlds in the Many Worlds theory, even though most of it may not affect any particles for most of the time.

"The pilot wave in your theory affects what the particles do, but the way that the individual particles actually move has no effect on the wave. This depends only on what particles *might* have done. There is no sym-

There are various "answers" to the measurement problem, but no one of them is universally accepted.

In practice, quantum mechanics is normally used to work out the amplitudes and hence the various probabilities for some physical system and then to use these to predict the behavior of large *ensembles* of simple atomic systems, without worrying too much about what would happen for a single system. The results for the ensembles can be compared with measurements, again without worrying too much about how the measurements could have been made.

The practical response to this problem is to "close your eyes and calculate." The interpretation of quantum mechanics may be difficult, but there is no denying that it works very well.

metry of action and reaction between the particles and the pilot wave. As a Classical Mechanic this must be a worry to you. You would not want to contradict Newton's Law that action and reaction are always equal, would you now?"

At this point the Quantum Mechanic, who had followed the Classical Mechanic into the room but remained quietly in the background, came forward and took his colleague by the arm. "Come along," he said. "You surely do not want to get involved in a charge of Classical Heresy by denouncing Newton's Laws. All this academic discussion of what electrons may or may not actually be doing is not for the likes of us. We are Mechanics. As a Mechanic, my main concern is that the Quantum laws do work and work well. When I calculate an amplitude for some process, this tells me what is likely to happen. It gives me the probabilities of different results and it does it accurately and reliably. It is not my job to worry about what the electrons are doing when I do not look at them, as long as I can tell what they are likely to be doing when I do look. That is what people pay me to do."

He led his subdued colleague quietly to one side, then turned to Alice and asked, "Have you learned as much as you want to know about observers and measurements?"

"Well," began Alice, "to tell you the truth I feel more confused than I was before I came here."

"Right," interrupted the Quantum Mechanic emphatically. "I thought as much. You have learned quite as much as you want to. Come along with me now and see some of the *results* of quantum theory. Let me show you some of the features of Quantumland."

Notes

1. The "measurement problem" is that the selection of one single possibility and the reduction of all the other amplitudes is quite unlike other quantum behavior, and it is not obvious how it can occur. The problem is stated most simply in the form: How can you ever actually measure *anything*? The conventional view of quantum mechanics is that, when there are several possibilities, there will be present an amplitude for each one, and the overall amplitude for the system is the sum, or superposition, of all of them together. For example, if there are several slits through which a particle might pass, then the overall amplitude for the system contains an amplitude for each slit, and you can have interference between the individual amplitudes. If the system is left to itself, then the amplitudes will change in a smooth and predictable way. When you make a measurement on a system which has a sum of amplitudes corresponding to different possible values of the quantity measured, then the theory says that you will, with some probability, observe one or another of these values. Immediately after the measurement the value is a known quantity (because you have just measured it), so the sum of eigenstates (see box on p. 76) reduces to a single one, the one for the actual value you have just measured.

2. The orthodox description of a measurement in quantum mechanics has the drawback that the process of making a measurement does not seem to be at all compatible with the rest of quantum theory. If the quantum theory is the true theory of atoms, as seems to be the case, and if the whole world is made of atoms, then presumably quantum theory should apply to the whole world and everything that is in it. That includes measuring instruments. Where a quantum system can give various values, its amplitude is a sum of states corresponding to each possible value. When the measuring device is itself a quantum system and there are various values which it *could* measure, it has no right to select just one of them. It ought to be in a state which is a sum of the amplitudes for all the possible results it *might* measure, and no unique observation could be made.

The conclusion you would draw from the above would seem to be either:

(a) We never actually observe anything

or

(b) Quantum theory is all nonsense.

Neither conclusion is really tenable (however tempting conclusion (b) might appear to be). We *know* perfectly well that we do observe things, but we cannot deny that quantum theory has an unbroken success rate at successfully describing all observations, while no alternative theory does as well. We cannot lightly abandon it.

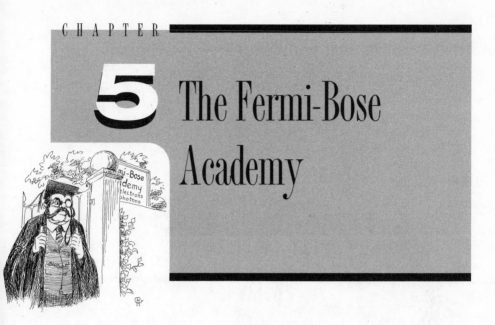

5 The Fermi-Bose Academy

Alice walked with the Quantum Mechanic along the path away from the school. As they traveled the path grew wider and gradually changed to a well-surfaced road.

"I think the most curious thing you have shown me," remarked Alice, "was the way that you got those interference effects even when there was only one electron present. Is it true then that it makes no difference whether there are many electrons or only one?"

"It is certainly true that you may observe interference whether you have many electrons or only one at a time. However you cannot say that it makes no difference. There are some effects which you only see when you have many electrons. Take the Pauli Principle, for example. . . ."

"Oh, I have heard of that," interrupted Alice. "I heard the electrons talk about it when I first came here. Would you tell me what it is, please?"

"It is a rule which applies when you have a lot of particles which are all the same—completely identical in every respect. If you would like to know more about it, it would be best if we were to call in here, since we happen to be passing, and they are very experienced in many-particle behavior."

Alice looked around at these words and found that, as they had been talking, they had come to a tall stone wall which ran along one side of the road. Immediately opposite them was a wide gateway. Impressive wrought-iron gates stood open between massive stone pillars with a

coat-of-arms painted in the center of each. To the right of the gateway, visible above the wall, Alice saw a wooden board which carried the message:

Fermi–Bose Academy
For electrons and photons

In the center of the gateway stood an imposing figure, a large and exceedingly well-built man made even more massive in appearance by the flowing academic gown and the mortarboard which he wore. His round, florid face was copiously adorned with a bushy mustache and side whiskers. Firmly fastened in one screwed-up eye he wore a monocle on a wide black ribbon.

"That is the Principal," whispered the Mechanic into Alice's nearest ear.

"Do you mean the Pauli principle?" asked Alice rather wildly. She had been taken off-guard by his sudden appearance.

"No, no," hissed the Mechanic, "he is the Principal of the Academy. Though of course Pauli's principle is the principal principle of the Academy, he is its Principal." Alice wished that she had not asked.

They crossed the road and went up to this imposing personage. "Excuse me sir," began the Mechanic. "Would you be so kind as to tell my young friend here something about many-particle systems?"

"Of course, of course," boomed the Principal. "We have no shortage of particles here, dear me no. I shall be most happy to show you around."

He turned around with a billow of his flowing gown and led the way toward the Academy. As they walked up the drive Alice saw small figures dodging in and out among the shrubbery. At one point a figure popped above a bush and made a face at them. At least Alice thought it had. As usual it was very difficult to make out any detail. "Ignore him," growled the Principal. "That is only Electron Minor."

They arrived at the door of the Academy, which was housed in a dignified old house of vaguely Tudor appearance. Without pausing the Principal led them through the main door into a vaulted entrance hall and up a wide carved staircase. As they walked through the building, Alice could see small figures hiding behind the banister, dodging in and out of rooms, and running off down side corridors as they approached. "Ignore him," remarked the Principal again. "It is just Electron Minor. Particles will be particles!"

"But it cannot be Electron Minor if we saw him on the drive," protested Alice. "Surely it cannot be the one particle in both places. Are we talking about something like the case when an electron managed to go through both holes in your double-slit experiment?" she asked the Quantum Mechanic.

"No, it is not that; they do have many electrons here. But don't you see, the electrons are all exactly the same. They are completely identical to one another. There is no way to tell them apart, so naturally they are *all* Electron Minor."

"That is right," confirmed the Principal emphatically as he led them into his study, "and it is a problem, let me tell you. You may know how difficult it can be for teachers when they have two identical twins in their school and are unable to tell them apart. Well I have hundreds of completely identical particles. It makes checking the register a nightmare, I can tell you.

"The electrons are not so bad," he went on. "We just count them and see whether we have the correct total. At least the *number* of electrons is conserved, so we know how many we ought to have, but for the photons even that does not work. The photons are bosons, so they are not conserved you see. We may begin a class with thirty, say, and have fifty or more at the end of it. Or the number may drop to less than twenty—it is hard to predict. This all makes it very difficult for the staff."

Alice had spotted a new word in that remark. "Do you think you might explain that?" she asked hopefully. "Would you please tell me what a 'boson' is?"

The Principal turned an even deeper red than he was before and spoke to the Mechanic. "I think it would be best if you took her to the beginners' Facts of Symmetry lesson, don't you? That should explain all about the Bosons and the Fermions."

"Right you are," replied the Mechanic. "Come along, Alice, I believe I can remember the way."

They walked down a corridor to a classroom and went in just as a lesson was beginning.

"Attention please," said the teacher. "Now as you well know, all you electrons are identical to one another and so are all you photons. This means that no one can tell when any two of you have changed places. As far as any observer could tell you *might* have changed places and so of course you will have to some degree. You all know that you have an associated wave function, or amplitude, and that this amplitude will be a superposition of all the things which you *might* be doing. Where there is no way of telling which things you are doing, then, as you know, you are doing all of them, or at any rate have an *amplitude* for every one of them. So you see, for any group of you it is impossible to tell when any two have changed places and this means that your overall wave function will be a superposition of all the amplitudes for which a different pair has swapped over. I hope that you have all made a note of that."
See end-of-chapter note 1

"Now the probability of making any observation is given by the *square* of your wave function, that is, the wave function multiplied by itself. As you are completely identical, it is obvious that when any two of you change places it can make no observable difference, so the square of your wave function cannot change. It might look as if there can be no change at all. Can anyone tell me what might change?"

One of the electrons put his hand up, or at least Alice assumed that was what had happened. She was not able to see at all clearly. "Please sir, the sign might change."

"Very good, that is an excellent answer. I would make a note in your record that you had answered so well, except that unfortunately I cannot tell you apart from the others. Yes, as you know your amplitudes do not have to be positive. They may be either positive or negative, so that two amplitudes may cancel one another out when you have interference.

This means that there are two cases in which the square of your amplitude would not be changed. It may be that the amplitude does not alter at all when two of you change places. In such a case the particles are bosons, like you photons. However, there is another possibility. When two of you exchange places, the amplitude may *reverse*. It changes between positive and negative. In this case the square is still positive and the probability distribution is unchanged, because multiplying the amplitude by itself will give *two* reversals, resulting in no change at all. This is what happens with fermions, such as you electrons. All particles fall into one or other of these two classes: They are either fermions or bosons.

"Now you may think that it does not matter much whether your amplitude reverses or not, especially as the probability distribution remains unchanged, but in fact it is very important indeed, particularly for fermions. The point is that if any two of you are in exactly the same state—that is, in the same place and doing the same thing—then if you exchange places, it is not only an unobservable change; it really is *no change at all*. In this case neither the probability distribution *nor* the amplitude can change. This is no problem for bosons, but for fermions, which always have to reverse their amplitude, such a situation is not allowed. For such particles you get the Pauli exclusion principle, which says that no two identical fermions may be doing exactly the same thing. They all have to be in different states."

See end-of-chapter note 2

"For bosons," as I said, "it is not a problem. Their amplitudes do not have to change at all when two of them change places, so they may be in the same state. In fact I can go further; not only may they be in the same state, but they positively *like* to be in the same state. Normally when you have a superposition of different states and square the amplitude to give the probability of observation, the individual states in the mixture are squared separately and contribute much the same to the overall probability. If you have two bosons in the same state, then when you square the two you get four. The two have contributed, not twice as much as one, but four times as much. If you had three particles in the same state they would contribute even more. The probability is much higher when there is a large number of bosons in one state, so they tend to get into the same state if at all possible. This is known as *Bose condensation.*

"So, there you have the difference between fermions and bosons. Fermions are individualistic, no two will ever do exactly the same thing, while bosons are very gregarious. They love to go around in gangs where each one behaves in exactly the same way as the others. As you will see later, it is this behavior and the interaction between you two types of particles which are responsible for the nature of the world. In many ways you are the rulers of the world."

At this point the Quantum Mechanic led Alice out of the classroom. "There you are then," he said. "That is the Pauli principle. It rules that no two fermions of the same type can ever be doing the same thing, so you can have one and one only in each state. The principle applies to all fermions of whatever type, but not to bosons. This means, among other things, that the number of fermions must be conserved. Fermions cannot just appear and disappear in a casual fashion."

"I should think not!" Alice said. "That would be ridiculous."

"I do not think you can say that, you know, because bosons do appear and disappear. Their number is not conserved at all. You can argue that the number of fermions must be definite if there is one and only one in each state, since a particular number of occupied states implies that there is that particular number of fermions to occupy them. The argument does not hold for bosons, since you can have as many as you like in any state. In practice the number of bosons is not at all constant.

"If you just look out this window here," he said suddenly as they were passing, "you can see the difference between fermions and bosons quite well."

Alice gazed through the window and saw that a group of electrons and photons were being drilled on the Academy field. The photons were doing very well, wheeling and reversing in perfect synchronism with no differences between any of them. The group of electrons, however, were behaving in a manner which was obviously driving the drill sergeant to despair. Some were marching forward, but at different speeds. Some were marching to the right and to the left, or even backward. A few were jumping up and down or doing headstands and one was lying flat on his back, staring at the sky.

"He is in the ground state," said the Mechanic, looking over Alice's shoulder. "I expect the other electrons wish that they could join him there, but only one of them is allowed you see. Unless the other had an opposite direction of spin, of course—that would make a sufficient difference between them.

"You can clearly see the difference between the fermions and bosons here. The photons are bosons, so it is easy for them to do the same thing. Indeed, they positively like to be the same as one another, so they are very good at marching in step. The electrons, on the other hand, are fermions and so the Pauli exclusion principle stops any two of them from being in the same state. They have to behave differently from one another."

"You often talk about the electrons being in states," Alice remarked. "Would you please explain to me just what is a state?"

"Once again," responded the Mechanic, "the best way will be for you to sit in on one of the classes here. The Academy teaches world leaders, since it is the interaction of electrons and photons that rules the physical world, by and large. If they are to be world rulers, they have to go to Statecraft classes naturally. Come along and let us see one."

He led Alice down to a large low building at the back of the Academy. When they went inside Alice could see that it was some sort of workshop. A number of electrons were working away at different benches. Alice went over to watch one group, who were busily erecting a set of fences around the edge of the bench. Alice could see there were various structures on the bench, and as the students moved the fences around, these structures all changed.

"What are they doing?" Alice asked her companion.

"They are setting up the boundary conditions for the states. States are controlled largely by the constraints which hedge them in. In general, what you can do is governed by what you cannot do and the restrictions serve to define the possible states. It is very much like the notes you can get from an organ pipe. For a pipe of a given length you can produce only a limited number of notes. If you change the length of the organ pipe, then you will change the notes. Quantum states are given by the amplitude or wave function which the system can have, and this is much like the sound wave in an organ pipe.

"As you have already discovered, you usually cannot say what an electron is really doing, because if you observe it, to check you will select out one particular amplitude and reduce the amplitudes to that one alone. The only time when you can be really certain about your electron is when it has a single amplitude instead of a superposition and when your observation can give but one value. In that case the probability of your seeing that value from your measurement is 100 percent and for any other result the probability is zero—it won't happen. When you make the observation, then you will see the expected result. In such a case, the reduction of the amplitude to that for your observed result has made no difference at all, as you were already in such a state. The state is not changed by the observation, and it is called a *stationary state*. In this class electrons are setting up stationary states."

Alice walked around the table, looking at the states which the electrons were crafting. They looked to her like a series of boxes, eight in all. There was one very large one, one slightly smaller than the large one, and six tiny ones of much the same size. She turned a corner of the table and was surprised to see that the states had changed completely. Now they

A *state* describes the condition of a physical system. It is the basic concept in quantum theory—the best description of the real world which can be given. In general the amplitude for a state gives the probability for various possible results of any observation. For some states there may be only one possible outcome to a particular measurement. When a system is in one of these so-called stationary states any measurement of that quantity will give one and only one possible result. Repeated measurements will give the same result every time. Hence the name *stationary state*, or the frequently used German equivalent *eigenstate*.

had the appearance of a number of stands, rather like cake stands, on tall pedestals. There were two which were much wider than the others; four of the same widths, but with successively taller pedestals; and two small ones. She walked quickly around another corner of the table. Now she saw that the center of the table was occupied with a large board to which were fastened a number of coat hooks. There were two rows of three and isolated single hooks top and bottom. "Goodness, whatever is happening?" she asked her companion. "I keep seeing the states quite differently when I look at them from different directions."

"Well, of course you do," replied the Quantum Mechanic. "You are seeing different *representations* of the states. The nature of a state depends on how you observe it. The very existence of a stationary state relies upon some observation for which it always produces a definite result, but a state cannot give definite results for *all* observations you can make. For example, the Heisenberg relations prevent you from seeing the position and the momentum of an electron at the same time, so a stationary state for one observation will not be a stationary state for the other. The observations which you use to describe the states are called its representation.

"The nature of a state may be very different, depending on how you observe it. Indeed the very identity of the different states can change. The states that you see in one representation may not be the same as the ones in another representation. As you may have noticed just now, the one thing which must remain constant is the *number* of the states. If you can

> **T**here are certain quantities which cannot share the same stationary state; position and momentum are two examples. If you have an eigenstate which gives a definite value for the position of a particle, then a measurement of its momentum could give any result. This leads to the Heisenberg uncertainty relations. If you have a mixture of states which correspond to different values for the position, then a measurement of position might give any one of the appropriate values. The position has become "uncertain," though now the spread of momentum values may be reduced.
>
> This spread is not caused by poor measurement technique; it is inherent in the physical state. The indefinite value of some physical quantity which may be inherent in a given state will allow such effects as barrier penetration, the exchange of heavy particles in nuclei, photons in electrical interactions, and indeed the existence of virtual particles in general. Virtual particles and particle exchange will be discussed in Chapters 6 and 8.

put one of the electrons in each state then you must always have the same number of states to contain them all, even though the individual states may have changed."

"That seems very vague to me," complained Alice. "It sounds as if you cannot be at all sure what is really there."

"Right!" replied the Mechanic gaily. "Hadn't you noticed? We can talk quite confidently about *observations,* but what is really there to be observed, now that is quite another matter.

"Come along, though. It is time for the evening assembly of the Academy. You should find that quite interesting."

The Mechanic led her back into the main building and ushered her through the entrance hall into a huge room with a high vaulted roof. The great tiled floor was completely filled with a crowd of electrons, packed in as tightly as possible. Overhead, a wide ornate balcony ran around the edge of the vast hall, and on it Alice could see the vague distant figures of a few electrons hurrying to an exit. There was just one tiny space on the floor near the doorway by which they had entered, and an electron which had been following them darted in to it and immediately came to a halt,

wedged in on every side by the dense crowd so that there was no room to move any farther.
See end-of-chapter note 3

"Why is it so crowded here?" cried Alice, overcome by the scale of the scene before her. "This is the valence level," answered one of the helpful electrons. "All the spaces on the valence level are full because the valence level is always full of electrons. None of us can move at all, as there are no states free to move into, you see."

"That is terrible," cried Alice. "How can any of you possibly move across the floor to get out if it is so crowded?" "We can't," said the electron with cheerful resignation. "But you can if you want to. You can go anywhere on the floor because there are no other Alices here, so there are plenty of Alice states free for you to move into. You will have no Pauli Exclusion problems at all." This still sounded very strange to Alice, but she tried to push her way into the tightly packed crowd and found, just as she had when she had tried to get into the full railway compartment earlier, that somehow she could move through without any trouble.

Alice made her way through the crowd of electrons toward a raised platform at the far end of the hall. On it stood the Principal, impressive as always in his gown and mortarboard. As she came closer Alice could hear his mellow voice booming out over the packed room.

"I know that you have all had a busy day today, but I trust that I do not need to remind you what an important role you must be prepared to play in the world. You electrons, each taking your place in your proper state, form the very fabric of everything we know. Some of you will be bound in atoms and will have to work away in your various levels, controlling all the details of chemical processes. Some of you may find your place within a crystalline solid. There you will be relatively free of attachment to any particular atom and may move around as far as the Pauli Principle and your fellow electrons allow. You may be in a conduction band, where you can move freely, and it will then be your task to rush around carrying your electric charges as part of an electric current. On the other hand, you may be in a valence band within a solid. Perhaps you will feel trapped there as there will be no states free for you to enter. Do not become discouraged. Not every electron may be in the highest energy states, and remember that the lowest levels must also be filled."
See end-of-chapter note 4

"As for you photons, you are the movers and shakers. Left to themselves the electrons would stay complacently in their proper states, and nothing would ever be done. It is your task to interact with the electrons at all times and to produce the transitions between states, the changes which make things happen."

At this point in the Principal's address, Alice became aware of the bright shapes of photons rushing though the crowd of electrons and of occasional flashes from different parts of the room. She turned around to see what was happening. It was difficult for her to see very far, because she was closely surrounded by so many electrons.

"This is really too bad!" Alice could not help exclaiming as she looked at all the captive figures, held fixed in position by the crush around them. "Is there no way in which anyone can move at all?"

"Only if we should get excited to the higher level," a voice answered. Alice could not see who had spoken. "But it doesn't really matter," she thought to herself. "Since they are all the same, then the same one as always must have spoken, I suppose." Just then there was a flash nearby and Alice saw that a photon had come rushing through the crowd and crashed into an electron. The electron soared upward and landed on the balcony, where he began running furiously toward the exit.

Alice was staring so hard at the retreating electron that she did not observe another photon rushing in her direction. There was a brilliant flash and she felt herself rising in the air. When she looked around she saw that she was now standing on the balcony also, looking down on the mass of electrons below. "This must be what the electron meant by being excited to the higher level. It doesn't seem all that exciting to me, but at least there is a lot more room here." She looked over the edge of the bal-

cony at the floor beneath and could see occasional little flashes here and there, each one followed by an electron floating up from the floor and landing on the balcony, where he or she immediately began to run at high speed toward the exit. One of them landed on the balcony close to where Alice was standing.

She looked down and could see a little electron-shaped hole where that electron had been a moment before. It was clearly visible, as the contrasting color of the tiled floor stood out sharply against the uniform background of closely packed electrons which covered the surface everywhere else. As she watched this space another electron nearby stepped smartly onto the gap which had just been created, although it could then move no further. Where this electron had been standing, however, there was now a hole and a more recently arrived electron stepped into that. "What a curious thing!" Alice said to herself. "I have become used to seeing electrons, but I did not expect to be able to see the presence of *no electron* quite so clearly!" She watched with interest as the movement along the balcony of the electron which had risen up to make the original hole was balanced by the movement of the electron-shaped hole as it progressed steadily across the floor in the other direction, toward the wide door by which she had originally come in.
See end-of-chapter note 5

When both electron and hole were out of sight, Alice herself walked along the balcony to the exit. She felt she had heard quite enough of the Principal's talk. She passed through the small door and found herself in a long corridor. Waiting for her by the door was the Quantum Mechanic. "How did you enjoy that?" he asked.

"Very well, thank you," replied Alice politely. She felt that it was expected of her. "It was most interesting to hear the Principal conducting the assembly."

"You say that," began the Mechanic, "but of course it was really the electrons which were doing the conducting, once they had been excited up to the conduction level. All electrons have an electric charge you know, so when they move around they cause an electric current to flow. The charge they carry happens to be negative, so the current flows in the opposite direction to the movement of the electrons, but that is a minor point. If all the states which any electron might reach are already full of electrons, as in the valence level, then there can be no movement and you have an electrical insulator. All the electrons and their charges are fixed in position in that case so there can be no electric current. In the present case you can get a current only when electrons have been carried up to the empty conduction level where they have plenty of room and can move easily. In that case you can get a current produced both by the electrons and by the holes they leave behind."

"But how can a *hole* give a current?" protested Alice. "A hole is something which isn't even there."

"First, you will agree that when the electrons are all present in the lower valence level, they cannot move and there is no current?" asked the Mechanic. "The current is just the same as if there were no negatively charged electrons present."

"Well, yes," answered Alice. That sounded fair enough.

"Then you must admit that when there is one electron *less,* the current will look like that due to one less than *no* electrons. The hole in the valence level behaves as if it were a positive charge. You saw how the movement of the hole toward the door was actually due to a lot of electrons taking one step in the opposite direction. So the electric current produced by negatively charged electrons moving away from the door is the same as a positive charge moving toward the door would give. As I said, the photons produce a current both from the electrons they put into the conduction band and from the holes they leave behind."

"The photons seem to be rather a bother to the electrons," remarked Alice, deciding to change the subject.

"Well, they are certainly rather hyperactive, but then photons are naturally very bright. As the Principal says, particles will be particles. I expect that at the moment some of them are lasing electrons in the dorm."

"I am sorry," queried Alice, "but don't you mean *hazing*? I am sure that is the word that I have heard used to describe student pranks."

"No, it is definitely *lasing.* Come and see."

They walked on down the corridor to a door at the end. The Mechanic opened this door and they entered, closing the door behind them. They were now in a long room which was lined along both sides with bunk beds. Alice could see that many of the top bunks were occupied by electrons, but the lower bunks were for the most part empty. "You sometimes find them in the top bunks rather than the lower ones," remarked the Mechanic. "It is called *population inversion*. It is only when they are like that that lasing becomes practical."

It was not very long before a lone photon came running into the room. He rushed to one of the bunks and careened into the electron which occupied that elevated position. With a thump the electron plummeted down to the lower bunk, and Alice was startled to see that there were now *two* photons rushing together around the room. They moved in perfect unison so that they almost seemed as one. "That is an example of *stimulated emission,*" the Mechanic murmured in Alice's ear. "The photon has caused the electron to make a transition to a lower level, and the

energy released has created another photon. Now just watch and see how the lasing develops."

The two photons rushed up and down the long room. One collided with an electron, and then there were three photons and another electron in a lower level. As Alice watched, the photons interacted with more electrons, producing more photons. Occasionally she noticed a photon collide with an electron which had fallen to a lower bunk. When this happened the electron shot up to the higher bunk and the photon vanished, but as there were initially very few electrons in the lower bunks this did not happen often to begin with.

See end-of-chapter note 6

Soon the room was crowded with a horde of identical photons, all rushing to-and-fro in perfect synchronism. There were now almost as many electrons in the lower bunks as in the upper ones, so that collisions were as likely to excite an electron to a higher position, with the loss of one of the photons, as to create a new one. The stream of photons poured out through the door at the end of the dormitory and down the corridor as a tight coherent beam of light. Before they had gone halfway down the corridor they collided with the massive form of the Principal who was walking toward them.

He immediately stopped, drew himself up to his full height, and spread his thick black gown to either side, so that he presented a dense black body, effectively blocking the corridor. The photons struck the inky black material and vanished completely. The Principal stood there for a moment, looking both hot and bothered and mopping the perspiration from his ruddy face with a handkerchief.

"I will not tolerate this sort of behavior," he puffed. "I have warned them before that any photons who carry on in this way will be instantly absorbed. It is hot work, though, since the energy released has to go somewhere, and it usually ends up as heat."

"Excuse me," said Alice. "Could you tell me where all those photons have gone?"

"Why, they have not gone anywhere, my dear. They have been absorbed. They are no more."

"Oh dear, how tragic," cried Alice, who felt sorry for the poor little photons who had been so abruptly snuffed out.

"Not at all, not at all. It is all part of being a nonconserved particle. Photons are like that. Easy come, easy go. They are always being created and destroyed. It is nothing very serious."

"I am sure that it must be for the photon," cried Alice.

"Well, I am not even so sure about that. I do not think it makes much difference to a photon how long it seems to *us* that it exists. They travel at the speed of light, you see, as after all they *are* light. For anything traveling at that speed, time will actually stand still. So, however long they seem to us to survive, for them no time at all will pass. The entire history of the universe would pass in a flash for a photon. I suppose that is why they never seem to get bored.

"As I said in the assembly, photons have many important parts to play in exciting electrons from one state to another and indeed in creating the interactions which make the states in the first instance. In order to do this, it is necessary that they be created and destroyed very frequently; it is part of the job, you might say. Creating interactions is more the task of virtual photons, though. We do not deal much with them here. If you are interested in states and how one goes about moving from one to another, then you should visit the State Agent. Your friend there will show you the way."

The Principal escorted them out of the Academy and back down the drive to the gate. As they walked on down the road, Alice turned back once to wave to the Principal, who was standing solidly in the center of the gateway where she had first seen him.

Notes

1. If you have many particles you will have some sort of amplitude for each of them and an overall amplitude which will describe the whole system of particles. If the particles are all different from one another then you know (or *can* know) the state each is in. The overall amplitude is just the product of the amplitudes for each particle separately.

    If the particles are identical to one another, then things get more complicated. Electrons (or photons) are completely identical. There is *no way* to distinguish one from another. When you have seen one, you have seen them all. If two electrons were interchanged between the states they occupied, there is no way that you would ever be able to tell. The total amplitude is, as usual, a mixture of all the indistinguishable amplitudes, which now includes all those permutations in which particles have been interchanged between two states.

    Interchanging two identical particles makes no difference to what you can observe, which means it makes no difference to the probability distribution that you get when you multiply the amplitude by itself. This could mean that the amplitude itself does not change either, or it *could* mean that the amplitude changes sign, for example, going from positive to negative. This is equivalent to multiplying the amplitude by -1. When you multiply the amplitude by itself to get the probability amplitude, then this factor -1 is also multiplied by itself to give a factor of $+1$, which produces no change in the probability. The change in sign sounds like a trivial academic point, but it has amazing consequences.

2.   There is no obvious reason why an amplitude *should* change sign just because it cannot be shown that it may not, but Nature seems to follow the rule that anything not forbidden is compulsory and to take up all her options. There *are* particles for which the amplitude does change sign when two of them are interchanged. They are called *fermions,* and electrons provide an example. There are also particles for which the amplitude does not change in any way when two are interchanged. These are called *bosons,* and photons are of this type.

Does it really matter whether the sign of the amplitude for a system of particles does or does not change sign when two of them are interchanged between states? Surprisingly, it does. It matters a great deal.

You cannot have two fermions in the same state. If two bosons were in the same state and you happened to interchange them, then it really would make no difference at all—it could not give even a change of sign. Such amplitudes are not allowed for fermions. This is an example of the Pauli principle, which says that no two fermions may ever be in the same state. Fermions are the ultimate individualists; no two may conform completely.

The Pauli principle is extremely important and is vital for the existence of atoms and of matter as we know it. Bosons are not governed by the Pauli principle—quite the reverse, in fact.

If each particle is in a different state and you square the overall amplitude to calculate the probability distribution for the particles, then each particle separately contributes much the same amount to the total probability. If you have two particles in the same state and square that, you get four times the contribution from only two particles. Each has contributed proportionately more, so that having two particles in the same state is *more* probable than having each in a different state. Having three or four particles in the same state is even more probable, and so on. This increased probability for having many bosons in the same state gives the phenomenon of boson condensation: Bosons *like* to get together in the same state. Bosons are easily led; they are inherently gregarious.

Boson condensation is seen, for example, in the operation of a laser.

3.   Electrical forces involving electrons can operate to hold atoms together, as discussed in Chapter 7, but they do not give rise to any repulsion which would push the atoms apart; so why do atoms keep a fairly uniform distance from one another? Why are solids incompressible? Why are the atoms not pulled into one another, so that a

block of lead would end up as one very heavy object of atomic size? Once again it is a consequence of the Pauli principle, which says that two electrons cannot be in the same state.

Since the atoms of a given type are all the same, each has the same set of states. Does this not put the equivalent electrons in each atom into the same state, which is not allowed? Actually, as the atoms are in different positions, the states are slightly different. If you were to superimpose the atoms, then the states *would* be the same, and the Pauli principle forbids this. The atoms are kept apart by what is known as Fermi pressure, but which is really the intense refusal of the electrons in one atom to be the same as their neighbor. Matter is incompressible because of the extreme individualism of electrons.

4. In a solid the electron states from the individual atoms have combined together to form a large number of electron states which belong to the solid as a whole. These states are grouped into energy bands, within which the energy levels of the states are so close together as to be almost continuous. Corresponding to the larger energy-level separations in the individual atoms, there are gaps in the energy bands of the solid. The lower bands are full of electrons which have come from the lower levels in the atoms. The highest of these full bands is called the valence band, and above it, separated by a band gap which contains no states, is another band: the conduction band. This band is either completely empty or, at most, only partly full.

    In the valence band the electrons cannot move. Any electron movement requires that electrons should change from one state to another, and there are no empty states for the electrons to go into. If an electric potential was put across a material, it would apply a force to the electrons in the valence band, but they could not move. If there were no electrons in the conduction band, the material would act as an electrical insulator.

5. If an electron in the full valence band is given enough energy, either by collision with a photon or even by a chance concentration of thermal energy, then the electron may rise across the band gap into the higher conduction band. As there are plenty of empty states in this band the electron can now move around, and an electric potential will produce conduction. Further, there is now a free space in the valence level, where the electron used to be. Another electron may move into this hole, and so on. There will be a hole in the otherwise full valence band, and it will be moving in the opposite direction to

the electron movement. This hole behaves very much like a particle with positive charge.

The above describes the behavior of semiconductor materials: materials such as silicon, which is widely used in electronics. Electric current is carried both by electrons in the conduction level and by *holes* in the valence level.

6. When a photon which has the correct energy interacts with an electron in an atom, it may produce a transition from one energy level to another, as described in Chapter 6. In most cases the transition will be from a lower to a higher energy level, since usually the lower levels will all be full. The photon is equally capable of producing a transition from a higher to a lower level, if the lower level is empty.

Should a substance happen to have a lot of electrons in a higher level, and a lower level is mostly empty (a condition known as *population inversion*), then a photon can cause an electron to transfer from a higher state to a lower one. This change releases energy and creates a new photon, in addition to the one which caused the transfer. This photon can in turn induce more electrons to fall to a lower state.

In a laser the light produced is reflected back-and-forth from mirrors at either end of the cavity, causing more photon emissions each time it passes repeatedly through the material. A little of this light escapes through the mirrors, which are not perfect reflectors, and gives an intense narrow beam: laser light. As the photons were emitted in direct response to the photons already present, the light is all "in step," or *in phase*, and has unique properties for producing interference effects on a large scale, as may be seen in holograms. (Not all holograms need laser light, but it helps.)

CHAPTER

6 Virtual Reality

The Quantum Mechanic led Alice down the road and through a wrought iron gate into an attractive park. Beautiful flower beds, full with an abundance of assorted blooms, lined both sides of the path, giving a most pleasing effect as they strolled along in the warm summer weather. In the sky the sun shone brilliantly, pouring out its light upon the idyllic scene. Beside the path colorful butterflies flitted from bright blossom to bright blossom and a small stream burbled downhill over a bed of rounded pebbles, while here and there along its path the water poured over a miniature waterfall. Alice thought it all very pretty and was looking around her in delight when she saw another figure approaching on a converging path.

The newcomer was plainly another little girl, but there was something very peculiar about her. She looked somewhat like Alice herself, but she resembled rather more the figure that Alice had occasionally seen in the negatives of her snapshots. Alice was reminded of the anti-electrons which she had watched in the Bank. She noted to her surprise that, although the girl was coming toward her, she was facing in the opposite direction and walking along backward.

Alice was so absorbed in the remarkable appearance of this strange girl that she did not consider how quickly they were approaching one another. Before she had fully realized what was happening they had collided. There was a blinding flash which quite took her senses away.

When she recovered from this she found herself walking alone down the path along which the other girl had come. Looking back she could see that the reverse-girl was walking away, still backward, along Alice's original path. Now, however, she was accompanied by another negative figure which was companionably walking along backward beside her. This second figure resembled her previous companion, the Quantum Mechanic.

When she looked around her, Alice was startled to discover that her surroundings had altered every bit as dramatically. Everything seemed to be reversed. In the sky the sun glowered darkly, draining the light from the scene below. Beside the path, dull butterflies flitted backward from dark blossom to dark blossom and a small stream burbled uphill over a bed of round pebbles, while here and there the water soared to the top of a miniature waterfall. Alice had never experienced anything like this before.

She was so fascinated by this remarkable scene that she did not observe that once again a small girl was bearing rapidly down upon her in reverse. Alice looked around just as they collided, with yet another blinding flash. When she recovered from her shock, she saw the girl was backing away along the path by which Alice had just come. She noticed furthermore that the scenery had now returned to normal. "Curiouser and curiouser," said Alice to herself. "The first collision somehow managed to make the whole countryside reverse itself, while the second one has put it back to normal. I'm sure I do not see how that could possibly happen. How can my colliding with that girl, however violently, affect the stream and the sun? It does not make any sense at all." Alice continued to debate with herself about the meaning of her recent experience. It had been so very remarkable that she paid hardly any attention when she heard a sharp detonation to one side of her, while shortly after an extremely energetic photon rushed across the path.

Alice had not reached any satisfactory explanation of her recent experience when the path led her out of the park and onto a wide, level plain. This seemed devoid of any feature, apart from a large, utilitarian building which stood facing her a little way ahead.

When she got close she could see that the building had a name board mounted centrally on the front, a little above the level of Alice's head. At one end this sign bore the words "State Agent." at the other "Virtual Realtor." In the center of the vast blank frontage were a door and a small window, which was full of notices.

Genuine Amplitude Reductions for Quick Sale
Fine Periodic Features.
States situated in desirable Energy Band.
Attractively priced for early Transition.

As Alice could see no one outside, she opened the door and went in. Immediately inside the door was a short counter and behind this a huge

room, almost empty apart from what seemed to be tiers of shelves rising up into the shadows in the distance. In the center of the room, a single figure was visible sitting at a desk and talking into a telephone. When he saw Alice he rose and hurried over.

He rested his hands on the counter and smiled widely in a toothy and rather insincere manner. "Come in, come in," he said, ignoring the fact that Alice had already come in. "What may I have the pleasure of showing you? Perhaps you are planning to move into your very first state? I am confident that we will be able to give you every satisfaction."

"To tell you the truth," began Alice, not that she had been at all tempted to lie about it, "I am not actually looking for anything. I was told that you would be able to tell me something about how electrons and other particles move between states."

"Well, you have certainly come to the right place. We have long been established in the particle transition business. If you would care to come with me to one of our locations I will endeavor to clarify the situation to your complete satisfaction."

Alice understood this to mean that he would explain things, so she came around the counter and followed him to one of the sets of shelves, or whatever they were. Either they were a long way away and very large, or perhaps she and the State Agent shrank as they came closer, but however it happened Alice found as she drew near that they now looked much more like a tall block of apartment buildings. These bore a sign which read:

> Periodic Mansions

They were very open at the front, and she could see electrons moving around on each level.

"There you have a good example of quality states built on well-spaced energy levels. Each one is occupied by the permitted number of electrons, up to the highest occupied level. Above that there are many vacant states, but there is currently no room for any more electrons on the lower levels. When an electron is a sitting tenant in a state, there is of course no room for another electron. Usually if he is left to himself an electron has no inclination to move from a state once he has settled into it. However, if we wait for a little we may be lucky and see some forced movement."

Alice stood and looked at the edifice, and after a short wait she saw a photon rush into the front. There was a commotion, and one of the electrons in the lowest level soared up and out of sight. Alice looked around to see where the photon had come from. Parked nearby there was a small truck with, painted on its side, the slogan:

> **Photon Removals**
> We make Light work of
> Transitions.

"I am in luck," cried the State Agent joyfully. "A photon has given its energy to an electron in the lowest level and excited it right up to one of the vacant levels at the top. It is not so often that we get a removal from the ground state. That leaves a very attractive vacancy. I must see to it at once."

He rushed off and soon came back carrying a notice board on a post, which he planted in the ground. The notice read:

> **Vacant Possession!**
> Attractive State on Ground
> Level.

Hardly had he got the board in position than one of the electrons in the second level gave a short cry and toppled down to the empty state. Once there he settled in and carried on as if nothing untoward had happened. As he fell Alice saw a photon rush out. Since the electron had not fallen very far, the energy carried by this photon was much less than the energy carried by the photon which had released the original electron.

The State Agent sighed, picked up a paintbrush from a pot he had brought out when he fetched the sign, and proceeded to cross out the word "Ground" and write "Second" in its place. The paint was hardly dry when Alice heard another sharp cry. An electron in the third level had fallen into the empty place in the second. The State Agent cursed and altered his board again to read "Third." He slammed the brush back into the paint pot and glared at the building.

There was another sharp cry. An electron from still higher up had fallen into the third level. The State Agent tore his notice from its post, flung it on the ground and stamped on it.
See end-of-chapter note 1

"Excuse me," said Alice, rather hesitant at interrupting this display of passion. "I had thought you said that the electrons would stay in their

states indefinitely if they were left alone, but those ones appear to have fallen down quite spontaneously."

"So it may seem," replied the Agent, quite glad to be distracted from his momentary fit of temper. "Actually all of those electron transitions have in fact been stimulated by photons, but you did not notice them because they were *virtual* photons. Virtual photons play a very important part in all electron interactions. They not only give these apparently spontaneous transitions between states, but they also help to create the states themselves in the first instance. So you see, the very particles which keep an electron in its steady state are also the ones which force the electron to leave it.

"Before I tell you about virtual particles we ought to look at normal particles, the ones which are *not* virtual. They are commonly known as real particles. The distinctive thing about them is that there is a very strict relation between their particular masses and the energy and momentum that they can have. That is what the notice there is about."

The agent pointed to a small sticker, printed on fluorescent green paper, which was attached to the front of the building. It said:

> Real particles do it on the Mass Shell.

Electrons may be excited by photons to make transitions in either direction, giving stimulated absorption and stimulated emission. Electrons which have been excited into a state of higher energy eventually decay back to a lower state if one is available, even if there are apparently no photons present. This is called *spontaneous decay*. Quantum mechanics maintains that all transitions are driven by something; they do not just happen. The apparently spontaneous decays are in fact caused by photons, but not by real photons. They are stimulated by virtual photons: quantum fluctuations in the vacuum.

Around any electric charge there is a cloud of virtual photons whose interaction with other charged particles produces an electric field. As they constitute the electric field, these virtual photons are always present in an atom and can produce the apparently spontaneous decays of electron states.

"They are certainly very fond of notices here," Alice thought to herself. "That one does sound rather suggestive, though I must admit that I do not know what it means."

"The mass shell," continued the agent, as if answering her thoughts, "is the region where energy and momentum are related in the strict way required for real particles. It is the straight and narrow path followed by conventional, hidebound particles.

"If you want to be a force in the community and to push things around, then you have to be able to transfer momentum. If you want something to move from where it is, or if you want to keep something from moving away, then you must transfer momentum. In each case you are concerned with moving, and moving means momentum. Whether you want to start a movement or to stop one makes little difference. It is changes in momentum which deflect objects from their paths and make things change, and it is control of momentum which makes particles take a particular path, for that matter.

"On the mass shell, you cannot have momentum without supplying the appropriate kinetic energy that befits your mass. A really massive particle, one with a lot of energy already invested in its rest mass, does not need so much extra kinetic energy to provide it with a given amount of momentum as would a lighter particle. All real particles must have the appropriate amount of energy if they are to have momentum. This is true even for photons, which do not have any rest mass at all."

The Agent reached into his pocket and took out a number of legal-looking documents. "The conditions are quite precise. Provided that real particles obey them, they are free, free of any energy debt. They can move around as they wish, as far as they wish. They are quite free to come and go. You may have seen the rule: `What is not forbidden is compulsory,'" he remarked.

"Yes, I did," replied Alice, anxious to air her knowledge. "I saw that in the Heisenberg Bank, and the manager told me about momentum and. . . ."

"There is another rule," continued the agent enthusiastically, without actually stopping to listen to Alice's reply. "It says `What *is* forbidden had better be done pretty quickly.' This is the rule followed by the virtual particles. These are not usually discussed much in polite, classical society, but they have a very important part to play in the world. Virtual particles behave in ways that classical laws say are simply not allowed."

"How can that be?" asked Alice, a little naively. "Surely if something is not allowed then no particle will be able to do it."

The agent heard her then and answered her question. "It is the quantum fluctuations which permit it," he said. "If you have been to the Bank, you will remember that particles may have a loan of energy for a short time. The larger the amount of energy the shorter the time of course. You n quantum theory the concept of a particle is not so sharp as in classical physics. Particles carry and deliver energy in a quantized form, in discrete packets. In many cases they have definite masses which clearly distinguish them from other particles and can carry specific amounts of other quantities, such as electric charge. Photons have zero rest mass (which also is a definite value). Real particles, those with a long-term existence, have strict relations between the values of mass, energy, and momentum. Where particles may be created and destroyed and have but a fleeting existence, then they do not obey such strict rules and the quantum fluctuations in their energy may be large. This is particularly true for those particles which are exchanged to provide an interaction between other particles. The entire energy of such particles is a quantum fluctuation. They are created literally from nothing. The vacuum is not completely empty, but is a seething mass of these short-lived particles.

may have heard the expression `The difficult we do immediately, the impossible takes a little longer.' Well, in quantum mechanics the impossible does not take a little longer, but it does last a little shorter. Virtual particles can enjoy all the benefits of energy which they do not possess, on a short-term free trial. This includes being able to transfer momentum."

"It must be a rather short free trial," said Alice thoughtfully.

"Oh, it is; it is. But it is something for nothing you see, so they all want it. You will have a better appreciation of virtual particles once you have seen them."

"But I can't see them," complained Alice. "Surely that is the point."

"You cannot see them at the moment," the Agent replied sternly, "but you will when you put on my *virtual reality* helmet." He walked quickly away in the direction from which they had come, and Alice hoped that she had not offended him. She was relieved when he returned soon after, carrying a large and highly technical looking helmet. This had a transparent visor which entirely covered the front, and there was a long cable attached to a socket at the back. The cable snaked away along the path by which he had come until it was lost from sight in the distance.

"Here it is," he said triumphantly, "a marvel of modern technology. Just put this on, and you will see the world of virtual particles."

Alice felt a little nervous as she contemplated the helmet. It was large, and it looked *very* complicated and even, she felt, a little sinister. However, if this was going to reveal the virtual particles she had heard mentioned so often, she was prepared to try it. She put the helmet on her head. It was very heavy. The Agent reached across to the helmet and made some adjustment at the side of her head, where Alice was unable to see. The view through the visor clouded over with little sparkling dots and. . . .

◆ ❖ ◆

When her view through the visor cleared, it had dramatically changed. Alice could still see the electrons in their various levels, but now instead of their appearing to be within a tall building she saw them as enmeshed in a network of vivid lines which joined one electron to another, so that they looked as much as anything like flies caught in some great spider's web of shining strands. As she looked more carefully at these strands she could see that they were actually composed of photons, but photons distinctly different from the ones she had seen before at the Academy.

All the photons which she had met before had been moving very rapidly, but they had at least been moving in a normal fashion. They had started at one position and a little time later they were at a new position, even if their positions were never precisely defined, while in the intervening period they passed through all the points between the two positions. It had never occurred to Alice that it was possible to travel in any other way, but some of these virtual photons seemed to manage it. As she looked at them she found it very difficult to say in which direction they were moving, or indeed if they were moving at all in any normal fashion. A given strand in the net, which represented the behavior of one photon, would seem to appear at the same moment at the positions of both the electrons which it joined, without apparently moving in the normal way from one to another. This link would then fade while others appeared elsewhere in the great mesh of photons which coupled together the electric charges of all the electrons.

It was a really beautiful sight, if rather peculiar. The virtual photons were moving in every way conceivable, while some photons seemed to have mastered the art of traveling from one position to another without actually requiring any time to elapse between the two events.

As Alice was watching this strange scene with interest, the helmet emitted a whirring noise beside her ear, followed immediately by a loud "clunk." The view in front of her shimmered and returned to the mundane view she had seen before she put on the helmet. Alice exclaimed aloud in annoyance at losing the fascinating picture. "I am sorry," said the Agent. "I am afraid that there is a timer built into the mechanism. I had intended to make it coin-operated, you see."

Alice was still too much enthralled by the vision she had just been watching to pay much attention to the Agent's apology and tried to

P articles in quantum theory are found to show properties which classically are associated with continuous waves. In a corresponding way, classical force fields are found to be composed of particles. The electrical interaction between any two charged particles is caused by the exchange of photons between them. These photons have a brief existence, which means they are well localized in time and so are uncertain in energy. They are *virtual particles*, whose energy and momentum may fluctuate well away from the values which would be normal for a long-lived particle.

describe to him what she had seen. As with all the people she had met in this odd world, he immediately began a lengthy explanation.

"That is just another aspect of the way that virtual particles do things which real particles cannot. It is a bit like barrier penetration in a way. I expect you have seen some cases of barrier penetration by now."

"I was told that I had," answered Alice carefully. "I saw someone penetrate through a door when I first came here and I was told that he could do this because his wave function spread into and through the door, to give a small probability of his being observed on the other side."

"That is quite true. That part of his wave function allowed your friend to penetrate into a barrier which would have stopped a real classical particle. He did not have enough energy to cross the barrier, so when he was penetrating the door he was in a sort of virtual condition. There are very few particles, if any, which are entirely real. They almost all have some virtual aspects, though some are more virtual than others. The exchange photons you have just been looking at are almost totally virtual.

"It is the general rule that virtual particles do not obey the rules, even though they cannot manage to escape them for very long. This means that they can do things for which they do not actually have enough energy. These exchanged virtual particles, such as the photons you saw, produce interactions between other particles. They can penetrate through barriers which would stop a real classical particle, and this includes the barrier of time itself. They can move in a *spacelike* way, while real particles can only be *timelike*. This means that, although a real particle can sit at the same position while the time changes, it is unable to sit at the same time while its position changes. A virtual particle is able to do both. It can move sideways in time if it chooses."

"That sounds very curious indeed," said Alice. "I am not surprised that real particles are unable to do that and that they only move from the past to the future."

See end-of-chapter note 2

"Well, actually that is not quite true," said the Agent a little apologetically. "It is certainly true that most particles move forward in time, just as you suppose. However most particles become a little virtual on occasion, during collisions for example, so it is possible for a real particle to revert. One moment it is moving forward through time in a respectable, law-abiding way. The next moment it finds that it has been quite turned around and that it is moving backward toward the past. Though it might surprise you to hear me say so, this is an allowable way for a real particle to behave."

"Oh!" cried Alice suddenly, startling the Agent in the middle of his careful description. "I think that must have been what happened to me earlier. I could not imagine what had become of me when I was walking through the park and everything seemed to reverse around me, but now I see that it was not the stream and the butterflies which were going backward. It was I who was traveling backward in time!"

Alice told her companion all she could remember about the incident, and he agreed with her interpretation. "It certainly sounds to me like a clear case of antiparticle production," he said.

"Antiparticle!" exclaimed Alice. "I did not know that this had anything to do with antiparticles. I remember seeing them at the Heisenberg Bank, but I do not understand why they should have anything to do with the present case."

"I would have thought it was obvious," said the Agent, though Alice did not feel it was in the least obvious. "Why, don't you see that when a particle moves backward in time it appears to an onlooker to be something totally opposite, moving forward through time in the normal way. Take the case of an electron. It has a negative electric charge, so when it moves from the past to the future in the normal way it carries its negative charge into the future. If, on the other hand, it moves from the future into the past, then it carries this negative charge from the future to the past, which is like a *positive* charge going from the past, to the future. Either way it is making the overall charge in the future more positive. It looks to an outside observer like a positron, or antielectron.

"What happened to you would have appeared to the rest of the world as an unusually high-energy photon giving up its energy to create an Alice and an *anti-Alice*. The anti-Alice would travel along until it collided with an Alice and the two mutually annihilated one another, converting their energy back to photons."

"How can that be?" cried Alice in some dismay. "I do not see how this anti-Alice could ever have found a second Alice to collide with anyway. There is only one of me and I certainly haven't been annihilated," she concluded defiantly.

"Ah, but what I have just described is how it would appear *to the rest of the world*. How it would appear to you is quite different, quite different altogether. For you the annihilation would come before the creation of course."

"I do not see any `of course' about it," answered Alice rather sharply. "How can anything be destroyed *before* it is created?"

"Why, that is the natural order of things when you are going backward in time. Normally, when you move forward in time you expect creation to come before destruction don't you?"

"Yes, of course I do," replied Alice.

"Well in that case, if you move *backward* in time you naturally expect the creation to come after destruction from your point of view. You are experiencing events in reverse order after all. I would have expected you to see that for yourself.

"In this case you were walking quietly along with the Quantum Mechanic and suddenly you collided with the anti-Alice. As far as your companion was concerned you and the anti-Alice were both utterly destroyed and your mass energy was carried off by high-energy photons."

"Oh dear! The poor Mechanic," exclaimed Alice. "He must believe that I have been destroyed then! How can I find him to set his mind at rest?"

"I shouldn't worry too much about that," the Agent assured her. "Naturally, the Quantum Mechanic knows about antiparticle annihilation, so he will know that you have simply gone back in time. He will no doubt expect to bump into you later, or perhaps earlier, depending on how far back you went. Anyway, the annihilation process converted you to an anti-Alice and you traveled backward in time until you were created, together with an Alice, by a high-energy photon. That is how it would have seemed to any onlooker. To you it just appeared that suddenly you were no longer traveling back in time, but had started to move forward as normal. You would not have seen the photon which caused this. You couldn't because it ceased to exist at the instant when you reversed your passage through time, so both as Alice and anti-Alice you were in a future which it never reached.

"You see now that, although anyone looking on would say that for a time there were three of you, two Alices and one anti-Alice, in fact they were all the same *you*. Because you had gone back in time you were living through the same period that you had already lived through as you walked along with the Quantum Mechanic. When you were returned to normal by the pair-creation process, you lived through the same period for a third time, now once again moving forward in time.

"That part of your life was rather like a road which zigzags up the side of a hill, climbing first to the east, then doubling sharply back to run toward the west before turning to the east again. If you climbed due north up the side of such a hill you might think that you crossed three different roads, while in fact you would have crossed the same road three times. It is the same sort of thing with antiparticle production. The antiparticle is a section of the road which goes the other way."

At that point there was a faint buzzing from the helmet and a small green light glowed in the corner of the visor. "I think the helmet is sufficiently recharged for another demonstration," said the Agent. "If you look

carefully this time you should be able to make out some of the second-order effects."

He adjusted the side of the helmet, and once again the view clouded over. . . .

The view cleared again to reveal that the landscape everywhere was strung together by an all-pervading net of photon lines. When Alice looked more carefully at one particular region, she could see that a few of the bright links were interrupted. In the middle of a shining photon strand she could see a sort of loop, where the photon changed in mid-position into what she could just recognize as an electron and a positron, an antielectron. These two combined together again almost immediately to form a photon strand which went on to attach itself to a real electron.

Peering still more closely, Alice could see another photon faintly issuing from the electron in the loop. Partway along the path of this photon she could see the dim outline of another electron-positron loop. From this emerged even fainter photons, and, if she peered really closely, she could just make out faint electron-positron loops along these. As far as she was able to distinguish she could see photons creating closed electron-positron loops and electrons or positrons emitting photons which created more electron-positron pairs. This went on and on, in apparently infinite profusion, but becoming fainter and fainter with each extra stage of complexity. Alice was becoming quite dizzy as she strained her eyes to try and see some end to this sequence. Finally an end came. She heard the whir and clunk from the helmet, and the entire pattern vanished from her sight.

"I thought that you said that electrons were joined by *photon* exchange," she said in a rather accusing tone. "I am sure that I saw electrons among the virtual particles. Quite a lot of them in fact."

"Oh yes, you would. The original real electrons act as the sources of the electric field, though it is more correct to say that the electric charges carried by the electrons are what produce the field. Photons do not really care about anything but electric charge, but wherever there is such a charge you will always get a cloud of virtual photons hanging about it. If another charged particle comes by, these photons are available to be exchanged and to produce a force between the two particles. Exchanged particles have to be created in order to be exchanged and they are destroyed afterward when they have been captured. Their number is obviously not conserved, so they have to be bosons.

"The relationship between photons and charge works both ways. Just as charged particles produce photons, so photons would like to produce charged particles, but they cannot produce just one charged particle because the amount of electric charge present is not allowed to change. That is another of the rules, and one that does not allow any uncertainty. What the photons can do however is to produce both an *electron* and an *antielectron,* or positron, at the same time. Since one has a negative charge and the other a positive one, the *total* charge in the universe has not changed. That was what you saw. The virtual photons produce virtual electron-positron pairs, which annihilate one another and return to being a photon. During the brief life of the pair, however, since they are both charged particles, they may produce more photons; those photons may produce more electron-positron pairs, and so on."

Not only can photons be created, but also particles such as electrons, though these have to be produced in company with their antiparticles so that there is no change in the total electric charge. Energy is required to create the rest masses of two such particles, but the necessary energy may be available briefly as an energy fluctuation. Such a fluctuation may occur even if there is no energy present initially, and the particles may be created literally from nothing. "Empty space" is, in fact, a seething brew of particle-antiparticle pairs.

"My goodness," said Alice. "It does sound excessively complicated. Where does it all end?"

"Oh, it doesn't. It goes on like that forever, getting more and more complicated. But the probability of an electron producing a photon, or of the photon producing an electron-positron pair, is rather small. This means that the more complicated amplitudes are weaker and eventually they are too weak to be noticeable. You must have seen that."

"Well," said Alice, whose head was spinning as she tried to follow what she had just observed and been told, "all I can say is that I have seen nothing like it before."

"You may well have done so," returned the Agent. "What you have just seen is like Nothing anywhere else. Though I am a little surprised that you have managed to see Nothing before you came here."

"I am sure that I wouldn't say that," replied Alice indignantly. "I may not have traveled very much, but I have still seen something, I would have you know."

"I have no doubt that you have," said the State Agent. "I am sure that you came from a very desirable location, but it is relatively easy to see Something, you know. It is much more difficult to see Nothing. I do not know how you could have done it without the aid of my virtual reality helmet."

"Just a minute," interrupted Alice, who had begun to suspect that they were talking at cross-purposes. "Would you tell me please what you mean by Nothing?"

"Why yes, certainly. I mean Nothing: the complete absence of any real particles whatever. You know: the Vacuum, the Void, the oblivion of all things, whatever you like to call it."

Alice was quite taken aback by the extent of this negative concept. "Would that look any different through your helmet? I should have thought that Nothing would look like nothing however you looked at it."

"Why, of course it makes a difference. The void is not the best neighborhood, perhaps, but there is plenty of undercover activity. Come and see for yourself."

The Agent set off at a smart pace and Alice followed him across the floor of his office. It was becoming increasingly difficult for her to believe that they were still inside an office, or a building of any sort, for it seemed remarkably large. They walked for some time, with Alice struggling under the weight of the helmet and of the cable which was still stretching out behind her. "I wonder how long this connection can be," she said to herself. "I am sure I must come to the end of it soon."

The Periodic Mansions, in which she had watched the electron states, were soon out of sight behind them, and still they marched on.

Just as Alice was about to beg that they stop for a rest, she saw ahead of them what looked rather like the shore of a lake, or of a remarkably calm sea. As they came closer she could see that it was a very large lake, that is, if it was a lake. It stretched ahead of them as far as she was able to see, an apparently limitless expanse. But if it was the sea, it was the strangest seascape that Alice had ever seen. It was very calm, completely and utterly still apart from a faint, hardly seen, quivering around the surface. It was not blue, or green, or wine-dark, or any other color Alice had heard used to describe water. It was completely without color. It was like a deep, clear night but without the stars.

"What is that?" gasped Alice, overcome by the eye-devouring emptiness of the sight.

"Nothing," replied the Agent. "That is Nothing. It is the Void!"

"Come now," he continued. "Let me switch on the helmet, and you may observe the activity in the Void."

He reached out to the helmet and once again did whatever he had done before. Alice's view, her view of Nothing, clouded over. . . .

Her view cleared to reveal a scene very similar to the last one she had seen through the helmet. Once more she saw a mesh of glowing strands. This time, however, she did not see the strands ending on the real electrons, which before had seemed to be trapped in the web but were in reality its source. Now there were no real particles present, only the virtual ones. Photons created electron-positron pairs. Electrons and positrons produced more photons, just as she had seen before. Previously the network had originated from the real electrons, which were its source and anchor in the world of real particles. Where was its source now? The electron-positron pairs were produced by the photons; the photons were produced by the electron-positron pairs, which were produced by the photons. Alice tried to trace back along the lines of particles to find their source, but she found that she was going around and around in circles. She felt that she must have lost her thread and was trying again to follow the lines more carefully when she heard the familiar buzzing and the loud clunk, and the whole scene vanished.

Alice once more explained what she had seen to the Agent and told him how she had been unable to decide which particles had been creating which others. "I am not surprised," replied the Agent. "They all create one another, you know. It is a chicken and egg situation, with them all laying and hatching at the same time."

"How can that be?" asked Alice. "There must be a source. They cannot have come from nowhere."

"I am afraid that they can and they have," was the answer. "All that prevents particle-antiparticle production normally is the need to provide energy for the particles' rest mass, and virtual particles are not even inhibited by that. The whole thing is a great big quantum fluctuation."

"Is it real then?" asked Alice. "Are all those particles really there at all."

"Oh yes, they are quite real, even if not in the technical sense of *real particles*. They are just as vital a part of the world as anything else. I should think, though, you have now seen as much through the helmet as you need," he continued and lifted the heavy device from Alice's head. "We shall not be needing it any more, so I shall engage the cable rewind mechanism." He touched a button on its side and the helmet began to rewind itself along its cable, scuttling over the ground in the direction from which they had come, like a mechanical spider, until it vanished from sight.

Although the helmet had gone, Alice's head was still full of the remarkable sights she had seen and she turned them over in her mind as she walked in silence beside the State Agent, along the shore of the infinite Void.

Notes

1. Within atoms, the allowed states for electrons have widely spaced energy levels and electrons may occupy only these levels. An electron can only transfer from one of these states if it goes to another (empty) one and in so doing its energy changes by a definite amount, the difference in the energies of the two states. An atom in its normal, or *ground*, state has its lowest energy levels uniformly filled with electrons, but there are levels of higher energy which are normally empty. When an electron is excited from its initial position it will end up in one of these empty higher levels or leave the atom completely.

    An electron which has been excited to a higher level can decay back to a level of lower energy if there is an empty state available. As the electron transfers to a level of lower energy, it must rid itself of the surplus energy, which it does by emitting a photon. This is how atoms come to give off light. Since the electrons all occupy definite states within the atom, any photon that is emitted can only have an energy equal to the difference of that possessed by the initial and final states of the electron. This gives a large number of possibilities, but nevertheless imposes a restriction on the energy which a photon may have. The photon energy is proportional to the frequency of the light and thus to its color, so the spectrum of the light produced by an atom consists of a set of colored "lines" of specific frequencies. The spectrum for a given type of atom is completely distinctive.

Classical physics can give no explanation for these spectra.

2. Virtual particles have a distinct fuzziness, both in time and in energy. This fuzziness shows itself as energy fluctuations, in which the particles behave as if they had more (or less) energy than they should. It can equally appear as an uncertainty in time. In a quantum system particles seem to be able to be in two places at the same time (or at least they have amplitudes which are).

The particles can even turn time around. The physicist Richard Feynman explains antiparticles as being "particles traveling backward in time."* This explains the way in which the properties of antiparticles are opposite to those of the particles: A negative electric charge carried backward in time is equivalent to a positive charge moving toward the future. In both cases the positive charge in the future is being increased, and a negatively charged electron traveling toward the past is seen as a positively charged positron, which is its antiparticle.

All particles have their antiparticles, as is to be expected if they are in effect the same particle behaving in a different way.

---

*Richard Feynman, *QED: The Strange Theory of Light and Matter,* Penguin, New York.

7 Atoms in the Void

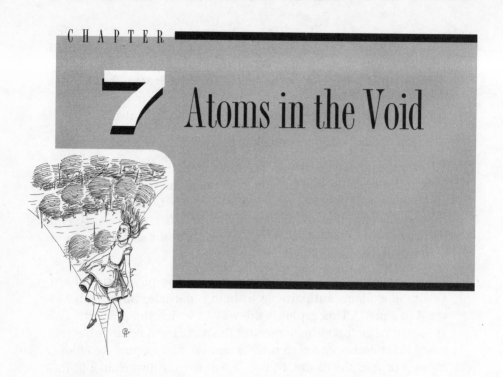

A lice walked with the State Agent along the edge of the Void, looking out over the shimmering tenuous surface which was continuously aboil with the activity of the virtual particles as they were born and died unnoticed.

A little way out from the shore Alice saw a disturbance in the surface, some sort of circular depression in the general uniform level. Further out she could see other pits, and many of them were clumped together into groups. Some of the groups were very small and contained just a couple of the circular objects. Other collections were more extensive. She could see one group which contained a ring made up of six of the objects arranged in a circle, while others were attached around the outside. In the distance she could see some enormous assemblies spread across the surface. The largest contained many hundreds of the circular things, whatever they might be.

As Alice watched, she saw photons soar intermittently from one or another of the shapes which were spread out before her. The brilliantly colored photons looked rather like flares fired from ships at sea.

The Agent followed the direction of her gaze. "I see that you are watching the atoms as they swim in the Vacuum. Atoms provide us with much of our work in the electron state business, one way or another. You can see from here the various molecular partnerships that they have set up between them. These range from small two-atom businesses to huge

organic conglomerates. Each different type of atom has its own distinctive spectrum of colors for the photons which it emits, so the photons act as signals which help you to identify the different types of atoms."
See end-of-chapter note 1

"I was wondering about all those things way out there," admitted Alice candidly. "I cannot see them very clearly from here. Is it possible to get any closer?"

"If you want to look closely at atoms we ought to go along to Mendeleev Moorings. There you will see every type of atom on display, with all the different elements laid out in a regular order."

The Agent led Alice along the shore until they came in sight of an extremely long, narrow jetty, which stretched far out over the Void. At the shore end there was an arched gate on top of which was a sign which read:

THE PERIODIC PIER
prop. D. I. Mendeleev.
Established 1869.

"There you are," announced the Agent. "That is where the atoms lie at dock before they set out to form their different chemical compounds. We usually call it the 'Mendeleev Marina' or the 'Atomic Pier,' though sometimes people talk of the 'Quay of the Universe.' You will find each different kind of atom represented here."

Together they walked beneath the sign and stepped onto the boards of the jetty. They strolled slowly out along the anchorage, while Alice looked at the long line of atoms moored in sequence to one side. Each of them appeared to her as a trumpet-shaped pit in the flat surface of the surrounding Void. The shape reminded her of the little whirlpool which she often saw forming over the drain whenever she emptied a bath, though these seemed to be quite still with no visible rotation. The surrounding surface of slick nothingness sloped down into each pit from the still flat level which stretched all around it. It sloped with almost imperceptible gradient at first, but ever more steeply as it funneled down toward the center. There were signs of activity taking place somewhere in the depths of the pit.

"Why is there such a deep hole?" Alice asked curiously. "As we are looking at Nothing, I would expect it to be all flat and featureless."

"That is a potential well," was the answer.

est. 1869

"What sort of well is that?" Alice continued curiously. "I know about garden wells which supply water and about oil wells, and I vaguely remember seeing something about a treacle well in a book I was reading recently, but what do you get in a *potential* well?"

"Why, the source of the potential, of course. You have to have a source to provide the water in a garden well. Here there is an electric charge as the source of electrical potential in the potential well. You should know by now what is in the well. It contains virtual photons. They provide the electrical attraction which would make the potential energy of a negative charge drop farther and farther below the surrounding vacuum level as it moves toward the potential source at the center of the atom. The potential source actually creates the well, you see."

The first pit was fairly shallow, but Alice could see that the others became successively deeper the farther they were positioned along the pier. The jetty stretched away ahead of her into the distance, with atom after atom moored along the side. Beside each one was a small notice to mark the mooring. The first of these read: $_1$H; the second, $_2$He; the third, $_3$Li. Each position had a different label. "Will these atoms all set out from here eventually to combine into groups like the ones already out on the surface of the Void?" asked Alice.

"Most of them will certainly, but there are a few which will not, like the one just here for example."

They paused beside an atom which carried the notice: $_{10}$Ne. "That is an atom of a Noble Gas element. They are an aristocratic lot and that means that they refuse to engage in commerce of any sort. They keep to themselves. They are perfectly satisfied with the way they are and will not mix with anyone else. They always travel around in splendid isolation. You never see them take part in any sort of compound."

They walked a little farther and the Agent explained that, even apart from the aloof Noble atoms, there was considerable variation in the enthusiasm with which different elements joined into compounds. "For example, this is a particularly active concern," he remarked, as they came to a notice which read: $_{17}$Cl.

Alice decided that it was time to examine one of these atoms more closely, so she tentatively extended one foot off the edge of the jetty. To her delight she did not sink. Her foot stood on a tiny dimple in the surface, rather like some pond-skating insects she had once watched. When she tried to walk toward the atom, however, she discovered that there was no friction in the void. The surface was *extremely* slippery, and she was quite unable to keep her footing. With a small cry she skidded down the increasingly steep slope and toppled into the deep pit.

As she fell, Alice found that she had plenty of time to look around her. The sides of the well became ever steeper as they closed in upon her, and she soon noticed that she was falling through the ghostly outline of a series of rooms which had low, closely spaced ceilings. The first few rooms were very low indeed, scarcely tall enough for a doll's house, but as she fell the rooms became steadily taller. Initially they were all completely empty and deserted, but then she came to a room which contained a large, round table surrounded by chairs. On the floor below this

The states which electrons may occupy within an atom tend to group into a set of levels which are separated by significant gaps in energy. If an atom has its outermost occupied level completely filled with all the electrons it can hold, then any extra electron added would have to go into a state of higher energy. It will usually have a lower energy than this if it stays in its original atomic state. Atoms of this type, whose outer shells are completely filled with electrons, form the noble gases and do not interact chemically with anything in the normal way.

she could see desks and filing cabinets, as if she was passing through some sort of office.

As the time passed she became increasingly amazed to find that she was still falling, without any sign of reaching the bottom. Down, down, down; would the fall *never* come to an end?

An atom is contained by the electrical field which is generated by the positive charge on its nucleus. This charge produces a *potential well* around the nucleus, which in turn defines the states available for the electrons to occupy. The selection of states available is a form of interference effect, akin to the range of notes you can get from an organ pipe or a violin string. One pipe can give only a few notes, for which the wavelength of the sound waves fits the pipe. In a similar way the allowed electron states fit within the potential well. The allowed states are grouped together into distinct energy levels. Any other wave function which does not correspond to one of these states is eliminated by destructive interference.

Alice gradually began to realize that her fall was *not* going to come to an end. She had not reached the bottom of the hole, but she was not getting any lower. She was floating quite unsupported in the center of the funnel, on a level with one of the shadowy rooms. She looked around her and noted that she was not alone. Near her were two electrons who were involved in a hectic flurry of activity. Around them she could discern the faint outline of an extremely tiny and cramped office. "Excuse me," she called. "Do you think that you could stop for a moment and tell me where I am?"

"No room, no room," they called.

"I beg your pardon, what do you mean?" cried Alice, to whom this reply did not seem particularly relevant.

"There is not enough room here for us to slow down at all, let alone *stop*," they answered her. "As you know, when the position of a particle is restricted, the Heisenberg relation forces its momentum to be large, and it is so cramped here that we have no choice but to keep moving. If we had as much room as they have on some of the higher levels we could afford to move in a more leisurely way, but not here. This is the lowest level, you see, so we must expect to be kept busy."

"Really?" inquired Alice. "What is it that you do that is so important?"

"We do not do anything in particular. No one is particularly interested in *what* the electrons in the ground state are doing, just as long as we keep moving."

"In that case, do you think you could tell me where I am, *without* stopping?" Alice asked. "For I do not know where I have come to. What is preventing any of us from falling farther down into the well."

"You are in the lowest level of a chlorine atom, as we have already told you. Here, we are so close to the potential source that there is very little room, so we have to move very quickly as our momentum is forced to be high. This means that our kinetic energy is also high. None of us is in a particularly virtual state, you see. Electrons have secure positions in atoms, with very good tenure. Most atoms have been around for a long time and the quantum energy fluctuations are small, so for us electrons the energy and momentum are properly related.

"You probably know that when an electron, or anything else, falls farther into a potential, it loses potential energy, and this will be converted to kinetic energy," they went on.

"Yes, that was explained to me when I visited the Heisenberg Bank," agreed Alice.

"Here in this potential well, though, when we get closer to the center, there is less and less room, so we need to have more kinetic energy. If we were to fall even closer we would need to have more kinetic energy than we can possibly get from converting potential energy, so we are unable to fall any further. In fact, paradoxically, we just do not have enough energy to be able to fall any lower and we cannot borrow the energy as a quantum fluctuation because we would need it for a long time.

"There are only two states on this level so there is only room for two electrons, one being in a spin-up and one in a spin-down state. There are more states available as you move up to higher energy levels, so you will find more electrons on the levels above. The next two levels can hold up to eight electrons on each level. In any atom, the lowest levels, the ones with the lowest potential energy, are the first to be filled. The Pauli princi-

ple allows only one electron in each state, so when all the states on one level already have an electron, any extra electrons have no choice but to move into levels higher up. The levels are filled from the bottom until all the electrons are accommodated. The highest level which contains any electrons is called the valence level. That is where the valence electrons live, though there are plenty of unoccupied states higher up in the attic. The valence electrons make all the decisions and control the compounds our atom can join. If you want to discover how an atom operates, your best plan would be to go up and talk to them."

See end-of-chapter note 2

"How shall I get up to that level from here?" asked Alice.

"Well, if you were an electron you would have to wait here until you were excited to the higher level by a photon which could give you the extra energy you would need. In your case, however, I expect that you can be carried up by the Ladder Operator."

"Don't you mean the elevator operator?" queried Alice. "I have been in an elevator in a large department store and it had an operator who took people from floor to floor, but I have never heard of a ladder that needed one."

When she looked around, however, she could see a sort of ladder with very widely separated rungs. Beside it stood a rather indistinct figure. "May I ask who you are?" said Alice curiously.

"I am the Ladder Operator. I am not a physical creature, but merely a mathematical construct. It is my job to transform a system from one state to a higher or lower one." He performed some complicated operation which Alice completely failed to understand but resulted in her being carried rung by rung up to the higher level.

In due course Alice arrived at the level on which she had seen the large, round table. This level contained more electrons than the first. She managed to count eight in all, though with some difficulty. As with all the electrons she had seen so far, they were moving energetically around. Several of them were circling around the table, some in one direction and some in the other. The others were not obviously rotating but were nonetheless in motion. None of them was sitting quietly on any of the chairs around the table, but they were leaping up and down, and some were stepping on and off the table. The electrons were never still, though on this level they were not moving quite so frantically as they had been on the lowest one.

"Hello Alice," they cried as she appeared. "Come, and let us show you how a reliable, medium-sized atom operates. The way in which Chlorine Corporation conducts its business is decided by us seven electrons in the valence level."

"But there are eight of you!" protested Alice.

"That is because we have entered into a partnership with another atom, Sodium Syndicate, to form a sodium chloride molecule. Working together in this way we like to think that we are the Salt of the Earth. An atom runs much more harmoniously when all of its levels that hold any electrons are filled completely. On our own we have only seven electrons in the valence level and Sodium has but the one, although there is room for eight. It helps both of us if the Sodium valence electron comes over here to sit on our valence level and give us a full board. This means of course that we now have an extra electron, and so we have a negative charge. The Sodium atom has an electron less than normal, which gives them a positive charge, and the electrical force between these opposite charges holds the two atoms together. That is known as *ionic bonding* between the atoms and is one of the common forms of corporate structure."

"That sounds very cooperative on both sides," agreed Alice tactfully. "Which of you is the electron that has come over from the Sodium atom then?" she asked.

"I am," they all cried, talking together. They paused for a moment and looked at one another. "No, he is the one," they now said, still speaking in perfect unison. Alice realized that there was absolutely no point in asking any question which tried to distinguish between the identical electrons.

"Could you explain to me, please, why you say that the sodium atom has a positive electric charge when it has lost one of its electrons," she asked instead. "Surely it still has quite a few electrons left, and they will presumably have negative charges also."

"That is quite true, all we electrons do have the same amount of negative charge, as we are all identical. Normally in an atom this charge is balanced and neutralized by an equal amount of positive charge carried by the Nucleus. Atoms are usually neutral, with no net electrical charge either way. So you see, when an atom has one electron more than usual, it will be negatively charged. It is known as a *negative ion*. If it has one electron fewer than normal, the positive charge on the Nucleus will dominate, and the atom becomes a *positive ion*."

"I see," said Alice thoughtfully, "but what is this Nucleus that you are talking about?"

"Every atom has one," was the evasive answer, "but you do not want to know too much about it. You really don't!" he added nervously.

At this point the conversation was interrupted by a faint cry which started somewhere below them, passed through the valence level close at hand, and finally stopped somewhere above. Alice looked up and saw that it was due to an electron which had apparently been excited by a photon from its position in a lower level and was now looking uncomfortably remote in one of the empty higher levels. The electron wandered rather slowly around the high wide level until eventually it gave a brief cry and toppled to the level below. As it did a photon rushed out of the atom, carrying away the energy released by the fall. Alice watched with interest as the electron fell in succession from one level to the next, in each case emitting a photon. As the lower energy levels were more widely separated than those above, each fall was farther than the one before, so the photons created had higher energy arising from each successive fall. As their energy increased, the color of the light moved farther toward the blue end of the spectrum.

Looking downward Alice saw that the space that had been left by the electron which was excited from the lower level had been filled and that one of her companions in the valence level was missing. Before long the electron falling from above had dropped to the valence level and filled the vacant place. The atom was now back to its original state. Two electrons had exchanged levels, but as they were identical that was no difference at all.

See end-of-chapter note 3

"You will have noticed all the different colors of the photons which I emitted," said one of the electrons proudly. This remark tended to suggest that it was the electron which had fallen that had just spoken, but Alice was now too experienced with the effects of electron identity to fall into that trap. "That is the way that atoms emit light you know: when electrons change from one level to another. All the photons were of different energy, and hence different color, because the levels are all different distances apart. They are very closely spaced at the top of the well but are farther and farther apart as you get lower down. This level spacing is different in atoms of different types, so the set of photon energies is completely distinctive for each type of atom—as distinctive as a human fingerprint."

Hardly had the eight electrons settled down, or got as settled as they could ever be while they were all in continuous frantic motion, when there was a tremor which seemed to run through the whole atom. "What was that!" cried Alice in some alarm.

"It was an interaction of some sort. We have been separated from our Sodium partner and are drifting through the void as a free negative ion. But do not worry. I do not anticipate that we shall drift about aimlessly for very long. We shall very soon be back in business if the Exchange is agreeable."

"What Exchange is that?" asked Alice. "Do you mean the Stock Exchange? I understand that controls business in my world."

"In our case we mean the Electron Exchange. All of our activities are governed by electron interactions of some sort, so it is electron exchange which is significant. Perhaps you would like to visit the Exchange?"

"Yes, I should think so," replied Alice. "How would I get there from here? Is it a long journey?"

"Oh no. Not really. In fact it is not really a journey at all. As you are in an interacting atom, you are already there in a sense; you just need a different representation. It is all a question of how you look at things. Just follow me."

As the electron had told her, they did not seem actually to go anywhere else, but still Alice found herself in company with an electron on the edge of a broad room. The floor was crowded with electrons which clustered around a large table in the middle of the room. It looked to Alice rather like one of the tables which she had seen in old war films, where commanders moved around various counters which represented aircraft, or ships, or armies. On this table also, she saw a great selection of counters which were being moved around into different groupings.

She looked more closely at some of these counters and saw that they bore the same labels as the atom moorings on the Periodic Pier. In fact, as she looked really closely, she was no longer so sure that they were merely counters. They looked like reduced versions of the atoms which she had seen along the side of that jetty. "Perhaps they are the same," she thought. "Maybe those are the same atoms which I am seeing differently. I suppose that instead of the Periodic Pier, that would make this the Periodic Table."

Around the side of the room, the walls carried rows of display screens on which she could see columns of numbers that changed as atoms were moved from group to group.

"Are those the prices for the various atoms?" asked Alice.

"Yes, after a fashion. Those numbers tell us the energies of the various electrons which are taking part in chemical combinations. They quote the *binding energies* of the electrons: the amount by which an electron's energy has been reduced below the value it would have if it were free. The larger is the value quoted, the lower is the potential energy that the electron has, and so the more stable and successful is the compound which it binds. The job of the Exchange is to make these binding energies as large as possible."

"And is this all done by moving electrons from one atom to another?" queried Alice, who remembered the explanation she had been given of ionic bonding in Sodium Chloride.

"Not always, no. Sometimes that is the most effective method and then the binding is done in that way. The Electron Exchange can get an advantage by moving electrons around because the electron states that are available within an atom are arranged in levels, or shells, with quite large gaps between. The binding energy for the last electron in a lower shell level is much greater than for the first electron that has to go into the next shell higher. This means that there is an easy method of improving the overall energy score for an atom which has only one electron in its highest shell. If this electron can move from its splendid but extravagant isolation into an almost full lower shell in some other atom, then there is almost certain to be an overall gain in binding energy.

"It is equally true that, when an atom has but one space remaining in its highest occupied shell, this state will have an unusually low energy, and any electron which transfers into it will be likely to improve its energy balance sheet. It is generally true that atoms with just one electron too many or too few are the most active—the most likely to take part in transactions and to form compounds. Atoms with two electrons alone in a high state and those with only two spaces left in a lower one may engage in similar electron transfers, but the gain in binding energy for the second electron is usually a lot less than for the first one, and it is much less effective."

"So what can an atom do if it has several electrons in its outer shell?" asked Alice, as this seemed to be expected of her.

I f an atom has just one electron in its outer level, while another atom is one electron short of a full level, the two can achieve a lower energy overall by transferring the isolated electron of the one to the nearly full valence level of the other atom. This is chemistry: The electrons in their various energy levels bind the atoms together. The details of chemistry can get rather complicated in practice, but that is the principle.

An atom contains that number of electrons which is needed to neutralize the positive charge on the nucleus. These electrons fill the states of lowest energy, with one electron in each state. If one atom has a single space left unfilled in its highest filled level and another atom has a single electron which has had to go into a higher level, then the overall energy may be reduced by transferring this electron into the space left in the other atom. Both atoms now have a net charge and the resultant electrical attraction binds them together to form a chemical compound.

"Such an atom has to change to a different kind of binding, one which is known as a *covalent bond*. An atom such as carbon, for instance, has four electrons in its outer shell. This means it has four electrons too many to be an empty shell and four electrons too few to be a full one. It is too nicely balanced to gain anything by actually transferring electrons to or from another atom, so instead it *shares* them. It turns out that if the electrons from two atoms are in a superposition of states such that each may be in either atom, then the energy of the two atoms may be lowered and this serves to bind them together.

"The ionic bond, in which an electron is completely moved from one atom to another, can only work between very different atoms, one of which has an electron too many, the other an electron too few. The covalent bond, on the other hand, can work when both atoms are of the same type. The most remarkable example is given by the covalent bonding of carbon atoms, the basis of the huge Organic Conglomerates." Alice could sense an atmosphere of awed respect emanating from the electron manipulators around the table as the Organics were mentioned.

"A carbon atom has four electrons in its outer, or valence, level. If each of these electrons is combined with electrons from other atoms, then all of the eight electron states contribute to the superposition and the shell is effectively filled. In this way a carbon atom can attach itself to as many as four other atoms, which may also be carbon. The carbon atom may also exchange two of its electrons with another carbon atom to give a *double bond*, in which case it will not be connected to so many other atoms, though the connection will be stronger.

"The ionic bond at its strongest connects but one atom to one other, so it does not produce large molecules. Where there are two electrons to transfer, things can get more complex. Even then the situation does not compare with that of carbon, where one atom may connect to four others and each of those may be connected to several others. Carbon-based compounds can form into enormous organic molecules of great complexity, which may contain hundreds of atoms in all."

"Do all of the different atom types that I can see there form compounds in the ways you describe?" Alice asked.

"Yes, apart from the noble gases. With the noble gases, the atoms have filled valence shells already and so do not stand to gain by any electron transfers. All of the others do form compounds to a degree, though some are more active than others and some are encountered much more often. The chlorine atom which you visited, for example, is very active. It will form compounds with the simplest atom, hydrogen, which employs only one electron in total, and also with the largest natural element, uranium. That is a very large establishment indeed. It employs almost a hundred electrons, but only the ones in the outer valence level really affect its chemical behavior. It is so large that there have been rumors that its Nucleus is unstable," he added confidentially.

"I wanted to ask about that," said Alice firmly. "You have mentioned the nucleus again. Please, would you tell me: What is the nucleus?"

All of the electrons looked somehow uncomfortable, but reluctantly answered. "The Nucleus is the hidden master of the atom. We electrons conduct all the business of forming chemical compounds and emitting light from the atom and so on, but it is the Nucleus which really controls the sort of atom we are. It makes the final policy decisions, and fixes the number of electrons that we can have and the levels that are available to put them in. The Nucleus contains the nuclear family, the hidden underground of Organized Charge."

Alarmed by this outburst of candor, the electrons around the room all tried to shrink unobtrusively into one corner, or at least as far as they were able without becoming too localized. Too late, the harm was done! Alice became aware of a new menacing presence nearby.

Amongst the scurrying electrons there was now a hulking shape, looming over Alice and her companions. She realized that it was a photon, but distinctly more massive than any she had seen before. Like all the photons she had seen it was glowing, but in a peculiarly dim and furtive way. Alice also noticed that, surprisingly for something which was

itself the essence of light, this photon was wearing a pair of very dark glasses.

"It is a heavy virtual photon," gasped the electrons, "Very heavy, a long way off its mass shell. It is one of the enforcers for the Nucleus. Photons such as him transmit the Nucleus's electrical control to its client electrons."

"I hear that someone is asking questions," said the photon, in a menacing tone. "The nucleons are the sort of particles that do not like to hear that questions are being asked by any other person whatever. I am taking that same person for a little ride to meet certain parties, or rather certain particles. They want to meet her very badly indeed."

This did not sound like a very promising start for a new encounter, and Alice was considering whether she might safely refuse. She could never quite make out, in thinking it over afterward, how it was that they began: All she could remember was that they were running side by side and the photon was continuously crying "faster," and Alice felt that she could not go faster, though she had no breath left to say so. They rushed across the surface of the table and dived into one of the atoms represented on its surface. It was one of the uranium atoms and it grew enormously as it rushed up to meet them.

The most curious part of the experience once they were inside the atom was that the things around them never changed in position: However fast they went they never seemed to pass anything. What Alice did note was that her surroundings, the busy electrons and the outlines of the levels which contained them seemed to be getting steadily *larger* as she ran.

"Is everything really growing, or am I shrinking?" thought poor puzzled Alice.

"Faster!" cried the photon. "Faster! Do not try to talk."

Alice felt that she would never be able to talk again; she was getting so out of breath: and still the photon cried "Faster! Faster!" and dragged her along.

"Are we nearly there?" Alice managed to pant out at last.

"Nearly there!" the photon repeated. "Why, we are there all the time and no other place, but we are not sufficiently *localized*, not hardly. Faster!" They ran on for a time in silence, going faster and faster while the surrounding scene ballooned in size around them, spreading upward and outward until everything she had seen before was too large to be readily appreciated.

"Now! Now!" cried the photon. "Faster! Faster! Your momentum is now nearly so large as will localize you within the Nucleus." They went so fast that they seemed to skim through the air, until suddenly, just as Alice was getting quite exhausted, she found herself standing in front of a tall, dark tower which rose smoothly in front of her, curving up from the ground and narrowing steadily as it rose. It was dark and featureless at the lower levels, though somewhere at the top Alice could see that it finished in a confusion of turrets and battlements. The overall effect Alice found extremely forbidding.

"There you see Castle Rutherford, the home of the Nuclear Family," said the heavy virtual photon.

Notes

1. Atoms had been found to contain light negative electrons and, later, were found to contain a positively charged nucleus. This suggested that they might be tiny versions of the solar system, with planetary electrons in orbit around a nuclear sun. The notion gave rise to fantasies in which the electrons were indeed miniature planets, with still more miniature people living on them, and so on *ad infinitum*. Unfortunately for such schemes, the "solar system" picture is clearly wrong.

    *   The only reason planets do not fall directly into the sun is because they are orbiting around it. There is definite evidence that many electrons do not have any rotation the nucleus.

    *   According to classical physics, the orbiting electrons within atoms should radiate energy and their motion should run down. Something as small as an atom would do this rather quickly, in less than a millionth of a second, and atoms do not collapse in this way. (The solar system is, in fact, running down, but rather slowly, on a time scale of millions of years.)

2. Because of the Pauli principle you can have only one electron in each state. As electrons are available in *spin-up* and *spin-down* versions, this effectively doubles the number of states. Electrons will fall into the atomic states because they have lower energy there and it is a general rule that things tend to fall to lower energy levels (as you may discover by holding a cup over a tiled floor and releasing it). Any atom has a large number of levels which could hold electrons; in fact, the number of states is infinite, though the upper ones are very close together in energy. An atom will continue to attract electrons into its levels only until it contains the correct number to compensate the positive charge of its nucleus, after which the atom no longer has any surplus positive charge with which to attract further electrons. When

an atom has reached its full complement of electrons, it will in almost all cases contain more than there are room for in the state of lowest energy. Some electrons must then be accommodated in states of higher energy.

3.   When people looked at the light given off by atoms of a single type, they found that the spectrum was not a uniform spread of colors like a rainbow, but a set of sharp lines, each of a distinct color. Every type of atom showed these *line spectra*, which were a complete mystery to classical physics.

The set of energy levels for the electrons is unique to any given type of atom. When electrons transfer from one level to another they emit photons which have an energy that corresponds to the difference in the energy of the two levels. As the energy of photons is proportional to the frequency and color of the light, this gives an optical line spectrum for the atoms which is as distinctive as a fingerprint.

The explanation of the existence of a line spectrum was the first major success of the developing quantum theory. The theory fitted the observed line frequencies and predicted other line spectra which had not been seen. These where all found in due course and showed that the quantum theory could not readily be dismissed.

CHAPTER

8 Castle Rutherford

Neutral
Particles
only

Alice stood gazing up at the dark heights of Castle Rutherford as it loomed overhead. "Where did that come from?" she asked her companion. "How did we get here from the atom's potential well?"

"I have to tell you that during no time are we *going* anywhere. We are remaining strictly in the vicinity of the atom, but are now somewhat localized in its center or indeed rather more than somewhat. What you see before you is the bottom of the same potential well. Do you not recognize that same item?"

"No, I certainly do not!" replied Alice emphatically. "The potential well was a *well;* it was a hole which went downward. This is a tower which goes upward. Quite a different thing."

"It is by no means so very different when you think about it," replied the photon. "The Nucleus is producing an electric field, and this same Nucleus gives a negative potential energy to any negative electrons which are in the locality. When you are mixing in such company, as with electrons and so on, you are naturally seeing the potential as a pit going downward. Nuclear particles like protons are such particles as carry a positive charge at all times, so if guys like these should come calling unexpectedly, they are liable to find their potential energy is rising more than somewhat as they approach the Nucleus. This will usually make characters like this keep a polite distance, and the field acts like a barrier.

In fact, it is for this reason it is called the coulomb barrier. The nucleons are apt to hate having uninvited visitors. If you are mixing with characters of this sort, you are seeing what they are seeing, which is a high potential wall around the Nucleus."

"How shall I get in then?" asked Alice. "I do not think I shall be able

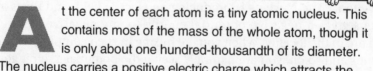

A t the center of each atom is a tiny atomic nucleus. This contains most of the mass of the whole atom, though it is only about one hundred-thousandth of its diameter. The nucleus carries a positive electric charge which attracts the negatively charged electrons and holds the atom together. This positive charge will, on the other hand, repel other positively charged particles and provide a barrier around the nucleus, the *coulomb barrier,* which keeps out protons and other nuclei.

to get over the wall. I am sure it will be very effective at making me keep a polite distance," she argued hopefully. She was still not at all sure that she wanted to meet the Nuclear Family.

"The coulomb barrier is acting to keep out only those same particles which have a positive electric charge. There are others that do not have any electric charge at all, and these particles can pass through easily. You are not carrying a charge at present, so you are liable to get in through the neutral particle entrance." He pointed toward a tall doorway in the bottom of the castle wall, which Alice had not noticed before. It was obligingly labeled: "Neutral Particles Only."

Alice and her escort went over to the door and knocked loudly. "What are the nuclear particles like?" asked Alice cautiously. "Are they much the same as the electrons I have already met?"

"They are commonly considered by one and all to be bigger than any electrons and are known to be about two thousand times more massive." This answer did nothing at all to reduce Alice's feeling of nervousness as she heard slow, ponderous footsteps approaching the door from within. These grew louder until she fancied she could feel the ground tremble slightly with each footfall. Finally they stopped and the tall door began to swing slowly inward. Alice looked up nervously to catch her first sight of

this monster which had summoned her. Finally the door was completely open and still she could see nothing. Were the nucleons invisible?

"Here I am," snapped an irritated voice, from somewhere below the level of Alice's knees. Startled, she looked down and there, standing in front of her, was a small figure. It looked not unlike the electrons she had seen before, except that somehow there was an aura of power about it and, like her companion, it was wearing dark glasses. However, when Alice remembered how far she had shrunk on her way to Castle Rutherford, she realized that this figure must be far, far smaller than the electrons had appeared to her before.

"I thought you told me that the nucleons were larger than the electrons!" she exclaimed, turning indignantly to the photon. She felt angry that she had been so deceived.

"Why, most informed citizens agree that they are indeed larger and I am sure you would not wish to question my word over so small a matter. Of course the nucleons are much heavier than the electrons and so they are inclined to be that much more localized. As they are two thousand times heavier, they naturally have two thousand times more rest mass energy, and it is widely accepted that they are in the region of two thousand times more localized, even when they are having the same energy as an electron type guy. This means that they are apt to occupy less space and so they may *seem* to be smaller than the electrons, but informed opinion is that they are in actual fact larger.

"Compared to the citizens of the Nucleus, the atomic electrons are such parties as have very little energy or momentum at all and are by no means well localized. They form considerable electron clouds which hang around in the vicinity of the nucleus and are very large indeed. They spread out over a volume which is hundreds of thousands of times farther across than the same nucleus." As Alice looked around she could see great gray clouds surrounding them, clouds which stretched away as far as the eye could see. It was strange to think that these were the electrons that she had seen so often before, but now seen from the viewpoint of a much more compact scale.

The neutron which had greeted them (for such it was) was becoming increasingly impatient with this exchange. "Don't just stand there, whoever you may be," it snapped querulously. "Come closer so that I can identify you."

"Why, he cannot see us," realized Alice. "I do believe that he is blind!"

"All neutrons are in such condition, as most people admit," replied her escort. "These parties are not such as have any interaction with photons, or hardly any, having no electric charge of their own. Neutrons are citizens who do not have much long-range interaction whatsoever, being only given to interactions of very short range indeed. Such a party is not much at recognizing others until they are close enough to touch."

They moved up close to the neutron until he bumped into them. "Ah, there you are!" he exclaimed sharply. "Come in and let me shut the door. It is much cosier inside." He ignored the photon, of whom he was largely unaware. Alice was interested to note that the photon simply faded into the castle's fortifications, which were after all composed of the virtual photons emitted by the charge of the Nucleus.

Alice followed the neutron into the Castle while he felt his way down a rough stone corridor. This passage was very narrow, but seemed obligingly to widen at their approach so that there was always just enough room to pass through. Alice found this behavior rather eerie, but she was never sufficiently sure that it was actually happening to make any comment. Now that she had met him, the nucleon whom she was following did not seem as threatening as she had feared. Impatient yes, but not in any way *sinister*. He reminded Alice of a distant uncle of hers.

Together they entered a tall vaulted central chamber of bare stone. The walls rose sheer on every side and vanished into the shadows of the ceiling. Around the walls overhead were arched openings leading to various higher levels, vaguely reminiscent of the electron energy levels that Alice had seen in the atom outside. The floor area was of moderate size and was crowded with as many particles as it could contain, but as Alice and her companion entered she clearly observed that the massive stone walls drew back slightly to create just the right amount of extra space needed to accommodate the new occupants.

Alice was quite sure of what she had seen this time and commented on the movement. "That is the effect of the self-consistent field within the castle," she was told.

"Like electrons and all other particles, we nucleons have to occupy quantum states, and the available states here are controlled by the local potential well. In the case of the electrons in the atom, that potential well is provided by *us*. The electron states are fixed by the electrical potential and we control that potential. The atom is our territory and the potential energy of the electrons within it is controlled by their distance from the positive electric charge of the protons in the central Nucleus. By means of the electrical potential produced by this charge, we in the Nucleus control the electron states, and the electrons must fit into them as best they can. In our own case the situation is different. We ourselves provide the potential for our own nuclear states."

"If you provide the potential in both cases, surely that makes the two cases the same," protested Alice.

"No, it makes the two cases quite different. You see, in the atom the potential is provided mostly by the Nucleus so that the Nucleus controls the states although the nucleons do not themselves make use of them. The potential controls the states which give the probability distributions for the electrons, but the electrons which use them have little effect on the potential. The atomic potential is much the same wherever the electrons may happen to be."

"For the Nucleus, on the other hand, the potential that we are now in is produced by the collective effort of all the nucleons within it. We have a very democratic system ourselves, though we rule the electrons autocratically. Our collective potential fixes the states which are available for us nucleons to occupy and so controls our probability distribution. This distribution subsequently controls the potential, as I said at the beginning. It is a vicious circle, as you might expect for the Nuclear Family, and you can see that the states we inhabit will naturally change as the distribution of nucleons changes."

"Is the nuclear potential produced by the same electric charge as the potential which holds the electrons?" asked Alice, who thought that she should get this point clear in her mind.

"Oh no, quite the reverse in fact. The electric charge in the nucleus is all carried by protons. You are bound to see some protons over there." He waved in the direction of the nearby particles. Alice glanced over and could see more neutrons, which looked just like her companion. Scattered among them were some other particles which looked distinctly more assertive. Where the neutron had been slightly irritable, these appeared to be in a state of barely suppressed fury. "The protons all carry positive charges, and particles which have the same sort of charge repel one another, you know. Protons are forever flying into a temper with each other and threatening to rush off. It is very difficult to keep them together, I can tell you."

"Don't the electrons have the same problem then? I should have thought that they would. If all electrons have a negative electric charge, then any two of them will have the same sort of charge and should repel one another."

"That is quite true; they do repel one another. However, you must realize that the electrons are relatively spread out and diffuse, and their charges are widely separated, so the repulsion they produce is fairly weak. The attractive force from the concentrated positive charge in the Nucleus is able to keep them in order. The protons in the Nucleus are crowded close together, so their repulsive force is very strong. The electrical forces threaten to tear the Nucleus apart."

See end-of-chapter note 1

"In that case what does keep you all together?" asked Alice reasonably.

"That is achieved by a completely different force, a strong force—in fact, the *strong nuclear interaction* is what it is called.

"The strong nuclear interaction is very powerful. It is able to overcome the disruptive electrical repulsion within the nucleus, even though it has no obvious effects outside the nucleus. It is a *short-range* force you see. Inside the Nucleus the nuclear forces are dominant, but outside there is little sign of them, and all that anyone sees is the electrical field due to the positive charges carried by the protons. We nucleons hold firmly onto our immediate neighbors when they are within reach, but we are not really aware of those farther away in the crowd and have very little effect on them."

Ever since she had entered the central hall of the castle, Alice had felt rather uncomfortable. Now she experienced a peculiarly eerie feeling and sensed that something was now in the chamber which had not been there just before. She looked around her and could see nothing. Then she looked upward toward the ceiling. She dimly preceived the great curved flank of some vast rounded shape which passed through the dim shadows of the soaring space above her head. It was obviously but a small part of some much larger object which looked vague and tenuous, like a ghost, and which was drifting through the surrounding walls as if they did not exist.

Alice exclaimed aloud, and then she had to describe what she was seeing to the neutron, who had not been able to see it of course. "Ah, that will be an electron," he said. "They fill the entire volume of the atom you know, which means that they pass through the Nucleus as well as elsewhere. Electrons are completely unaffected by the strong interaction, so they are not aware of us when they do pass through. The nucleus is a tiny part of the volume occupied by electrons, so we do not see very much of them here. Well, actually I do not see them at all, but you know what I mean."

"Is this strong interaction not caused by photons then?" Alice inquired. She had been told that photon exchange held atoms together, but she had understood that that was due to the interaction between electric charges, and she gathered that this was something quite different.

"You are right, it is nothing to do with photons. It *is* caused by particle exchange—all interactions are—but it involves a different sort of particle. The strong interaction is in fact caused by the exchange of many different particles, the most evident of these being called pions. These are of necessity bosons, as they are created and destroyed during the exchange process. Pions have much greater mass than photons. Indeed photons do not have any mass at all, which makes them quite inexpensive to create, in energy terms. Pions have a mass about three hundred times that of an electron. They may still be created using an energy fluctuation, as allowed by the Heisenberg relation, but the fluctuation must be very large to provide the rest-mass energy of the pion, so it cannot last for long. In the available time the pions cannot get far from their source, so they can only be exchanged with particles which are close at hand, almost touching in fact. The strong interaction is consequently of very short range."

At this point a disturbance broke out. Two of the protons had had a sudden and violent argument and were threatening to storm off in opposite directions. Neutrons rushed in to separate the contestants and keep them well apart, so diluting the strength of their mutual repulsion. While the neutrons crowded between the protons to increase their separation, they also grasped them firmly to hold them within the Nucleus.

"You see how we neutrons are necessary to hold the Nucleus together, particularly in the larger nuclei," remarked a neutron. "In a Nucleus every proton repels every other proton, not just those immediately next to them, as is the case for the strong interaction. The repulsion rises rapidly with the number of protons in the Nucleus, and this means that heavy nuclei, which have a large number of protons, need proportionately more neutrons to keep them well away from one another so that their repulsion does not overwhelm the attractive force exerted by their immediate neighbors.

"The Family of nucleons comes from two distinct clans, the protons and the neutrons. The lineage displayed on the wall over there shows how they combine." He indicated a large diagram hanging on the wall, among various other symbols and heraldic decorations. This showed a large and fanciful drawing of a proton and a neutron at the top two corners of the chart. Down the center were listed all the different nuclei in which the Family were involved. Alice saw that they were identified by the same labels that she had seen marking the different atoms at the Mendeleev Marina. On close examination she noted that the labels were slightly different: There was another number given for each one. Now the nuclei were given as, $_1H^1$, $_2He^4$, $_3Li^7$ and so on.

From the original proton and neutron at the top of the picture, lines were drawn to the various nuclei listed. There was one line from the proton to the $_1H^1$ nucleus and no line at all from the neutron. To the $_2He^4$ nucleus there were two lines from the proton and two from the neutron. Thereafter many nuclei had approximately equal numbers of lines from the proton and from the neutron. As Alice looked toward the bottom of the chart she saw that each nucleus depicted there had many more neutron lines than proton ones.

"That chart shows how the different nuclei are populated from the two distinct clans of nucleons. The first number tells you the number of protons involved. This is the same as the number of electrons which can be controlled and hence decides the chemical behavior of the atom. The second number gives the total number of nucleons which populate a given Nucleus.

"Lighter nuclei have the same numbers of protons as of neutrons. A carbon nucleus, for instance, contains six protons and six neutrons. The repulsion given by six protons, each repelled by every single one of the five other protons, is still not enough to overcome the attraction caused by the strong interaction. Here in our uranium Nucleus, on the other hand, we have 92 protons. The repulsive force between all the different pairs of protons is now very large, so a relatively large number of neutrons is needed to keep the protons apart and dilute their electrical repulsion. In our Nucleus we have all of 143 neutrons. The number of neutrons need not be quite the same in every uranium nucleus. For a given element the number of protons is always the same, since this fixes the number of electrons and hence the chemical behavior, but the number of neutrons does not have much effect on the chemistry of the atom and can vary slightly from one Nucleus to another. Nuclei of an element which have different numbers of neutrons are known as *isotopes*. We have 143 neutrons in this Nucleus, as I said, but many uranium nuclei have 146, which makes them a little more stable."

"I have heard of stability before," said Alice. "I thought that atoms were completely unvarying, and, although they might take part in different compounds, the atoms themselves last forever."

"Not entirely. The walls of the nuclear potential barrier serve to keep us inside, just as the coulomb barrier keeps other protons out. Occasionally, however, there is penetration, and the Nucleus is changed in some way. It works both ways; particles outside the Nucleus might break in, or some from among our complement may try to escape.

"The reason that protons and neutrons stay in the Nucleus is the same as the reason that electrons stay in the atom: because they require less energy where they are than they would if they were outside. The decrease in energy from the value they would have outside the Nucleus is called the nuclear *binding energy*, or BE. There are energy levels for nucleons within the Nucleus in much the same way as for electrons in the atom, and, as neutrons are not identical to protons, these levels may be filled with neutrons and with protons independently. Because the level-filling process is the same for neutrons and protons, stable nuclei tend to have equal numbers of the two types. For the heavier nuclei, which have larger numbers of protons, the proportion of neutrons is greater, as I have described already. For all nuclei there is a ratio of protons to neutrons which gives the most stable atom. An excess of either type will give a tendency to instability and some form of decay. I am forced to admit that, in uranium, the repulsion between the protons is so great that the Nucleus is barely stable at the best of times. Any disruption of the balance between protons and neutrons could well be disastrous."

Suddenly an alarm trumpet sounded and a strident voice echoed through the vaulted chamber. "Alert! Alert! Condition Alpha. We have an escape attempt in process."

I n large nuclei with many nucleons the repulsion between all the protons becomes proportionately stronger and the nuclei may be *unstable*. They may undergo *radioactive decay*, where the nucleus emits an α particle, which is a tightly bound group of two neutrons and two protons which penetrate through the coulomb barrier. Neutrons may also undergo β (beta) decay, where an electron is created within the nucleus and immediately escapes because electrons are not affected by the strong interaction. Nuclei may also emit Γ's (gammas), which are just high-energy photons.

Alice looked around to see if she could see any cause for this alarm. Everything looked much as before. There was considerable movement among the assembled nucleons, but then they, like other particles which she had encountered, were always in continual agitation, so that was nothing new. As she watched carefully she noticed that a small group of particles, two protons and two neutrons, were moving together through the crowd, holding tightly to one another. They would rush up to the wall, collide with it and bounce back, and rush across the chamber to collide with the opposite wall. Alice was strongly reminded of the person she had seen trying to penetrate his locked door when she first arrived in Quantumland.

She commented on this to her companion, and he replied, "That is alpha particle clustering that you are describing. An alpha particle is a group of two protons and two neutrons which will bind together so tightly that they act as one particle. As it contains two protons the alpha particle is repelled by the overall positive charge of the protons and is trying to escape, but is prevented by the wall around the nucleus. The group is trying to *tunnel* out. They are planning to escape by barrier penetration, and sooner or later, of course, they will succeed."

"How long is it likely to take them to manage it?" asked Alice curiously.

"Oh a few thousand years, I should think."

"Don't you think it is a bit premature to sound the alarm then?" inquired Alice. "It sounds to me as if you have plenty of time to deal with such an escape without having to panic!"

"Ah, but we cannot be sure of that. It will *probably* take them thousands of years to escape, but they *might* get out at any moment. There is no way of being sure; it is all a matter of probability."

"Are all escapes from the Nucleus by barrier penetration then?" asked Alice.

"Not at all. Alpha emission is by barrier penetration, as I have just stated. We also get beta and gamma emission, and neither of those requires barrier penetration."

"What are they, then?" asked Alice dutifully. She suspected that she was about to be told whether or not she asked, but it seemed more polite to inquire.

"Gamma emission is photon emission, much as you get from the electrons in the atom. When an electron has been excited to a high state and then drops back to a lower one, it will emit a photon to carry off the energy released. The same thing happens when an excitation of the nucleus rearranges the charged protons: A photon is emitted when the nucleus returns to the state of lower energy. Because the interaction energies in the nucleus are so much greater than in the atom generally, the gamma photons have much higher energy than those from the atomic electrons. Indeed they will have some hundred thousand times more energy, but they are still photons.

"Beta emission is the emission of an electron from the Nucleus," her informant continued.

"I thought you said that there were no electrons in the Nucleus," protested Alice. "You said that the electrons were not aware of the strong interaction and just drifted through occasionally."

"That is quite true. There are no electrons in the Nucleus."

"If the Nucleus cannot hold electrons and there are no electrons in the Nucleus," said Alice patiently, "how can one escape from it? That does not make any sense. It cannot escape unless it is there to begin with."

"It is because the Nucleus cannot hold electrons that they *do* escape from it so readily. The electrons are produced right inside the Nucleus in a weak interaction, and, of course, as the Nucleus cannot hold them they immediately escape. It is quite straightforward when you think about it," said the neutron kindly.

"That may be," said Alice, who felt that it was not *at all* clear yet, "but what is a weak interaction? How do the electrons. . . ?"

Once again a trumpet sounded and a herald somewhere in the top of the chamber cried out. "Attention everyone. The Castle is under attack! We are besieged by a hot plasma of charged particles."

"Oh dear!" cried Alice. "That sounds serious."

"No, it isn't really," replied a nearby neutron soothingly. "None of the charged particles in the plasma are likely to have enough energy to breach our defenses. Come and see."

He led Alice up through the various galleries and energy levels within the Castle until they came to a position from which Alice could view the outside. She saw other nuclear castles in the distance and, spread across the plain, a number of protons moving quickly around. "Those protons are from a hot hydrogen plasma," Alice's companion told her. "In a plasma the atoms have lost some of their electrons and become positive ions with an overall positive charge. The nucleus of hydrogen contains only a single proton, so when a hydrogen atom loses its electron there is nothing left but a proton. Plasmas can be made very hot, and then the protons rush about with a lot of energy, but not enough for them to break in here," he finished complacently.

Alice watched as some protons came running toward a nucleus and on up the curving base of its wall. As they rushed upward they moved more and more slowly as they lost their kinetic energy, eventually coming to a halt a short way up the wall. From that point they slipped back down again and rushed off in a different direction than that from which they had come.

"You will see, even if I can't, that they are having no success at all at actually getting inside," continued Alice's guide.

"Could they not get in by barrier penetration then?" asked Alice.

"Well, yes. They could in *principle,* but they spend so little time near the Nucleus that it is really most unlikely."

At this point Alice noticed a disturbance in the distance. Something was getting closer at a most remarkable speed. "What is that approaching?" she asked, rather anxiously.

"I have no idea," answered the neutron. "Is there something approaching?"

Alice realized that the neutron would naturally be unaware of the approach of the fast charged particle as it came galloping up, with plumes of scarcely seen virtual photons trailing from it in its whirlwind passage. As Alice was describing its appearance to the neutron, the newcomer arrived at a Castle in his path. With little apparent reduction in his mad onward rush, he ran up the barrier wall and over the top. A moment later Alice saw him galloping off into the distance, apparently little affected by his encounter. She could not say the same for the Nucleus he had entered. This had burst completely asunder, and large portions of it were flying off in different directions. Alice completed her description of the event.

"Ah, that would be a Cosmic Ray-der. We very occasionally get one passing by. They come from somewhere way outside our world and they have enormous energy. To them the energy needed to cross the coulomb barrier of a Nucleus is as nothing and it presents no barrier at all. We have no defense against them, but fortunately they are, as I said, very rare."

Looking down on the area outside Alice could make out a few unobtrusive figures moving quite slowly and stealthily about. "Oh look!" she cried, forgetting who her companion was. "There are some neutrons moving about out there."

"What?" cried the neutron by her side. "Are you sure? This is serious. Come, we must get down to the main hall at once."

He rushed Alice back down through the successive energy levels to the hall she had first entered, ignoring her protest that there had not been very many neutrons outside and that they had not had much energy at all, really.

Hardly had they arrived when, without warning, an invading neutron popped right through the wall and landed in the middle of the chamber, on top of all the other particles. This was not one of the normal occupants of the Nucleus, but one of the foreign neutrons which had come in from outside. Alice remembered that the virtual photon had told her how the coulomb barrier had no effect on neutral particles and how she herself had come in through the barrier without difficulty. In the same way this neutron had entered uninvited.

There was immediately a great bustle and panic among all the nucleons. They rushed to-and-fro in consternation, surging from one gallery to the next, calling out that the stability of the nucleus had been totally upset by the addition of this excess neutron. As they surged to-and-fro, Alice was much alarmed to discover that the whole room was shaking violently in sympathy. The massive stone walls were quivering like a vibrating drop of liquid. One moment the chamber would be square and compact, the next it would stretch out very long and thin. A narrow neck formed in the middle close to where Alice was standing, so that the room was almost separated in two. Back and forth the walls swung, and each

The electrical potential of the nucleus gives a coulomb barrier which repels positively charged particles. Protons of low energy are unable to pass above this barrier. They could in principle pass *through* it by barrier penetrtation, but the probability of this is low as they are only passing by and have only a fleeting interaction with the nucleus. Some particles in cosmic radiation have sufficient energy to overcome the barrier and can readily pass through, giving enough energy to the nucleus in passing to disrupt it completely.

Neutrons do not have any electrical charge, so for them the barrier does not exist. A neutron which happens to collide with a nucleus can pass straight through.

time the room became narrower and narrower at the midpoint. The room stretched out for a last time. Alice saw the far walls rushing away in opposite directions while the nearer walls came closing in as if they were going to crush her and the particles which were in her vicinity. Previously the movement had always reversed before the gap closed, but this time the walls clashed together, just where Alice was standing together with a few neutrons.

When the walls had moved through her, Alice found that she was back on the plain outside the Castle. She looked back toward it and saw that the tall, dark tower was split by a fissure which ran all the way down its middle. As she watched, the Castle was torn into two half towers, which slumped apart. Each one was shaking violently, its outer surface vibrating wildly like a bag full of jelly. High-energy photons soared from the two castles like some dramatic fireworks display as both of them shed their surplus energy. Gradually the shaking died down and both the irregular shapes flowed into the same tall soaring shape which she had first seen. Two smaller replicas of Castle Rutherford now stood before her, except that they did not stand but slid rapidly away from one another, driven by the positive charge which they had previously shared between them.

"Come, I am glad that is over. It was really rather frightening," Alice admitted to herself. As she looked around the now quiet landscape she could see a few neutrons which had been ejected with her from the Castle when it had split in two. They spread out over the plane, rushing off in random directions. As she watched, one arrived by chance at the distant shape of another nuclear castle and promptly dived into it through its side.

For a short time nothing appeared to happen. Then she could see this castle also begin to shake. The shaking increased until suddenly the castle split down the middle. "Oh no!" cried Alice in dismay as she saw the two halves thrust apart, spitting out energetic photons. Almost unnoticed, a fresh group of neutrons ran away from the scene of the catastrophe.

Before much time had passed, a couple of the neutrons which were now roaming aimlessly around the plain had chanced upon and entered other nuclei. Again the process repeated, ending once again with these nuclei splitting, more gamma photons pouring onto the scene, and more neutrons being ejected to roam around in confusion. Again and again the process was repeated. Soon there were four nuclei all in the pangs of separation, then ten, twenty, fifty. All around her, Alice could see nuclear castles falling apart in fiery fission, while overhead the scene blazed with the intense, vivid radiation of high-energy photons.

"This is terrible!" cried Alice in horror. "Whatever can be happening?"

"Do not worry, Alice," said a calm voice by her side. "It is only induced nuclear fission. A chain reaction, you know. It is nothing for you to worry about. It is just that you are standing in the middle of what, in your world, would be called a nuclear explosion."

Alice whirled around and saw the mild features of the Quantum Mechanic. "You do not have to worry," he said again. "The energies involved in a fission reaction are less than those you have already met within the Nucleus itself. The only problem is that they are no longer confined within the Nucleus. I have been looking for you," he continued, still calmly, "as I have an invitation to give you."

He presented Alice with a stiff, ornately engraved invitation card. "It is an invitation to the Particle MASSquerade, a party which is being held for all the elementary particles," he said.

> ome nuclei may split into two smaller and more stable nuclei, a process known as *nuclear fission*. This may be caused by the addition of an extra neutron, which is not kept out by the coulomb barrier and is the "last straw" for an already unstable nucleus. The fission may release several other neutrons, leading to a *chain reaction*.

Notes

1. Almost everything in the physical world may be seen as caused by the interplay between electrons and photons, virtual or otherwise. The properties of solids, of individual atoms, and of the chemical behavior which comes from the interplay between atoms, all reduce to an electrical interaction between electrons. As well as the electrons which interact with the rest of the world, there is within the atom a positively charged nucleus. The nucleus is not held together by electrical forces, quite the reverse in fact.

   The atomic nucleus contains neutrons, which have no electrical charge, and protons which are positively charged. Within the small space of the nucleus, whose radius is typically a hundred thousand times smaller than the overall size of an atom, the mutual repulsive force of the protons is enormous. This electrical force tends to tear the nucleus apart, so there must be an even stronger force which holds the nucleus together, one that, for some reason, is not evident elsewhere. Such a force exists, and it is called the *strong nuclear interaction*. Although it is strong, it has a very short range, so that its effects are not obvious outside the nucleus. This strong interaction is produced by the exchange of virtual particles, just as the electrical interaction is produced by photon exchange. Photons have no rest mass, but the exchanged particles in the strong interaction are relatively heavy. They must get their rest mass energy through a particularly large quantum fluctuation, which is only possible for a very short time. Such heavy virtual particles are very short-lived and unable to travel far from their source, so that the interaction they produce is consequently of short range.

9 The Particle MASSquerade

Clutching her invitation, Alice climbed up the broad stone steps which led to the tall polished door. She could not remember how she had come to be there, though she remembered being given the invitation. "So I expect this is probably the right place for the MASSquerade, whatever that may be," she said to herself encouragingly. "I always seem to end up somehow where people want me to be."

She stopped outside the door and examined it. Its paint was very smooth and glossy, deep red in color. It had a shiny brass doorknob and an equally shiny brass knocker in the shape of a grotesque face. It was also closed and locked. Cheerful candlelight streamed from the keyhole and Alice could hear loud music being played within.

How was she to enter? The answer seemed obvious enough, so she firmly grasped the knocker and hammered loudly.

"Ow! Do you mind!" exclaimed an anguished voice from close at hand. Literally at hand, in fact. Alice stared in surprise at the door, to meet the furious glare of an irate door knocker.

"That was my nose!" it exclaimed indignantly. "What do you want anyway?"

"I am really sorry," said Alice, "but I thought that, as you *are* a door knocker, I might use you to knock at the door. How am I to get in if I do not knock?" she asked plaintively.

"There's no use in knocking," said the knocker huffily. "They are

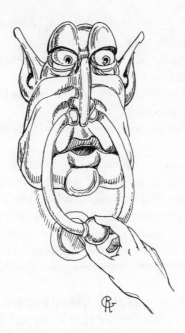

making such a noise inside no one could possibly hear you." And certainly there was a great deal of noise going on within: a buzz of conversation, a louder voice speaking above the rest, but still not quite audible through the door, and, above it all, the sound of the music.

"How am I to get in then?" asked Alice, in some frustration.

"Are you to get in at all?" said the door knocker. "That's the first question, you know."

It was, no doubt, but Alice did not like to be told so. "It is really dreadful," she muttered to herself, "the way everyone will argue so." Raising her voice she addressed the knocker, though she felt a little self-conscious in talking to a door knocker at all. "I have an invitation," she said, holding it up in front of his face.

"So I see," replied the knocker. "That is an invitation to the Particle MASSquerade, which is a function for particles only. Are you a particle?"

"I am sure that I don't know," declared Alice. "I did not think that I was, but with all the things that have happened to me, I am beginning to feel that I must be."

"Well, let me see if you meet the requirements," said the knocker, rather more agreeably now that its nose felt recovered. "Let me just look at my notes for a moment." Alice did not see how a door knocker could keep notes, let alone look at them, but after a short pause the knocker continued. "Ah, yes. Here we are. The list of specifications to define a particle."

"One," it read out. "Whenever you are observed, are you invariably observed in a reasonably well-defined position?"

"Yes, I think so, as far as I know," answered Alice.

"That's fine," said the knocker encouragingly.

"Two. Do you have a unique and well-defined mass—apart from the normal fluctuations, of course."

"Well, yes. My weight has not changed very much for some time." That was what Alice believed, at any rate.

"Good, that is a very important requirement. All the different particles have their particular masses. It is one of their most distinctive features and very useful when it comes to telling one particle from another." Alice was rather taken with the notion that people might be identified by weighing them rather than looking at their faces, but she realized that particles did not on the whole have anything very definite in the way of faces.

"Three. Are you stable?"

"I beg your pardon?" said Alice, feeling distinctly affronted.

"I said, 'Are you stable?' It is a simple enough question. Or at least it ought to be: The requirement has become increasingly blurred recently. It used to mean quite simply, 'Do you decay to something else?' If you were likely to decay at any time in the future, then you were unstable, and that was that. But that wasn't good enough! People started to say, `We cannot be sure that anything lives forever, so a distinct state that exists for a long enough time can be classed as a particle.' Then the question is, 'What counts as long enough?' Is it years, or seconds, or what? At the moment they accept lifetimes of less than a hundred million millionth of a second as being stable," he finished disgustedly. "So, I must now ask

There are many strongly interacting particles as well as the proton and the neutron. It is not all that easy to distinguish one type of particle from another. Some have different electric charges, but there are many with the same charge. Particles are usually distinguished in practice by measuring their masses, which are fairly distinctive. Most of the particles are to some degree unstable: a heavier particle decaying into lighter ones. Outside of a nucleus even the neutron is unstable, with a mean lifetime of about 20 minutes.

you: Do you expect to survive for longer than a hundred million millionth of a second?"

"Oh yes, I should think so," answered Alice confidently.

"Good, then I can count you as a stable particle. You had better go inside. You may not have anything better to do than stand about out here, but I have," grumbled the door knocker. There was a click and the door swung open. Alice lost no time in passing through it.

Inside she walked through an elegant entrance hall, with pale paneled walls, chandeliers, and alcoves containing statues. As they were all statues of notable particles, it was rather difficult for Alice to make out much detail. She thought it was rather clever the way the sculptor had managed to make the features of a statue appear so vague and unlocalized. In fact, to the uninitiated, they looked much like shapeless pieces of stone.

Beyond the entrance hall Alice entered a large room, which seemed to be a main ballroom or salon. It was lit by ornamental chandeliers hanging from the ceiling, but somehow they did not give much light and the room was mostly in shadow. The shadows were made more intense by contrast with a few bright spotlights which spun around the room. One came to rest in a circle of light immediately in front of Alice. Into the center of this circle leaped a figure clad much like the joker in a pack of cards. His ridiculously cut costume was gaily striped in red, blue, and green. On closer examination, Alice saw that it was also striped in *antired, antiblue,* and *antigreen.* Alice had never seen such colors before. (Unfortunately this book does not have colored illustrations, so you cannot see what these colors look like.) His fantastic appearance was completed by a mask, which was set in an unbelievably wide permanent smile.

He addressed Alice. "Bon soir, mademoiselle. Guten Abend, Fraulein. Good evening, young lady. Willkommen. Bienvenue. Welcome. Welcome to the MASSquerade."

"Thank you," replied Alice. "But who are you and what is a MASSquerade?"

"I am the Master of Ceremonies for this MASSquerade," he replied, "which is the Masked Dance of the Particles. An evening of Revelry and Revelation. An Exploration of the Mystery behind the Mask. The particles all come here to whirl about in joyous dance and, at suitable occasions, they unmask. Your mask, if I may say so, is particularly inspired," he added.

"I am not wearing a mask," said Alice coldly.

"Ah, but can you be certain of that? We all wear masks of some sort. Why, tonight we have had two unmaskings already."

"I do not see how that can be," challenged Alice. "You can only unmask once. You are either wearing a mask or you are not, surely."

"Why, it depends how many masks you wear. Particles can wear many masks. At the beginning of the evening we had a group of atoms, and then these unmasked to reveal themselves as a crowd of electrons and a number of nuclei. Later in the evening we had a further time of unmasking and the nuclei shed their disguise to show that they were actually neutrons and protons, with a sprinkling of pions among them. I confidently anticipate that there will be further revelations before the night is over.

"But now," he cried out in a voice suddenly loud enough to carry throughout the whole room, "on with the festivities! Mesdames et Messieurs, Damen und Herren, Ladies and Gentlemen, I pray you to step lively in a Collider-Dance."

There was a bustle of movement and Alice saw that the assembled particles were beginning to circulate around the room. She could not truly say that they were dancing, but they were certainly going around and around, with ever-increasing speed. The main problem was that there seemed to be no general agreement on the *direction* in which they were to circulate, so some went round one way and some the other.

Faster and faster the circulating bunches of particles rushed through one another. Before long the inevitable happened and two particles collided with a great crash. Alice looked over in concern to see if they had been hurt in the collision. She could not really determine whether they had been hurt, but they were certainly not the same after their interaction. She saw several small pions rush away from the collision, which she did not believe had been there before, and the colliding particles themselves were changed into something quite new. They were larger and somewhat more exotic particles than they had been—definitely they were not the same.

The dance continued and there were further collisions, more and more as time went on. With every one which took place, relatively familiar nuclear particles changed to something new and strange. Soon there was a bewildering variety of different particles present—far more types than Alice had seen before or than she had imagined to exist.

"A marvelous sight, is it not?" inquired a voice by Alice's ear. It was the Master of Ceremonies, his grinning mask a mere arm's length away. "Such a fine hadronic assembly of particulate revelers. Such a splendor of baryonic variety. Why, I do believe that there are now no two of them the same!"

Alice did not understand many of the words he had used and felt that it was wisest not to ask about them. She wanted to know, as simply as possible, just what had been happening. "Where have all these new types of particles come from?" she asked.

Particles may be created in collision processes, the kinetic energy of the colliding particles being converted to produce the rest mass energy of the new particles. Very many such particles were discovered and were classified in various *symmetry groups,* but they are now known to be different combinations of quarks, rather as atoms are combinations of electrons with protons and neutrons in their nucleus. The fermions, or *baryons,* contain three quarks, while the bosons, or *mesons,* contain a quark and an antiquark.

"They have been created in the collisions, of course. As you saw, the particles were all circulating very quickly indeed, so they each had a large amount of kinetic energy. When they collided, this kinetic energy could be converted to rest mass energy, so that particles of higher mass could be created. In the different collisions which took place, different particles were produced. Each one has its own distinctive rest mass which serves conveniently to identify it, though there are also other, more subtle, differences. I expect that by now there are no two strongly interacting particles present here which have the same mass. That is what happens at a MASSquerade."

Once again his voice became loud as he addressed the whole room. "The dance is now finished. Please assemble in your appropriate multiplets."

At his request the assembled particles began to gather together into separate little clumps, scattered around the room. Alice saw that mostly they gathered into groups of eight particles, six arranged in the form of a hexagon around the outside and two together in the middle. A few groups contained ten particles in a triangular layout which had four of the particles spaced across its base.

"There you see the particles gathered into their symmetry groups," the Master of Ceremonies said quietly to Alice. "These groups are collections of particles which all have the same values for some property, such as spin. You can see that there is a striking regularity in all the different groupings. This provides an indication of an underlying similarity beneath the skin, or rather beneath the mask. You may recognize some of the members of that nearest group," he added.

Alice looked at the eight particles nearby and saw that the two on the top edge of the six-sided pattern were a proton and a neutron. The others, however, were unknown to her.

"That is a group of baryons which all have a spin of one-half," she was told. That meant nothing at all to Alice, but for the moment she was quite prepared to believe it.

"The neutron and proton I believe you have already met. In the next row you have the sigma particle, which can manifest with both positive and negative electric charge and also with no charge at all. It consequently appears as if it were three different particles. In the center of the pattern you have the lambda, which is a single particle with no charge. These are all strange particles," he added.
See end-of-chapter note 1

"They all seem very strange to me," agreed Alice, as she came over to look at them more closely.

"No, no. *Strangeness* is simply a property possessed by certain parti-

cles and which happens to have been given the *name* of strangeness. Rather like electric charge, you know—except that it is totally different," he added unhelpfully. "The remaining two particles are both the Cascade. It comes in two different charge states, so there are two of it," he explained. "It is doubly strange, of course."

"Of course," echoed poor Alice.

"And now the time is upon us," he called out suddenly, speaking loudly and clearly so that his voice carried through the whole room. "Now is the time for the final unmasking of the evening. Mesdames et Messieurs, Damen und Herren, Ladies and Gentlemen, I bid you all. . . unmask!"

Just how it was done Alice could never quite decide, but all around her the aspect of the particles was changed. She looked at the particle nearest to her, which was the one the Master of Ceremonies had called the lambda. It no longer looked like a particle, but like a sort of bag, within which she could see three shapes. She came closer, to try to make them out more clearly, and felt herself being pulled within the enclosure. She tried to pull away, but despite her efforts she found herself sucked in.

Once inside, Alice found that there was not sufficient room for her to stand. She tried kneeling down on the floor, but the container still pressed in on her so closely that she tried lying down with one elbow on the floor and the other arm curled around her head.

In this awkward position she looked around and stared at the three small figures which she had dimly glimpsed from outside. Now that she could see them, she observed that they were different from any of the particles which she had so far encountered. Each one was colored in a distinctive shade. One was red, one was green, and one was blue. She noted that they were attached to one another with lengths of multicolored cable of some sort. It was variously striped in the three colors, together with the three anticolors which she had seen on the costume of the Master of Ceremonies.

Alice was so engrossed in studying these odd new particles that she was quite startled to hear a voice coming from one of them.

"If you think we are moving pictures," he said, "you ought to pay, you know. Moving pictures are not made to be looked at for nothing. No how!"

"Contrariwise," he added, "if you think that we are alive, you ought to say hello and shake hands."

"I am sorry," exclaimed Alice contritely as, with some difficulty, she held out her hand. She was not quite sure how it happened, but somehow she found that, instead of a hand, she was holding the large rubber bulb of an old motor horn. When she pressed it there was a loud honking noise.

"Well, who are you then?" she asked, a little irritated by this foolery.

"We need no introduction, so I shall perform it. We are the Three Quark Brothers," answered the spokesperson (spokesparticle?), wiggling heavy eyebrows at her. "I am Uppo, this is Downo, and that is Strangeo over there." Uppo was green, Downo red, and Strangeo blue.

"I hope you do not mind if I join you," said Alice, trying to make light of her awkward position.

"Why? We are not going to come apart," answered Uppo and they all laughed uproariously.

Alice was not amused; she had not found the joke very funny. In fact, on further consideration she was not sure that she had found it at all humorous. She looked at the three brothers in irritation and was struck by the fact that now Uppo was red and Downo was green.

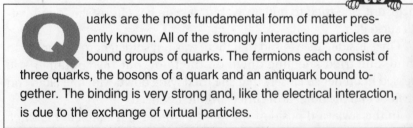

uarks are the most fundamental form of matter presently known. All of the strongly interacting particles are bound groups of quarks. The fermions each consist of three quarks, the bosons of a quark and an antiquark bound together. The binding is very strong and, like the electrical interaction, is due to the exchange of virtual particles.

"You have changed color," she announced in a tone that was almost accusing.

"Naturally," replied Uppo calmly, "we are usually off color. When I started I was quite green, then I felt a little blue, but now I am beginning to see red. You know that particles which have electric charge exchange photons?" he said abruptly.

"Yes, I was told about that," replied Alice.

"Well, we Quarks are colorful characters. We stick together by exchanging gluons. Through thick and thin, or rather through red, green, and blue. The gluons stick around when they see the color of our money; they monitor our color. Particles which have color all exchange gluons. The gluons hold them together in much the same way as photons do for particles with charge."

"But why do you change color?" asked Alice. "Charged particles do not change their electric charge when they exchange photons."

"No, but photons do not carry charge. There is no charge on a photon, which is why they are so popular. Gluons do carry color. When a colored gluon escapes from its source, then that hue is transferred to the Quark that catches it. It is a regular who's hue, I can tell you." As Uppo was speaking, Downo changed his hue to blue, and Strangeo became red, his curly hair taking on a particularly vivid shade.
See end-of-chapter note 2

Uppo indicated Strangeo. "There," he said, "that is a source of a different color!

"It is because we have such colorful gluons that we can never be separated. One for all and all for nothing. United we stand and divided we remain inseparable."

"I am afraid that I do not see what you mean at all," protested Alice.

"Well, we all know that opposite electric charges attract, but you can separate particles which are suffering from that sort of attraction. They are held together by photon exchange, but the photons have no charge."

"If there is-a no charge on photons then-a they free. They go wherever they want," said Downo suddenly.

"Right, because photons have no charge they are free, free to spread out as far as they want. They do not exchange other photons between themselves."

"If there is-a no change and no charge, then there is-a no transaction," added Downo. "These photons, they no do-a business together."

"Without charge the virtual photons have no business with one another, so they do not attract one another. No one gets a charge out of them. So they just spread out all over the place. The farther apart the source charges get, the more place there is for the photons to spread out over. The photons are spread out thin. They have a thin time of it, with less momentum to transfer."

"Last job, I get-a transfer," cut in Downo helpfully. "They say they going to give-a me a little momentum, but all they give-a me was the boot."

"And you felt the force of their argument," replied Uppo. "But with less momentum to give, the force gets weaker. You pull charges far apart, they lose touch, the attraction gets weaker and weaker, and eventually they get so out of touch that they don't even remember to write. Give them enough energy and you can pull them anywhere. They can get so far apart there is no attraction left to speak of. The charges are then quite independent. I expect you know what someone means by an `independent charge,' or what I charge someone with independent means, for that matter?" he added.

"But enough about electric charges, we are here to talk about Quark charges."

"What's a Quark charge?" asked Alice curiously, always anxious to get as much clear as she could manage.

"Double rate on weekends and for up-Quarks," answered Downo. "But we-a very cheap. Our charge only a third of other particles' charge."

"One thing I do not understand," said Alice to Downo. (That was an understatement, as there were many things that she did not understand by now.) "Why do you try to talk as if you were Italian? I do not believe that you are."

Many particles have electrical charges, and it is a striking fact that the observed particles have charges which are all of the same size. Some particles have positive charge and some negative, but the *amount* is the same in each case. This amount is usually called the *charge on the electron,* simply because electrons were the first particles to be discovered. Estimates of the charges possessed by quarks require that they be different. A quark may have a positive charge which is *two-thirds* the size of the electron's charge, or it may have a negative charge which is *one-third* as large as an electron has. As quarks cannot escape from their bound groups, these fractional charges cannot be observed directly, but there is strong evidence that they are correct.

"It is because he is a fermion," replied Uppo on his behalf. "Enrico Fermi was Italian."

"But aren't you all fermions?" protested Alice.

"Certainly, one for all and all for Pauli. Which nobody can deny." All three Quarks stood to attention and saluted.

"We are one group indivisible. A Quark cannot escape from inside a proton or from any other particle. This is all because of the red, green, and blue. There's Old Glory for you."

"Pardon me," began Alice.

"Gesundheit!" answered Uppo, but Alice continued determinedly.

"I don't know what you mean by *glory*."

"Of course you don't—until I tell you. I meant, 'There's a nice knock-down argument for you!'"

"But *glory* doesn't mean that!" protested Alice.

"When I use a word, it means just what I choose it to mean, neither more nor less. The question is, which is to be master—that's all. But it is another question with gluons," he added gloomily. "There is no mastering them, they never let go—not like the photons. The trouble is that the gluons are all colored. And color creates gluons like charge creates photons, so all the gluons emit other gluons, and those gluons emit more gluons. You start with just one or two, and you end up with hundreds. It's like having the wife's family stay. And because they are all exchanging gluons, they all stick together, just like the wife's family. Instead of spreading out in a wide fuzzy cloud like photons, they bunch up to form the tight, colored strings of virtual gluons that you see here. Because they are bunched up they are not free to spread out like the photons. There is no such thing as a free bunch."

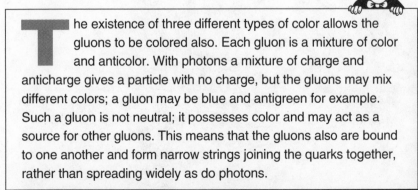

The existence of three different types of color allows the gluons to be colored also. Each gluon is a mixture of color and anticolor. With photons a mixture of charge and anticharge gives a particle with no charge, but the gluons may mix different colors; a gluon may be blue and antigreen for example. Such a gluon is not neutral; it possesses color and may act as a source for other gluons. This means that the gluons also are bound to one another and form narrow strings joining the quarks together, rather than spreading widely as do photons.

"When one Quark moves away he soon comes to the end of his tether. If we have more energy, the gluons will give us more rope, but we are still hanging on the end of it. However far we roam, the gluon attraction pulls us home. We cannot break free, but we can still escape with a little help from our friends."

At that singularly appropriate moment, a photon of very high energy crashed into the little group of Quarks. Alice had had no warning as she had not seen it coming. In fact, as she now realized, photons move so fast that she had never yet seen one coming before it arrived. This photon collided with Strangeo, exciting him to a quite manic frenzy, and he rushed off, honking loudly on a horn. Behind him his tethering rope stretched out farther and farther. Alice could see that, however far it stretched, the rope was not becoming in any way thinner or weaker. It was obvious that it could go on stretching indefinitely and that the escaping Quark would soon run out of energy, with no chance of breaking free. But no sooner had Alice reached this conclusion than . . . the rope broke!

Where a moment before there had been one long and steadily lengthening cord which was soaking up all the energy that the photon had delivered, now there were *two* very short lengths with a large and steadily widening gap between them. On either side of this break had appeared a new Quark, each anchoring one of the fractured ends of the ropes. On the end of the rope which was still attached to the two Quarks who had remained with Alice was a Quark who looked exactly like Downo, apart from being a different color. The rapidly receding Strangeo was trailing his own short length of rope, to which was attached a

reversed version of Downo. This Alice assumed correctly to be an anti-Quark. "Whatever has happened?" asked Alice in some confusion.

"You have just seen a Quark escape with the help of friends in low places. In the vacuum, in fact, and you can't get much lower than that. You cannot detach a gluon rope once it has seen the color of a Quark, so we have to fool it with something which looks just like a Quark."

"And what is that?" asked Alice.

"Another Quark, of course. When the gluon string has stretched long enough so that it now contains enough energy to create the rest masses of two Quarks, then we cut the string and work the switch. One end gets a new Quark, the other Not."

"There is-a knot in the string?" asked Downo (one of the Downos).

"That's right, there's a Quark at one end, Not-a-Quark at the other."

"What's a Not-a-Quark?" asked Alice.

"An anti-Quark. And if you believe that, you should see my uncle. Part of the original string has vanished rapidly into the distance, carrying off the energy and connecting the absent Strangeo to the new anti-Quark. So, you see, absence makes the part go yonder."

"He may have escaped, but he still isn't free," protested Alice.

"With a bound he was free. He is free of us now, but he is still bound. With his anti-Quark he is bound into a boson. That's like a pion, but pions can be deceiving and in this case they have formed a kaon instead. You do not see a free Quark—or even free a Quark sea, but that's another kettle of fish."

"Are the fish in-a the Quark sea?" asked Downo.

"No, there's nothing fishy about the Quark sea. Its sole purpose is to hold virtual Quark-Antiquark pairs."

"The sole I understand, and the porpoise I understand, but why the pears in-a the sea?" argued Downo.

"Forget the sea," replied Uppo, "or we will all be at sea. The point is that you will never find a Quark on his own."

See end-of-chapter note 3

"Does that mean that you have to stay here forever with no chance of a change?" asked Alice in concern.

"Oh, we can have a change all right. They say a change is as good as a rest, but I feel quite at liberty to discuss the weak interaction."

"I heard that mentioned when I was visiting the Nucleus. I believe it had something to do with beta decay of nuclei, whatever that may be."

"It is the same thing. In fact it is a far, far beta thing. What happens is that a neutron inside the Nucleus changes into a proton and an electron, together with another particle called a neutrino. This neutrino has no charge, no mass, and no strong interaction. It doesn't do much of anything really, like most of the folk I know. Anyway, that's the story we tell. What *really* happens is that a down Quark inside the neutron changes to an up Quark, an electron, and a neutrino. When the down Quark changes to an up Quark, then everything is on the up. It puts the charge up, the neutron becomes a proton, and there you are. Hang around, and you might get lucky."

Hardly had he spoken when, by a most convenient coincidence, one of the two Downos became blurred and began to change and lose his identity. After a fleeting moment of transition, Downo was no longer there and in his place stood a duplicate of Uppo. As he moved aside, Alice saw an electron rush away from the same spot. This was followed by yet another particle. Alice caught only the briefest glimpse of this one, something barely perceived and very hard to see at all. This she assumed to have been the neutrino, performing its usual role of ignoring and being ignored by everything and everybody.

The group of three Quarks now consisted of one Downo and two identical Uppos. Identical, that is, but for the fact that one was currently green and the other blue. "My," said Alice. "That was a most remarkable thing altogether."

Obediently the two Uppos replied, in perfect unison, "That was a most remarkable thing."

"But what can you expect," they added, "when the particles exchanged in an interaction have an electric charge? Photons don't have a charge, but this isn't the Charge of the Light Brigade. When a source emits one of these charged particles, it has to share the charge. There are no fluctuations allowed there, you know. When the particle's electric charge has changed then it counts as a different particle. You must have heard of charge accounts. That is how we Quarks get to change," they added.

"But where does the electron come from?" asked Alice, who felt that the explanation was a little lacking.

"The particles exchanged in the weak interaction are called W," began Uppo rather inconsequentially.

"What?" responded Alice, temporarily forgetting her manners. "Not 'What,' just W. It is not much of a name, but it is all they have, poor things. There are two of them, you know: One is W Plus, and one is W Minus. No one has ever asked them what the W stands for," he finished thoughtfully. "Anyhow," he continued, "these W's, as their friends call them, are very friendly types. They will mix with anyone. They interact with both leptons and hadrons, with electrons as well as strongly interacting particles. So when a down Quark decides it is time to change into an up Quark, it gets all charged up. The electric charge of the Quark has increased, so it gives out a W Minus particle to balance the books. This W in turn plays it by the book and interacts with a passing neutrino, which has no electric charge at all, turning it into an electron, which does have an electric charge. The electron finds itself in company with a lot of strongly interacting particles, where it has no right to be, and leaves as rapidly as it can."

See end-of-chapter note 4

"But where does the W find a neutrino which it can change into an electron?" asked Alice in some puzzlement. "I didn't think there had been a neutrino there before. I thought that it was emitted after the decay, along with the electron."

"Ah, that is where it fools you. You thought it should be there *before*, but instead it was there *after*. You are expecting it to arrive from the past, so it sneaks up on you, back from the future, and still arrives just when it is needed. Of course, because it came from the future, it is still around afterward, on its way to arrive. In this way it gets to be both the neutrino converted by the W and the one emitted after the decay. That cuts down on the overheads."

"But how can it arrive from the future?" asked Alice. As she spoke she had a distinct feeling that she already knew the answer to this question.

"It is an antineutrino, of course. One of my favorite anti's. Every particle has its antiparticle, which travels backward in time and so is opposite in every way. That's the great principle of antiparticles—`Whatever it is, I'm agin it.'"

"And is there no way any of you can ever get free?" asked Alice, to be quite sure on this point.

"No, no way at all," they assured her.

"Does that mean that I cannot escape either?" asked Alice in dismay, as she did not really wish to be trapped with them forever. "Not at all. You have no color so the gluons don't hold you. You are one of the most colorless people we have ever met, so there is nothing to keep you; you can leave whenever you wish. We won't even notice. You can get up and walk away. Just don't forget to leave a tip."

This sounded much too simple, but Alice tried it anyway. She stood up and found that indeed there had been nothing to prevent her from leaving the group at any time. She stretched after her cramped confinement in such a small space, looked around her, and found that she was standing face to mask with the Master of Ceremonies. His grinning mask was just a few feet from her face. She stared at him, hypnotized by his wide, fixed grin and the dark eyeholes above. Deep within their black depths where his eyes should have been, she thought she could see an intense blue spark, like a distant star on a clear, frosty night.

"And how did you enjoy your meeting with the Quarks?" he asked her merrily.

"It was very interesting," she replied truthfully. "They were most colorful characters, but I did find them rather changeable.

"Was that the last unmasking that will take place tonight," Alice continued, "or are there further layers to be stripped away before I can see what is really there?"

"Who can say?" he replied. "How can you ever know if you are finally looking on the naked face of Nature or if you are simply looking at yet another mask? Tonight however there is but one more unmasking to come. I have yet to remove my own mask."

As he was speaking, the bright spotlight which had followed him all through the evening began to dim, and the light from the chandeliers overhead became even fainter than it had been before. As the light died, the Master of Ceremonies lifted both hands to his face and slowly pulled off his mask.

In the rapidly fading light Alice looked at the face behind the mask. She could see nothing but a smooth oval, a total blank with no features of any sort discernible. She stared in astonishment at this enigmatic visage, and, as the last gleam of light died, she saw the *mask* wink at her.

Notes

1. The protons and neutrons which inhabit the nucleus (known collectively as nucleons) are examples of strongly interacting particles, also known as *hadrons*. There are many other hadrons, though not all particles interact strongly. The class of particles which are known as *leptons* do not feel the strong interaction at all. Electrons belong to this class and so are not bound inside the nucleus together with the nucleons. They are aware of the nucleus only as a positive electric charge which holds them loosely bound within the atom.

    Experiments in high-energy physics have discovered hundreds of strongly interacting particles. This situation presents a fairly familiar scenario in physics. Whenever a class contains a very large number of members, they usually turn out to be built up as composites of something more basic. The various chemical compounds identified are all composed of atoms. There are 92 naturally occurring varieties of atoms that are stable, and they are all composed of electrons arranged in differing numbers around a central nucleus. Nuclei in turn are composed of neutrons and protons bound by the ex-change of pions. These are mentioned in the previous chapter. Now the neutron and proton are found to be just two members of a class with hundreds of others: K, ρ, ω, Λ, Σ, Ξ, Ω, Δ, and so on. These particles have now been shown all to be composed of quarks.

2. Quarks are held together by forces like, but yet unlike, the electrical interaction. These forces do not act on electrical charge, but on something else which is called *color charge* or just color. This has nothing to do with color as we normally understand it; it is just a name which has been given to something completely new. The fact that the word *color* is already in use is perhaps unfortunate, though it is not the first time that a word has had two different meanings.

    The interaction between two electrically charged particles is due to the exchange of virtual photons. The interaction between quarks is caused by the exchange of a new class of particles which have been named *gluons*. There are differences between the interactions: Electrical charges come in only two forms, positive and negative, or charge and anticharge. The photons which are exchanged between electrical charges are themselves electrically neutral; they carry no charge and so do not emit more virtual photons in their own right.

The gluons exchanged between quarks are emitted by a form of charge carried by the quarks, but completely different from the normal electrical charge. It is called color charge, though it has nothing whatsoever to do with the colors that we can see. While there is only one form of electrical charge, together with its opposite, or anticharge, there are three different forms of color charge, given the names *blue*, *green*, and *red*. Again it should be stressed that these names are just conventions and have nothing to do with ordinary color. Associated with each color charge there is an anticolor, and there are two ways of producing color-neutral objects. With electrical charge you can only produce an electrically neutral object by combining charge and anticharge (positive and negative charge). There are two ways to produce color-neutral particles: a combination of color and anticolor (as in bosons) or a combination of all three colors of quarks (as in fermions).

3. When particles are bound together by the electrical interaction, the potential energy in the binding decreases rapidly as they move far apart. If a particle is given enough energy, it can break free completely, as a rocket which has reached escape velocity has then enough energy to break free of the earth's potential. When a gluon string has already been stretched, however, it takes just as much energy to stretch it a little farther as it did initially. It is like stretching an elastic string; it does not get any easier the farther you stretch it. It is also like an elastic string in that, when you stretch it, it can break.

   The gluon string is capable of absorbing more and more energy as the quarks separate and the string stretches. Eventually the energy in the string is more than is necessary to create a quark-antiquark pair. The string breaks and its broken ends are terminated on the color charges of the new quark and antiquark. In place of the original bound system of three quarks there are now two separate systems, one of three quarks and one of a quark and an antiquark. Instead of releasing a free quark the energy has created a new particle, a boson. This will always happen and free quarks are never produced.

4. Though the quarks cannot escape from the "particles" within which they are bound, they can change from one type to another. This is caused by a peculiar process called the *weak interaction*. The weak interaction is a broad-minded process which will interact with virtually everything. The electromagnetic interaction affects only particles that have electric charge. The strong interaction affects only the strongly interacting particles (or hadrons) and not leptons. The weak

interaction will affect them all, though the effect is rather slow and weak because it is a *weak* interaction.

The weak interaction is peculiar in that it can change quarks. It can change either a *down* quark or a *strange* quark to an *up* quark. In the process, the electric charge of the quark is changed, with the surplus charge carried off by the "W boson," the type of particle which is exchanged in the weak interaction. This charge may then be handed over to newly created leptons, an electron and a massless electrically neutral lepton known as an antineutrino. This happens in the process of nuclear β decay, in which a radioactive nucleus emits a fast electron. This process had been known for many years, but was odd in that it was quite clear that there were not any electrons available within the nucleus to be so emitted. The electron is created in the decay process and, as it is not bound, leaves the nucleus immediately.

10 The Experimental Physics Phun Phair

T he darkness slowly cleared from about Alice. The shadows lifted from her eyes, which were immediately dazzled by a chaos of bright lights and colors. At the same time her ears were assaulted by an assertive cacophony of sounds. She looked around her and found that she was in the midst of a merry and diverse throng of people. There appeared to be all manner of folk present, in every kind of dress. She could see that some of them were wearing white coats, such as one imagines scientists to wear in their laboratories, while others in the crowd were dressed in very casual clothes or in formal suits. She could see costumes from countries all over the world and indeed from many different times in the past.

There were men in Victorian frock coats, with impressive bushy side whiskers, and others in burnooses, or traditional Chinese costume, with wide flowing sleeves and long pigtails. She saw one particularly hairy-looking individual who staggered past dressed in untreated animal skins and carrying what looked rather like a roughly formed wheel, which appeared to have been chipped out of stone. One the side of the wheel the words *Patent applied for* had been carefully chiseled. One man in particular caught her attention for some reason. She sensed some special quality about him, without being able to pin down exactly what it might be. He had a pale, intense face and was dressed in the breeches, waistcoat, and wide frock coat of the seventeenth century.

He was walking along absentmindedly taking a large bite out of a bright red apple.

"Where am I?" she asked herself, speaking aloud but hardly expecting to be noticed in the hubbub which arose all around her.

"You are in the Experimental Physics Phun Phair," came the unexpected response. Alice looked to see who had spoken, and found that, once again, she was accompanied by the Quantum Mechanic, who was walking quietly by her side. He indicated a banner stretched across a gateway by which they had, apparently, just entered. It bore the slogan:

> Experimental Physics Phun Phair

"It does seem to be spelled rather strangely," commented Alice, this being the first thing which struck her about it.

"Well, what do you expect? They are all scientists here, you know. This is the great carnival of empirical observation. Here you will find many demonstrations of physical phenomena and sideshows of experimental results."

Alice gazed around her and saw a splendid variety of tents and stalls, with here and there a more substantial looking building. They all carried large, brightly colored posters which vied with one another for the attention of the crowd. She read a few of them:

> Enjoy the thrill of particle
> collisions.

> Hunt the neutrino.

> Knock out a quark and win a
> Nobel Prize.

There was a disturbance of some sort in the crowd nearby. Alice looked across and saw a balding bearded man wrapped in what appeared to be a large white bath towel. He was shouldering his way through the crowd, hampered by the fact that he was carrying a large posterboard in one hand and an incredibly long pole or lever of some sort in the other. She peered carefully at the notice he was carrying. At the top, roughly painted out, she could just make out the words:

> Feel the Earth Move!

Below the erased words she read the modified message:

> See Me Move the World!

"Who is that," asked Alice, "and what is he planning to do?"

"Oh, that is a well-known Greek philosopher. He is obviously intending to go into his old `Moving the World' routine."

"Really?" exclaimed Alice. "Does he often move the world then?"

"Oh no, he never does. He can never find a fixed place to stand while he uses his lever, you see."

As this did not appear to offer much immediate entertainment, Alice looked around for something more promising. Her attention was attracted by a stall nearby which bore the name "Photoelectric Canon." There was a sort of stylized gun from which the player could direct a beam of light onto a metal surface. The light caused electrons to be emitted where it struck, and the idea, as explained by the stall's occupant, was to get the electrons to move a little distance to a sort of bucket, where they would be collected. This seemed easy enough to Alice, even when it was explained that, to make things a little more interesting, there was a weak electric field which resisted the passage of the electrons and turned them back just before they reached the collector. After all, as the stall owner explained, there was a control which would allow Alice to increase the intensity of the beam of light to many times its present value.

However hard she tried, though, she found that she *could not* get any of the electrons to travel that last little distance. She turned the intensity of the light higher and higher. More and more electrons came streaming out, but every one was turned back at the last minute by the electric field.

"This is really too bad!" exclaimed Alice in frustration.

"I am afraid that it is what you have to expect," replied her compan-

The quantum description of the world is hardly what we would have expected. The reason for believing in it is that its predictions agree with experimental results. It is the only theory that can give any sort of explanation for the behavior of matter on the atomic scale, and it does it remarkably well.

ion sadly. "You see, you have only been given control of the *intensity* of the light and not its *color*. If light were a classical wave, then you would expect that as you increased its intensity, the associated disturbance would increase, and it would give more energy to the electrons that were emitted from the surface of the metal target. In fact, it is the color, or frequency, of the light that decides the energy of the individual photons which compose it. As you are not provided with any way to alter this, you cannot change the energy of the photons or for that matter the energy of the electrons which those photons will knock out of the metal surface. The stall has, of course, carefully been set up so that this energy is not *quite* enough to get through the retarding electric field. When you increased the intensity of the light, you directed more photons onto the surface, and these produced more electrons, but they all had the same energy, and in every case it was not quite enough for the electron to make it to the collector. I am afraid that you cannot win."

Alice felt somewhat cheated by her experience with this stall and looked around for something different to occupy her. Nearby was a small tent with a sign which read:

ROLL UP! ROLL UP!
See the largest collection of quarks in captivity.

The central features of quantum behavior are the detection of discrete particles and the observation of interference. The observation of quanta is shown by the photoelectric effect: the production of electrons by light falling on a metal surface. The only result of increasing the light intensity is to increase the number of photons present and consequently the number of electrons. Each photon still interacts on its own, so if the *frequency* of the light is unchanged as the *intensity* is altered, then each photon will still have the same energy, and the energies of any electrons produced will be the same, whatever the intensity of the light. This is quite different from the behavior expected for a classical wave, where a greater intensity should mean more energy delivered.

Alice and her companion slipped inside the tent, where the exhibitor was telling a small crowd how fortunate they were to see all six quarks captured and displayed for their entertainment. Alice looked at the exhibits. None of the quarks were present singly, of course. They were all assembled in pairs, each one bound unbreakably to his antiquark. Alice realized that this was as close to a collection of isolated quarks as it was possible to get. "And after all," she thought, "he did say that they were in captivity."

Alice looked at the assembled quark pairs. They were assembled on a platform which had various levels, with the heavier quark combinations standing on a higher energy level. She saw an *up* quark, wiggling his heavy eyebrows at her as before, a *down* quark, and, slightly higher up, a *strange* quark with flaming red curly hair.

As well as these three types, which she had already met at the MASSquerade, there were two others higher still. One projected a captivating personality, and she saw a quick flash of light as from a display of incredibly white teeth. "That is a *charm* quark," the Quantum Mechanic

murmured in her ear. The other new quark was heavier yet. This one was placed quite high up, and Alice saw him even less clearly than other particles she had met, but she got the strangest impression that this one had a donkey's head. "That is a *bottom* quark," her companion informed her.

Alice now looked higher to see the sixth quark. There was the position on the platform, but it was empty. There was no sign of the sixth quark, which, she was informed, would be the *top* quark.

Other members of the audience also had noticed the absence of this sixth quark and were protesting noisily. "All right, all right!" said the exhibitor soothingly. "I know that he is in here somewhere. The top quark is the heaviest of them all, so we have to look for him at high energies, but he must be there." He picked up a large butterfly net which was leaning against a pole, climbed up a stepladder, and began making wide random sweeps of the net close beneath the roof of the tent.

During this, his audience became increasingly restless, with uncomplimentary remarks being made on all sides. Gradually, the mood of the crowd grew distinctly ugly and people began to slip away in order to write critical letters to their favorite technical journals. "Come away," said the Quantum Mechanic to Alice. "This is no place for us now."

They went outside and Alice's attention was caught by another stall at which people were clustered, throwing balls at various prizes which they would win if only they were able to knock them from their stands. It looked very much like the sort of fairground stall which she had seen near her home, except that in this one there was a sort of fence of thin, uniformly spaced wires between the competitors and their targets.

Alice watched for a while and noted that, as soon as a ball was thrown, it became quite blurred and it was impossible to see exactly where it had gone until it hit some point along the back wall of the booth.

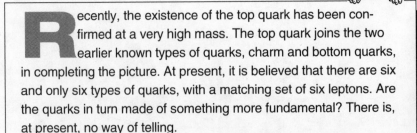

Recently, the existence of the top quark has been confirmed at a very high mass. The top quark joins the two earlier known types of quarks, charm and bottom quarks, in completing the picture. At present, it is believed that there are six and only six types of quarks, with a matching set of six leptons. Are the quarks in turn made of something more fundamental? There is, at present, no way of telling.

She saw that most of the balls did just that; they hit the wall rather than any of the prizes. Gradually piles of balls built up where they had hit, and Alice could see that these piles were positioned neatly in the spaces between the prizes.

"Exactly right," said a voice by her ear, echoing her thoughts. "The regularly spaced wires are acting to produce an interference pattern, with much greater probability for the balls being observed in some places than in others. Naturally the minima, the places where the probability of finding a ball is at its lowest, are arranged to fall where the prizes are."

"That doesn't seem very fair," remarked Alice.

"Well no, perhaps not, but in the Phun Phair you do not expect it to be fair. After all, the man who runs the stall has to make a living, so he does not want to give out prizes too often. Of course, there is still *probability* of the ball being observed even at the minima, so *some* prizes do get won, but not too many."

Alice still felt that it was not quite right somehow, but before she said anything further, her attention was caught by a large pavilion a little way off. It was surmounted by a huge glowing sign which said:

THE GREAT PARADOXUS
Spooky action at a distance!

Below the sign there were a number of large posters strung across the front of the building:

Extraordinarily astounding!
Paradoxically incomprehensible!
Rather surprising on the whole!

Alice and her associate made their way to this exhibit and joined the crowd which was streaming in through the doorway. Within was a long, high-ceilinged enclosure with a raised platform in the center. On either side were a pair of short ramps leading up to doors at the two ends of the building. On each ramp stood a short metal cylinder with a pointed nose and stubby fins at the back.

On the central platform stood the Great Paradoxus, a tall figure with glossy black hair, a spiky waxed moustache, and a flowing black cloak. "Good evening, Ladies and Gentlemen," he greeted them. "Tonight I plan

The central features of quantum behavior are the detection of discrete particles and the observation of interference. Particles, or quanta, are observed at one place rather than spread over a wide region like a classical wave. Despite this, particles seem to behave like waves, in that they show interference effects between the different amplitudes which describe all the things that a particle might be doing. Interference may be demonstrated by scattering electrons from a regular grid, as illustrated by the arrangement of atoms in a crystal, and may be performed at such a low intensity that only one electron will be present at a time.

to conduct a little experiment on the reduction of amplitudes, which you may find to be of some interest. Here on the platform beside me," he continued, "you see a source of transitions, transitions which will release two photons in exactly opposite directions. As you know, if you were to measure the spin of the photons along some direction of your choice, you would find them to have a spin either *up* or *down,* with no intermediate choice possible." Alice had not known that, though she had heard talk of spin-up and spin-down electrons, but all the other people present were nodding their heads wisely, so she assumed it must be all right.

"As I say, *if* you were to measure the spin, you would find it to be spin-up or spin-down, but if you do *not* measure it, then there will be a mixture, a superposition of states having different directions for the spin. Only when you make a measurement of the spin will the amplitudes reduce. One will be selected, and one will no longer be present. Now," he said abruptly, "the source which you see here makes its transitions from states which have no spin at all, so the *total* spin of the two particles produced must also be zero. That means," he explained kindly, "that the spins of the two photons must be opposite: If one has spin-up then the other *must* have spin-down. But, mark you, the spin direction of the photons is only selected from the superposition of states when a measurement is made—that is the usual understanding. Thus, you see that when a measurement is made on one photon with the discovery, let us say, that it has spin-up, then the amplitude superposition for that photon will reduce to the appropriate state.

"However," Paradoxus continued, drawing himself up to his full height, "at the same time, the superposition for the *other* photon must

reduce also, because we know that this photon must have the opposite spin. This must happen however far apart the photons might be at the time, even if they had arrived at different stars in the sky. In this demonstration we will not be making our measurements quite so far apart as that," he smiled at his audience. "I now call for two volunteers, two trusty and reliable experimenters who will agree to travel to the opposite ends of Quantumland and make our observations for us."
See end-of-chapter note 1

There was a buzz of argument and discussion from the assembled crowd. Eventually, two people were pushed forward. Both were dressed in long frock coats and narrow trousers, and both had bushy side whiskers. Both wore waistcoats, each with a gold chain attached to a watch which its owner had obviously checked recently against a reliable standard clock. These two were not actually identical to one another, since only particles are completely identical, but they were certainly very much alike. They were obviously both honorable, honest, and reliable as well as being competent and conscientious observers. If they were to say that they had seen something, no one would dream of disputing it.

Paradoxus handed to each a polarimeter, a device with which they might measure the spin directions of the particles. With military precision each stripped the instrument he had been given, examined it to make sure that it had no unusual features, and quickly reassembled it.

The showman then summoned two attractive female assistants, who escorted the volunteers to the metal cylinders and opened a door in the side of each. For some reason each of the two observers then put on a tall top hat before he squeezed himself into the cramped space within. The assistants closed the doors, lit a fuse at the rear of each cylinder, and stepped hastily back. With a roar the stubby rockets rushed up the ramps, through the doors at the end of the pavilion and arced over the horizon, making for the opposite ends of Quantumland.

"And now we wait for them to arrive," remarked the impresario. "As soon as each is in position he will send a message through his telegraph line." He indicated two bells which stood on small tables at either end of the platform. Everyone looked at them, waiting for them to ring as a signal that the show might proceed. It was a long wait.

"Everyone seems very patient," remarked Alice, who was beginning to feel a bit restless herself.

"They have to be," replied the Quantum Mechanic. "Experimental scientists all learn to develop patience."

Finally the bells rang, first one and then shortly after the other. This indicated that both observers were in position, and, with a dramatic gesture, Paradoxus opened windows at either end of his photon source. Two by two the photons went rushing off in opposite directions.

After some time he closed the windows again and there was another long pause. "I wonder what we are waiting for now," thought Alice, who felt that the entertainment might move just a *little* faster. There was a flapping of wings and through the door at one end of the building flew a carrier pigeon, which was expertly caught by one of the assistants. Not long after, a pigeon arrived through the other door, and the messages carried by both could be compared. Paradoxus displayed the two messages, which showed a perfect correlation, with a spin-up photon going to one side invariably accompanied by a spin-down version detected at the other, even though the two detectors were too far apart to have had time to exchange any information.

"That's no mystery!" shouted someone from the other side of the large room. The voice had come from a tall figure whom Alice could not see very clearly, but he did look *rather* like the Classical Mechanic. "It is obvious," he went on, "that the photons are not in fact completely uncertain whether they are really spin-up or spin-down when they leave the source. In some way they know which they will be, and they know also that the two of them must be opposite. It does not matter then how long they wait before they are detected; they will be found to have that spin direction which was already decided when they were emitted."

"That sounds a very reasonable argument, does it not?" beamed the impresario, looking not in the least dismayed. "We shall have to extend

our demonstration a little. You say that it has been decided at the time of emission whether the photons shall have spin upward or downward, and they carry this information with them as they travel. What would happen if our two observers were to measure the spin in other directions, say to the left and to the right, or at some other angle in between? And what would happen if our observers were to rotate their polarimeters as and when they felt like it, without referring back to us or collaborating with one another? Would it be possible for the source to know in advance what information it should transmit along with the particles such that their spins would match properly *whatever* the angles our friends chose for their measurements? I think not!"

Quickly he wrote out new instructions for the observers, bound the notes to the pigeons' legs, and sent them back to whence they had come. After a pause the telegraph bells rang once more to show that the messages had been received and understood. Again, with a flourish, he opened the windows on the central source and let the photons stream outward. After a suitable period, he closed the windows again, and then it was back to waiting. Alice was feeling distinctly tired of waiting for something to happen when, at last, there was a rushing noise from both sides. This grew steadily louder, and then the two rockets came arcing back down through the doors at the two ends of the building and landed back on the ramps from which they had departed.

As the stubby cylinders sat smoking gently, the doors opened and from each vehicle descended one of the observers, still wearing his tall formal hat. They both marched over to the impresario, removed their hats, bowed, and presented him with their notes. As far as Alice could tell, everyone in the audience apart from herself immediately crowded around to try to be the first to get a glimpse of the results. There was a tremendous hubbub of discussion and argument, and they all began to make their own calculations. Alice saw people with tiny portable computers, with electronic calculators, and with slide rules. She also saw someone with a strange mechanical calculator which had scores of tiny cogwheels. The Chinese folk whom she had noticed earlier had each produced an abacus, and their deft fingers slipped the beads to-and-fro along the wires more quickly than her eyes could follow. Even the hairy gentleman in the animal skins was involved. He had abandoned his wheel and was going through some complicated procedure with several little piles of bleached knucklebones.

Finally the arguing groups quieted down and came together in a common conclusion. It was true, they said, that there was a quite inexplicable agreement between the spin directions of the two photons. Even when arbitrary changes were made in the directions along which the two spins were measured, the observed correlations were greater than could be explained by any information sent out along with the particles. It was all quite clear, they agreed; in fact, it was clear as a bell. It didn't look all that clear to Alice, but if everyone was agreed, she supposed it must be correct.

"That is a very interesting result," remarked the Quantum Mechanic as he returned from the middle of the crowd. Most of the other people present were still arguing excitedly, despite the fact that they were apparently all in agreement. "It shows that the behavior of the wave function in different places cannot be caused by messages passed from one position to the other. There simply isn't enough time for that. It presents a whole new Aspect of quantum nature."
See end-of-chapter note 2

Interesting it might be, but Alice felt that there had been too much sitting and waiting and that she would like more action, so they left the pavilion and went off to investigate the rides.

"You will have to behave as a charged particle if you want to take any of the rides," remarked the Quantum Mechanic. "The rides all operate by electrical acceleration, so they will only work for charged particles. As you are just a sort of honorary particle, I see no reason why you should not be a charged particle just as readily as an uncharged one."

They had come to a very long, narrow building on which was a sign reading:

> **RIDE THE WAVE!**
> Ride the electromagnetic wave for mile after mile.
> (That makes two miles; count them: 2.)

There was an excited line of electrons outside, waiting their turn to get on, but Alice did not think that this was quite the ride she wanted at the moment. She would rather go on something like the Big Wheel, which she had ridden in a fair near her home.

She mentioned this to her companion, who said that he would take her to the circular machines. As they headed in that direction they were passed by a parade. There was a succession of little carts, on each of which was balanced a huge apparatus built around a vast iron magnet with copper coils wound around it and various intriguing devices embedded in its center. From all of these snaked great bundles of wires and cables.

"How do those little carts manage to support all that weight?" asked Alice. "Surely they should be squashed flat under such huge masses of metal?"

"Oh, they would be if the pieces of equipment were *real*, but this is the Experiment Funding Parade, so each one is only a proposal. They are rather like the experiments we did in our *gedanken* room. They are just ideas at the moment, not real at all, so they are not very heavy. In fact, most of them carry very little weight indeed."

Alice looked at the procession and observed that the second cart carried an apparatus exactly like the first, the third carried another exactly the same, so did the fourth, the fifth, the sixth, and on and on as far as the parade was visible. "There doesn't seem to be a great deal of variety," remarked Alice.

"That is because multiple copies of each proposal must be submitted," replied her associate. "There will be a different one along in due course."

As they watched the floats go by the air was filled with a snowstorm of irregular scraps of paper. "Torn-up unsuccessful grant applications," said the Quantum Mechanic before Alice could ask. "Come along, we had better find your ride."

They passed a succession of Big Wheels. They were all lying on their side instead of being upright as they would be in a normal fair, and Alice's companion told her that in the Phun Phair they called them rings rather than wheels. There was the Big Ring, the Much Bigger Ring, and the CERN Truly Enormous Ring. Alice decided that she would like to ride on this last one.

She lined up with a jostling bunch of protons, and soon she was entering the machine and being seated, or "injected" as they termed it, in a *beam bucket*. This was a sort of electrical enclosure which Alice shared with a large crowd of protons who were milling around excitedly in all directions. They moved off, accelerated forward by strong fields which pulled on their electric charges. As they gathered speed, the protons quieted down, and they all rushed forward together.

Faster and faster they went, guided around and around by magnetic fields. After some time, Alice began to notice that their speed was no longer increasing much, though she could still feel an acceleration. She asked one of the protons about this and was told that they were now going almost as fast as photons did and nothing could go much faster than that, but that their *kinetic energy* was still rising. This seemed odd to Alice and she was about to argue, when there was a sudden wrench and she felt herself flung out of the ring together with the protons.

Through the air she rushed at what now seemed an incredible veloc-

t is part of the paradox of quantum physics that measurements on very small objects must be made with enormously large particle accelerators. Because of the *Heisenberg relation*, a small size is coupled with a large momentum, and it requires a large machine to accelerate particles up to the enormous energies needed. Most of the very high energy accelerators are circular and the particles travel around and around many times as they are accelerated. There are a few large linear accelerators, in which electrons are accelerated along a single straight path, such as the 2-mile-long "linac" at Stanford in California.

ity. Looking ahead, she was terrified to see a wall directly in front of her and to realize that she and the protons were headed straight toward it! She tensed for the collision as the wall rushed closer, but to her amazement the wall stopped her no more than would a fog or a dream.

She looked around her and saw that, although the wall had had little effect on her, the reverse was far from true. She passed an atom some way off and it burst asunder, the electrons spilling out and the nucleus cast free to drift on its own. All around her she trailed a deadly train of virtual photons. These tore at the atoms she passed, which seemed as gossamer, ripped apart by the distant effect of her passage. She came close to a nucleus and it too was shattered, the protons and neutrons scattered in every direction. In dismay she recollected the Cosmic Ray-der whom she had seen from Castle Rutherford and who had so effortlessly destroyed a nuclear castle. Now she was horrified to realize that she had become as he, leaving a wide swath of destruction among the atoms and nuclei which she passed!

She saw a neutron straight ahead for just a moment before she plowed into it. Briefly she glimpsed its three quarks, who were thrown into a panic by her passage. They were not cast individually out of the neutron, because they were too firmly bound to one another, but their chains stretched and broke, stretched and broke, with the creation of a host of quark-antiquark pairs. Where previously the neutron had been standing was now a great jet of mesons carried forward by the wake of Alice's own enormous momentum.

Alice hid her eyes to blot out the image of the chaos about her, lest she should see some even more violent catastrophe. She had a brief sensation of falling and felt a slight bump.

The high-energy particles produced by accelerators can penetrate for considerable distances through normal matter. They have such high energy compared with that in the electronic bonds between atoms that these have little effect in slowing them. Such particles leave a swath of ionization and broken bonds along their path. If they pass close to the nucleus of an atom, they will split this apart also. Eventually these fast particles will lose all their energy through such processes, but they can go a long way.

Alice quickly opened her eyes, to find that she had fallen off the couch in her own front room and was lying on the floor. She got up quickly and looked around. The sun was shining cheerfully in through the window and the rain had cleared away. She turned to look at the television, which was still operating. The screen showed a group of rather serious folk sitting around a studio, arranged carefully on either side of a commentator, who informed Alice that they were about to have a studio discussion on the future of scientific planning in the country.

"Boring," said Alice. She switched off the television firmly and went outside into the sunshine.

Notes

1. There have been many attempts to set up an experiment which would contradict the more extreme predictions of quantum theory, but so far quantum mechanics has always been vindicated.

    An example is the Aspect experiment to investigate a form of the *Einstein-Podolsky-Rosen* (EPR) paradox. There are various forms of this paradox, which involves measurements of particle *spin*, the strange quantized rotation posssessed by elementary particles such as electrons and also photons. The paradox treats the case of a system which has no spin but which emits two particles that do have spin and which travel directly away from one another. The restrictions of quantum theory tell us that a measure of the spin of either particle can give only one of two values: *spin-up* or *spin-down*. If the original system has no spin, then the spins of the two particles must compensate; that is, if one is spin-up, the other *must* be spin-down, so that the sum of the two gives a total spin of zero. If no measurement is made of the particle spins, then quantum mechanics says that they will be in a superposition of spin-up and spin-down states. When a measurement is made on the spin of one, then *at that point* its spin will be definite, either up or down. But at the same time, the spin of the *other* particle becomes definite also, as the two must be opposite. This would be true no matter how far apart the particles have moved since they separated. This is in essence the EPR paradox.

2. It would seem reasonable to explain the EPR paradox by saying that in some way the spins were predetermined from the start: that, in some way, the particles *knew* which would be spin-up and which spin-down when they set out. In that case it would not matter how far they had traveled as they would bring the information with them.

The limits of the information which it would be possible for the particles to fix in advance are considered in *Bell's theorem,* which treats what happens if the spin measurements are not made along one predetermined direction, but at a selection of different angles for the two particles. The calculation is rather subtle, but the outcome is that, in some cases, quantum mechanics predicts a greater correlation between the measurements on the two particles than could be arranged by any advance information which could be sent with the particles without prior knowledge of the directions along which the spins would be measured. Alain Aspect in Paris has measured this effect and found, as usual, that quantum mechanics appears to be correct. It seems to involve some sort of information which travels more quickly than the speed of light.

The Aspect result does not directly contradict the normal understanding of Einstein's special theory of relativity. This says that no information, no message, may travel more quickly than the speed of light. The effect considered in the EPR paradox cannot be used to send messages. If you could decide whether you would measure the particle spin as up or down, then the opposite spin of the other particle would convey information in a sort of Morse code, but you cannot do this. You have no control whatsoever of the result of a measurement on a superposition of quantum states; the result is completely random and no signal can be forced onto it.

Index

NIMMO, Dan D. and Robert L. Savage. Candidates and their images: concepts, methods, and findings. Goodyear, 1976. 250p bibl index 75-21173. 6.95 pa. ISBN 0-87620-154-0
The book is somewhat misnamed or, at least, mis-subtitled. There are few concepts and the findings do not seem particularly substantial or satisfying. There is, however, some discussion of Q-sort and quantitative analysis. The authors, indeed, recognize the limits of their effort but conclude that their studies move the discipline forward or, at least, point it in a fruitful direction. These limits include an ill-defined conceptual base and methods that measure unsurely or posit that which is to be determined. Perhaps the most interesting aspect to this book is that it offers the following sort of "noteworthy" observations with a straight face: (1) "[A]mong party identifiers loyal to their party's candidate, the proportion positive about their choice is high . . ."; (2) "[P]arty identifiers who defect generally have more positive images of the opposition's candidate than of their own party's"; and (3) "Independents voting Democratic have expressed positive images of the Democratic candidates." Professor and student will find value, however, in the annotated bibliographies concluding several chapters.

Candidates and Their Images
Concepts, Methods, and Findings

Candidates and Their Images
Concepts, Methods, and Findings

Dan Nimmo
University of Tennessee—Knoxville

Robert L. Savage
University of Arkansas—Fayetteville

Goodyear Publishing Company, Inc., Pacific Palisades, California

Library of Congress Cataloging in Publication Data

Nimmo, Dan D.
   Candidates and their images.

   Includes index.
   1. Public relations and politics   2. Electioneering
3. Imagery (Psychology)   I. Savage, Robert L., joint
author.   II. Title.
JF2112.P8N55   659.2'9'32901   75-21173
ISBN 0-87620-154-0

Current printing (last digit):
10 9 8 7 6 5 4 3 2 1
ISBN: 0-87620-154-0
Library of Congress Catalog Card Number: 75-21173
Y-1540-7
Printed in the United States of America

Editor: Victoria Pasternack

TO NIETZSCHE

who wrote:

A great man, did you say?
All I ever see is the actor creating his
own ideal image.

Contents

Preface

For a number of years both authors undertook a variety of research projects seeking to understand how citizens see, think, and feel about candidates for public office. That research grew from a general interest in political opinion and voting patterns but more particularly, from the conviction that election campaigns, contrary to the findings of early voting studies, were playing an increasingly important role in how voters view competing candidates and thereby fashion electoral choices. Out of these concerns came published works pertaining to the technology of persuasion in political campaigns, the character of popular beliefs, values, and expectations about politics, and the measurement of political perceptions. Then through our collaboration we focused on what has become a fashionable way of talking about candidates for public office; in other words, we began research into the "images" of political candidates. We have pursued that research from numerous perspectives and employing diverse methodologies and procedures. *Candidates and Their Images* is the product of our accumulated studies into the images of political candidates.

In certain respects this volume departs from what in recent years has been the conventional way of reporting the results of political inquiry. Readers will quickly note two of these departures, both of which we believe afford us a greater opportunity than do conventional approaches to integrate the findings of published and unpublished research pertaining to candidate images. First is our conscious decision to concentrate on a series of coordinated studies concerning a limited set of problems rather than to reduce our findings to a single statement of broad scope and generality, as is more typical in the social sciences. Thus we return to the rewarding tradition of

Irving Janis, Muzafer and Carolyn Sherif, and William Stephenson. We maintain the conviction that by emphasizing a limited set of problems relating to candidate images, such as that of the stimulus- versus perceiver-determined controversy or the effects of victory or defeat on the perceptions of political figures, we can explore the relative merits of various explanations for a phenomenon as complex as personal imagery.

Second we have used multiple methods and techniques to develop the research reported in this volume, thus departing from the view that meaningful research must be wedded to a single, prized technique. Readers will find that alongside secondary analysis of data derived from large-scale, nationwide probability samples are findings based on studies employing small, restricted, purposive samples; intensive investigations which we deem as vital as extensive research; both survey and quasi-experimental techniques which provide instructive insights; and both R- and Q-methodological designs. We think that such pluralist inquiry not only enriches our substantive findings but also provides an informative introduction to lesser-known research techniques. For example, although it may be more widely known among psychologists and market researchers, a growing number of political scientists employ Q-technique to good advantage. Q-techniques offer considerable promise in testing the general application of propositions suggested by a single study, for they permit repetition and replication with strategic variations. For readers especially interested in this facet of image research we have described in the appendixes the most basic features of Q-technique.

Many people have assisted us in the research reported in the following pages; we single out a few for special acknowledgment. We are especially appreciative of the time and effort devoted to this project by Michael W. Mansfield of Baylor University. He was involved as a co-researcher in conducting the Q-studies reported herein and in compiling the data for measuring the effects of campaigns on perceptions of presidential candidates, which we report in Chapter 6. We wish to thank Harry Howard of the University of Tennessee for his assistance in the difficult task of secondary analysis of the 1972 election studies reported in Chapters 6 and 8. In addition we wish to acknowledge the full support of the University of Tennessee—Knoxville, the University of Missouri—Columbia, and Auburn University for making computer time and assistance available. We owe special gratitude to Rondal Downing, who provided assistance for our initial Q-studies through the facilities of the then-existent Research Center, School of Business and Public Administration, University of Missouri—Columbia. And, above all, we are indebted to Roy Pfautch, president, Civic Service, Inc., for his enthusiasm and genuine concern in sponsoring a study of the differences between old and new political campaign styles, his considerable contribution to the project, and his willingness to make the resulting data available for publication in this book. To Mary Noblitt we owe acknowledgment for her patience in typing the manuscript and, more especially, the tabular data. Victoria Pasternack and Gail Weingart contributed incisive and invaluable editorial supervision. Finally we wish to thank David Grady of Goodyear Publishing Company for taking an interest in this project and permitting the authors to approach it in an unfettered way.

Chapter One

Introduction: The Image in Politics

At 7:30 in the evening of October 15, 1973, the thirty-ninth vice-president of the United States, Spiro T. Agnew, bade farewell to the American people in a nationally televised address. Five days earlier he had resigned his high office preparatory to pleading "no contest" to a charge of tax evasion. He thus became the second vice-president in history to resign from the office and the first to do so under charges of wrongdoing. In his farewell, Mr. Agnew made one statement of particular interest to readers of this book: "In this technological age," stated the former vice-president, "image becomes dominant; appearance supersedes reality. An appearance of wrongdoing, whether true or false in fact, is damaging to any man, but—more important— it is fatal to a man who must be ready at any moment to step into the presidency."

Mr. Agnew thus voiced a proposition widely shared by observers of contemporary American politics; that is, that public figures are known primarily for their "images," images that somehow mask the "true" or "real" nature of things. Indeed the many faces, or images, of Mr. Agnew himself long obscured the "reality" of the man. Seeking the vice-presidency in 1958, Mr. Agnew admitted that he was an unknown quantity to most Americans, a man whose name was hardly "a household word." Once in office the vice-president seemed simultaneously to be Hero, Villain, and Fool:[1] a hero to those who rallied around his frequent attacks on antiwar protesters, against the major national news media, or against "effete, intellectual snobs"; a villain to those he attacked; and to people who scoffed at his overusage of alliterative language or laughed at comedian Bob Hope's repeated portrayals of his friend's inepti-

1

tude on the golf course, the vice-president acted the fool. Yet if none of these images captured the substance, was the appearance of a man "on the take" or a "tax evader" closer to reality? Or was the "reality" that had been super-seded by the "image" (as the former vice-president himself suggested) one of Ted Agnew as a victim of "scurrilous and inaccurate reports" and of accusers who were nothing more than "self-confessed bribe brokers, extortionists, and conspirators"?

The case of Spiro T. Agnew illustrates a significant problem confronting people wishing to make sense out of American politics, particularly the poli-tics of office seeking and holding in the 1970s. In a democracy, the assump-tion is that self-government requires popular control of policy makers, exerted in part by choosing public officials through popular elections. But in a far-flung republic consisting, as did America in 1972, of more than 136 million people of voting age assigned the responsibility of electing a multitude of federal, state, and local officials, firsthand knowledge of even the major candidates, let alone of the less well-known ones, is impossible. Instead voters must respond to contenders on the basis of such subjective knowledge as they are able to assemble from secondhand accounts—from news reports, a few candidates' personal and televised appearances the observations of friends, and so on. In short, voters take an interest in selected candidates, form images of them, and on election day those images help them make up their minds.

But how, as citizens, are we to judge and make decisions about poli-ticians on the basis of their images? What is the nature of those images? Do they supersede reality, as Mr. Agnew and others have argued? Or are they the substance of political reality, at least as we *perceive* it? Do our images of candidates determine how we vote, or are they just one among many sets of influences over what we do? Can images be manipulated, as some observers have warned, by politicians bent on selfish gain using highly sophisticated per-suasive techniques,[2] or are our images derived in such complicated ways that the candidates and their technical advisers simply cannot make "a silk purse out of a sow's ear"?

These are only a few of the questions that surround the topic of polit-ical candidates and their images. In this book we shall return to these ques-tions and raise several others as well. In some instances we attempt answers. In others we suggest that the questions may be unanswerable because they rest on a misunderstanding of the role of images in influencing behavior. In still other instances we admit that political scientists do not yet know enough about certain problems to permit even grounded speculation, much less firm answers. This is a book about the part played by the images of political can-didates in contemporary electoral politics. We endeavor to clarify what it means to refer to a candidate's "image"; discuss various dimensions, typolo-gies, and traits associated with candidate images; look at how election cam-paigns affect images; and come to an understanding of how the images of candidates influence voting. To begin the task we have assigned ourselves, in the remainder of this introductory chapter we first provide a few brief illus-trations of how the notion of images is used in selected areas; then we advance a tentative definition of the concept and outline our major concerns with it.

Contemporary Uses of the Image Concept

People use the word *image* in a variety of ways. Politics is only one area in which the term has considerable currency. In the visual arts, for instance, the concept enters into discussions of how people can look at an abstract art object (say, a painting or a piece of sculpture) yet think they see a tree, a reclining human, or some other recognizable form. Psychologists of perception suggest that when we see a physical object such as a tree, we retain a memory of it after that object has been removed from our sight. This memory, or "mental image," we can then recall when we see a figure whose outlines seem similar; thus, our mental image of a tree can help (or some artists might say, hinder) in interpreting a piece of abstract art.[3]

In a later chapter we consider in detail the relationship between mental imagery and perception, so we need not dwell on it here. We mention it merely to urge the reader to recognize that the notion of image refers to relatively complex psychological phenomena. Discussions of the images of political candidates frequently overlook the psychological aspects of image perception; rather, our sketchy understanding of images in politics is more apt to stem from the term's usage in other areas—public relations, popular culture, and campaign politics.

The Image in Public Relations

In November 1973, an Associated Press wire story reported that the United States Chamber of Commerce had announced "a massive campaign to counter critics and spruce up the corporate image in the minds of Americans." In a four-pronged program budgeted at $500,000, the chamber declared it would use education, politics, the courts, and communications media to "push the good points about business." The chief executive officer of the chamber emphasized that, "We're not trying to propagandize. We're trying to find a way to talk facts."[4]

The Chamber of Commerce's announcement typifies the usage of the term *image* in the commercially oriented field of public relations. One student of public relations, Albert J. Sullivan, has written that, "Images are the proper function of public relations and warrant the most careful understanding and usage."[5] Sullivan distinguishes between images and reality, but notes a close relationship: images are *reflections* of reality that carry information about things that lie outside the mind. According to Sullivan, the mind does not, indeed cannot, confront reality directly; instead it operates with images as indirect reproductions of the external world. But reflections of reality vary in accuracy, from close correspondence to partial, colored inaccuracies. The responsibility of public relations, Sullivan writes, is to present truthful images: "The whole image process is its responsibility; when it sets information out on its journey, it is not enough that 'some' of it arrives 'somehow' and does 'something.' It must wholeheartedly investigate what its images turn into;

evaluate far more effectively the service its images perform for the whole truth; and above all, it must learn to understate the information it projects."[6] Sullivan concludes that in dealing with images, public relations must, like Caesar's wife, be above suspicion in its regard for truth.

Closely associated with public relations, although not strictly identical to it, is the field of advertising. Advertisers also employ the term *image* in their efforts to sell products. Even though advertisers want to guide the buyer's behavior, as John Downing has pointed out, their notion of what is a brand image is not merely manipulative. A brand image, writes Downing, is "a constellation of feelings, ideas, and beliefs associated with a brand by its users and nonusers mainly as a result of experience of its advertising and performance." In light of such a definition, many advertisers would argue that they, like public relations men, simply offer a view of a product, and that such an image seldom misrepresents it.[7]

Critics, however, rarely exonerate public relations and advertising as above suspicion. Image builders have been charged with engaging in a needless, misleading enterprise, whether engaged in public relations to promote a product or an institution. Take as an example a widely sold book of the 1950s, *The Hidden Persuaders*. Its author, Vance Packard, levelled his criticism against image building in product selling. Noting that there were so many products of a similar type on the market (such as countless brands of beer) that people could not discriminate among them reasonably, Packard argued that marketers decided that consumers "should be assisted in discriminating *unreasonably*, in some easy, warm, emotional way." Hence, salespeople began exploring the molding of product images, "the creation of distinctible, highly appealing 'personalities' for products which were essentially undistinctive." These images strike a responsive chord in buyers "at the mere mention of the product's name, once we have been properly conditioned."[8] For Packard (and for similar critics of public relations and advertising) the imagery associated with such latter-day slogans as that of a best-selling beer, "You only go around once in life," is but another name for a Pavlovian process of conditioning people to behave in a mindless way.

There is also criticism of the use of images to promote an institution, the practice of institutional advertising. Whereas product advertising helps sell specific goods and services, institutional advertising strives to build "goodwill" for large corporations confronting a potentially hostile public.[9] When competitive products offered for purchase are so similar that consumers cannot discern the merits of alternative goods, the public relations specialist seeks to assure prospective buyers that distinctions can be made on the basis of the integrity and competence of the manufacturer. He works with the "corporate image" rather than a "brand image."[10] For example in the 1970s, with the advent of America's "energy crisis" and its attendant shortages of gasoline and heating fuels, one major oil company ceased advertising that its multiple grades of gasoline were superior to others'; rather, televised commercials and printed advertisements stressed how the corporation was stepping up exploration of domestic oil reserves to "meet the crisis" and implied that purchasing the company's products would contribute to a self-sufficient America. The principal purpose of institutional image making is to serve the interests of the corporation by convincing people that specific qualities of competing gasolines, drugs, deodorants, automobiles, and thousands of other items are less important in making rational purchases than the public reputations of the companies that "stand behind" the products. Thus, the public relations posture is that images are summary reflections of things consisting as much of what we *feel* about a product or company as what we *know*.[11]

The Image in Popular Culture

Images pervade popular culture as well as public relations. Popular entertainers are celebrities celebrated as much for the impressions audiences have of their styles and selected attributes as for the tangible qualities of their performances. Well-known athletes have images, ranging from Joe Namath's "swinger" to Johnny Bench as the "All-American professional." The images of such popular figures parallel product images supported by brand (the New York Jets or Cincinnati Reds) and corporate images (the National Football League or Major League Baseball).

Various writers have described the images in popular culture,[12] but Daniel Boorstin in his work *The Image: Or What Happened to the American Dream*[13] offers the most succinct summary. The term *image,* Boorstin says, stems from the Latin *imago* and *imitari,* "to imitate." Thus, he defines an image as "an artificial imitation or representation of the external form of any object, especially of a person," and then provides a detailed characterization of images: they are synthetic, believable, passive, vivid, simplified, and ambiguous.[14] They are *synthetic* in that they are "created especially to serve a purpose, to make a certain kind of impression."[15] A trademark, brand name, jingle, slogan, or television talk show host's reputation for spontaneous wit and congenial chatter—all are contrivances to command attention, attract audiences, and sell products. An image is *believable* for, in the absence of direct contact with the institution or person it stands for, it *is* that institution or person. Hence, the National Football League is bone-crushing violence, "Ford Has a Better Idea," Bobby Riggs is a hustler, and Billy Jean King is liberated. An image is *passive* in that it does not change; rather a person lives up to it. (Perhaps Jack Benny was not a miserly skinflint, but the image was there and he strove to fit into it in public performances.) An image is *vivid* in its portrayal of reality in concrete ways. When actor Burt Reynolds offers the image of manliness, then poses nude for the centerfold of a popular women's magazine, the image is seemingly real. An image is *simplified* in that it is not the object or person it represents; rather it is a far simpler depiction. (The cocky, garrulous Howard Cosell is simple and distinctive enough to be remembered.) Finally, an image is *ambiguous* in that it is partial, incomplete, an outline that we are left free to fill in with a combination of our senses and imagination. As with the abstract painting, the image provides the hint of a tree or perhaps of something else; it remains for us to complete the picture.

This encapsulates Boorstin's argument about the nature and characteristics of images in popular culture. What disturbs him is not that images are artificial imitations or representations of something else but that people are prone to regard images as the real thing. Images begin as ideas that represent something and are thus ways or means to visualize objects, persons, or goals. However, they seem to take on a life of their own and no longer serve as means to an end, but become the end itself. Boorstin makes the point best by relating a conversation between a mother of a young child and a friend: "My, that's a beautiful baby you have there," says the friend. Replies the mother, "Oh, that's nothing—you should see his photograph!" As another example, in the conviction that the Good Life consists of retirement from a lifelong occupation, many people begin striving to retire as an end in itself; but alas, they all too often face the disillusionment that their retired status is much less than they had hoped for. The magic of images lies not in their truthfulness— that is, how closely they correspond to what they purportedly represent—but their believability.

The Image in Campaign Politics

Rarely do we hear as much talk of images as during elections. Since images of political candidates comprise the subject matter of this book, we will have a great deal to say in the chapters that follow about the use of images in campaigning. At this point, however, we want merely to indicate how a few political observers use the term. As in public relations and popular culture, discussions of campaign images frequently have a pejorative quality. Images in campaigns imply something contrived and artificial and are creations of the mass media: "It is difficult to say what the politician 'really' is as distinguished from the 'persona' which emerges from the media to be projected to the voting public. Thus the media have from time to time given us 'country lawyer' Sam Ervin, 'ruthless' Bob Kennedy, 'idealistic' George McGovern, and other variations on these themes, which may or may not get us closer to the respective 'realities' of the politician (as distinguished from the person)."[16]

Most accounts of contemporary American political campaigns couple "image campaigning" with the advent and growth of the electronic means of mass communication. A prime illustration comes from Joe McGinniss's *The Selling of the President 1968*, which details how Richard Nixon's stable of television advisers portrayed the image of their candidate as a calm, cool, thoughtful choice for voters alarmed by the turbulence of American politics in the sixties.[17] And a self-designated "image specialist" has provided a revealing account of how he, a radio-television-film producer and writer, successfully assisted the campaigns of "image candidates."[18]

It is a mistake, however, to think that politicians have started to worry about their images only in our era of electronic communication. One need only read Niccolò Machiavelli's "How a Prince Must Act in Order to Gain Reputation," written more than four centuries ago, to recognize that the concern with imagery in politics has a long history.[19] (Of course, one can find even earlier examples.) In America, the distinct style of campaigning con-

cerned largely with images dates from at least the 1920s. Historian Richard
Jensen argues that prior to the 1920s, election campaigns were militarist in
nature; that is, campaign organizations consisted of structures along military
lines with precinct "captains," party "loyalists," "drilling" of voters, and so
on. Around 1916 the trend shifted toward a different campaign mode, one
more mercantilist than militarist. Advertising became a basic campaign technique; candidates directed their appeals less to party "loyalists" than to the
undecided and wavering voters of both parties.[20] Election campaigns assumed
the tone of sales campaigns rather than military campaigns. The way was
opened for the eventual transfer of the notion of image from public relations
and advertising arenas to politics.[21]

   Not only did a style congenial to image campaigns predate the electronic media, but even today image campaigns are not restricted to film,
radio, and television. The newspaper is one case in point. Image advertising—
combining pictures and messages to highlight selected qualities of a candidate
—is as commonplace in newspapers as on television. In 1972 for instance,
advertisements in newspapers depicting George McGovern wearing a protec-
tive "hard hat" and speaking to laborers exemplified imagery.[22] Newspaper
accounts of campaigns also emphasize the candidates' images. A study of
campaign coverage in leading U.S. newspapers in 1968 revealed a focus on the
personal qualities of competing candidates; newspapers furnished the raw
material for constructing a variety of images of competing candidates or of a
single candidate.[23] Generally, contemporary image campaigns are carefully
orchestrated combinations of advertising and news accounts furnished through
all available media—radio, television, newspapers, campaign brochures and
mailings, and billboards, among others.

A Working Definition of Image

Our review of the uses of the term *image* in public relations, popular culture,
and campaign politics indicates its popularity in recent decades. Although the
way was prepared for what is now called "image politics" by the shift from
militarist to mercantilist campaign styles in the 1920s, the word *image* did
not really become fashionable until relatively recently. It has had notable
forerunners, however. For example in 1922, political journalist Walter Lippmann wrote that the world is too subtle, complex, and fleeting for direct acquaintance. Unequipped as we are to deal with so much variety, we recon-
struct our environment using simpler, manageable mental models. These
reconstructions he called "stereotypes," and he argued that public opinion—
about government, issues, and politicians—was a function of these "pictures
in our heads."[24] Similarly, such terms as "charisma," "personality," "political appeal," and "personal appeal" found their way into discussions of campaign politics to suggest that people respond to candidates, as to products, on
the basis of how they *feel* as well as what they *know* about them.[25] By the
mid-1950s and early 1960s, articles appearing in scholarly books and journals
began to employ the concept of image with increasing frequency in political

analyses.[26] Today there is much talk of product images, brand images, corporate images, images of celebrities, of ideals, and of political figures. Whether
viewed from the perspective of the practitioner endeavoring to sell products,
win audiences, or win elections, or from that of the scientist trying to explain
these processes, we live in an era in which images are thought to be important,
at least sufficiently important to know more about. But, what is an image?

From our discussion thus far, it should be clear to the reader that there
is no apparent consensus on a definition of the term *image*. Some writers
speak of it as a mental construct, that is, the way a person represents reality
to himself. Others imply that images are sets of visible attributes of a product,
object, or person that are "projected" or "transmitted" to the consciousness
of an audience.

A book written almost two decades ago by Kenneth Boulding, entitled
The Image,[27] will help us derive a definition of the concept "image" that will
assist us in discussing the images of political candidates. Its thesis is that each
of us possesses an image of the world; that is, each person possesses knowledge of the world that he or she *believes* to be true. Thus defined, *an image is
a person's stock of subjective knowledge.* An individual's image, says Boulding, governs that person's behavior. Where do our images come from? Generally, from our experience. More specifically, our images develop as we
confront our environment. Those contacts between a person and his surroundings constitute what Boulding calls the *message-image relationship.* Boulding
is careful to distinguish between images and the messages that reach those
images. Messages contain information about the world and such information
may result in changes in the image: *"The meaning of a message is the change
which it produces in the image."*[28]

The barest outlines of Boulding's detailed and provocative discussion
consist of three key points. First, images are unique to the persons who hold
them; that is, they are mental representations of the world as it is perceived
by the person holding a particular image. Second, as we have seen before,
images are reflections of reality, not definitive knowledge of it: "Knowledge
has an implication of validity, of truth. What I am talking about is what I
believe to be true; my subjective knowledge."[29] Third, although images are
unique and specific for each individual, there is a close relationship between
a person's image and messages that reach it; through the message-image transaction, a person's image meets with sources of information affecting his image.

What does this imply about the images of political candidates? We shall
begin by defining an image as *a human construct imposed on an array of perceived attributes projected by an object, event, or person.* Such a definition
summarizes our view that an image is (1) a subjective, mental construct
(2) affecting how things are perceived but also (3) influenced by projected
messages. Thus, our focus is largely on the message-image relationship;
that is, jointly on what attributes a candidate publicizes to appeal to the
voter and the attributes the voter sees in the candidate, which comprise
the voter's image of the candidate. For our purposes, therefore, the candidate's image consists of how he is perceived by voters, based on both the

subjective knowledge possessed by voters and the messages projected by the candidate.

As we explore the subjective knowledge of voters that, in transaction with candidates' messages, comprises the substance of our concern in this book, we will find it helpful to think of images as consisting of what a person knows, feels, and proposes to do. Thus we try to capture in this notion of image the thought of "a constellation of feelings, ideas and beliefs" associated with a politician, which Downing speaks of in his definition of brand image.[30] Put more technically, the content of our images has *cognitive, affective,* and *conative* aspects. The cognitive aspect consists of what we know about the candidates—such things as their age, experience, background, and stands on issues. How we feel about the candidates (whether we like them or not, and how strongly) is what comprises the affective aspect. What a voter proposes to do about a candidate, his behavioral intentions (for example, to vote for, against, or not at all) relates to the conative aspect. Thus, in contrast to the emphasis in public relations and popular culture, where images consist of how people feel about products, corporations, or celebrities, we regard as important the cognitive and conative aspects as well as the affective component.

Having said what we think images to be, we want also to indicate what we think they are not. For example, our definition is nonpejorative; that is, we will not be attacking images for superseding reality, for replacing ideals, or for misrepresenting the truth. We begin with the assumption that we all have subjective knowledge of things, which varies greatly in its accuracy or validity. We are interested in discussing what the images of candidates are and how they relate to voting behavior. Moreover, we treat images as associated with a timeless problem of political leadership, namely, that of mobilizing support in a diverse constituency. Even candidates who avoid television appearances and use only personal-contact campaigning have images. Thus we do not treat images as products of a "television age." Finally, we wish to avoid the view that somehow an "image candidate" is insincere, whereas some other type of candidate is, by contrast, certain to be a man of principle. To be known as a "man of principle" is also to have an image. At issue is whether any specific image is a mere pose or is reinforced by substantial actions in support of certain causes. Perhaps by learning more about the character of images of political candidates we can learn to make informed judgments rather than simply rejecting all images out of hand.

Plan and Themes of the Book

In the chapters that follow, we shall be examining in detail a variety of questions that bear on the role of candidate images in electoral politics. To our knowledge, this text is the only existing effort to synthesize the numerous studies previously published that concern the images of political candidates, and also to combine our synthesis with an attempt to deal with a broad range of issues implicit in those studies. For that reason we deem it particularly important to establish the context of our work in Chapter 2 by reviewing the studies, techniques, and theories of previously published research and to note that systematic studies of voting behavior reveal the crucial part played by candidate images in voting, whether implicitly or explicitly.

In Chapter 3 we take the view that candidate images are multifaceted phenomena with several key dimensions and identifiable modes of orientation, requiring alternative approaches to understanding. In Chapter 4 we consider the various determinants of candidate images and look at a major controversy surrounding the derivation of those images. Our theme will be that a candidate's image is as much a function of what people project to him as of his efforts to project to them. "Image campaigns" involve differing modes and styles of campaigning. In Chapter 5 we look at these styles with the view that voters respond to different campaign styles in ways that have implications for planning campaign strategies.

A point of controversy has been whether or not election campaigns and exposure to mass media make any difference in how people respond to candidates. Do people in fact change their images of candidates across the course of campaigns, or are campaigns simply a waste of time? In Chapter 6 we detail the effects of election campaigns and media exposure on candidate images. Winning or losing an election can influence how people perceive contenders, particularly winners moving into positions of authority or losers hoping to live to fight another day. In Chapter 7 we explore the effects of victory or defeat on the images of public figures. Finally, in Chapter 8 we turn to a crucial question for anyone interested in candidate images—that is, what, if any, is the independent effect of candidate images on voting behavior?

Writing in 1956, Boulding argued for scientific study of the message-image relationship in all phases of behavior; he labelled such a science "eiconics." We are under no illusions that we have progressed very far toward that goal, but we share the view that the road to such a goal "will be paved not only by descriptive surveys and historical reports, but by cumulative, relatively microscopic studies which have undertaken the tedious task of identifying and validating the conceptual variables that enable us to explain uniformities and deviations of human behavior in politics."[31] We therefore derive much of the material in our discussion either from original small-sample research projects conducted by the authors or from extensive secondary analyses of data provided from the large-scale election studies of the Survey Research Center, University of Michigan (now called the Center for Political Studies).[32] Reporting that research necessarily requires a discussion of the technicalities of data gathering and analysis. Many readers may find such discussions too esoteric for their tastes. Consequently, we confine the body of each chapter only to those matters pertaining directly to the substantive question at issue. We place technical discussions related to each chapter in separate appendixes at the end of the book. In this volume we explore the barest essentials of the message-image relationship that we refer to as "candidate imagery" in the hope that such an exploration will stimulate a greater appreciation for the role of imagery in politics.

Notes

1. Such images are discussed by Orrin E. Klapp in *Heroes, Villains, and Fools: The Changing American Character* (Englewood Cliffs, N.J.: Prentice-Hall, 1962); and in *Symbolic Leaders: Public Dramas and Public Men* (Chicago: Minerva Press, 1964).

2. For a discussion of the alleged dangers underlying the manipulation of images, see Herbert I. Shiller, *The Mind Managers* (Boston: Beacon Press, 1973).

3. See Rudolf Arnheim, *Visual Thinking* (London: Faber and Faber, 1969), chap. 6; and Mardi Jon Horowitz, *Image Formation and Cognition* (New York: Appleton-Century-Crofts, 1971).

4. Associated Press wire story, November 12, 1973.

5. Albert J. Sullivan, "Toward a Philosophy of Public Relations: Images," in Otto Lerbinger and Albert J. Sullivan, eds., *Information, Influence and Communication* (New York: Basic Books, 1965), p. 247.

6. Ibid., p. 249.

7. John Downing, "What Is a Brand Image?" *Advertising Quarterly* 1 (Winter 1964-65): 13-19.

8. Vance Packard, *The Hidden Persuaders* (New York: Pocket Books, 1957), p. 38.

9. Leonard I. Pearlin and Morris Rosenberg, "Propaganda Techniques in Institutional Advertising," in Daniel Katz et al., *Public Opinion and Propaganda* (New York: Henry Holt, 1954), pp. 478-90.

10. Otto Lerbinger, "The Social Function of Public Relations," in Lerbinger and Sullivan, *Information, Influence, and Communication*, pp. 53-60.

11. William Stephenson, "Methodology of 'Imagery' Measurement: MR Redefined" (paper delivered at the Sixth Annual Public Utilities Seminar, American Marketing Association, St. Louis, Missouri, February 18, 1960).

12. See, for example, Shiller, *The Mind Managers*; and Orrin E. Klapp, *Collective Search for Identity* (New York: Holt, Rinehart and Winston, 1969).

13. Daniel J. Boorstin, *The Image: Or What Happened to the American Dream* (New York: Atheneum, 1962).

14. Ibid., p. 197.

15. Ibid., p. 185.

16. Richard D. Cudahy, "Politics and the Press," in *Two Perspectives on—Politics and the Press* (Washington, D.C.: The American Institute for Political Communication, 1973), p. 1.

17. Joe McGinniss, *The Selling of the President 1968* (New York: Trident Press, 1969).

18. Gene Wyckoff, *The Image Candidates* (New York: Macmillan, 1968); see also Joseph Napolitan, *The Election Game* (Garden City, N.Y.: Doubleday, 1972).

19. Niccolò Machiavelli, *The Prince* (New York: The New American Library, 1952), pp. 120-23.

20. Richard Jensen, "American Election Campaigns: A Theoretical and Historical Typology" (paper delivered at the Annual Meeting of the Midwest Political Science Association, Chicago, May 2-4, 1968).

21. For a discussion of how public relations techniques were transferred to political campaigns, see Stanley Kelley, Jr., *Professional Public Relations and Political Power* (Baltimore: Johns Hopkins Press, 1956).

22. James J. Mullen, "Newspaper Advertising in the Kennedy-Nixon Campaign," *Journalism Quarterly* 40 (Winter 1963): 3-11.

23. Doris A. Graber, "Personal Qualities in Presidential Images: The Contribution of the Press," *Midwest Journal of Political Science* 16 (February 1972): 46-76; "The Press as Opinion Resource during the 1968 Presidential Campaign," *Public Opinion Quarterly* 35 (Summer 1971): 168-82; "Press Coverage Patterns of Campaign News: The 1968 Presidential Race," *Journalism Quarterly* 48 (Autumn 1971): 502-12.

24. Walter Lippmann, *Public Opinion* (New York: Macmillan, 1922).

25. See, for example, Herbert Hyman and Paul B. Sheatsley, "The Political Appeal of President Eisenhower," *Public Opinion Quarterly* 17 (Winter 1953-54): 459; James C. Davies, "Charisma in the 1952 Campaign," *American Political Science Review* 48 (December 1954): 1083-1102.

26. Ithiel de Sola Pool, "TV: A New Dimension in Politics," in Eugene Burdick and Arthur J. Brodbeck, eds., *American Voting Behavior* (Glencoe, Ill., The Free Press, 1959), pp. 236-61; Leo P. Crespi, "Some Observations on the Concept of Image," *Public Opinion Quarterly* 25 (Spring 1961): 115-20; Sidney Kraus, ed., *The Great Debates* (Bloomington, Ind.: Indiana University Press, 1962), especially chaps. 11-18.

27. Kenneth E. Boulding, *The Image* (Ann Arbor, Mich.: University of Michigan Press, 1956); see also James W. Fox, "The Concept of Image and Adoption in Relation to Interpersonal Behavior," *Journal of Communication* 17 (June 1967): 147-51; Dennis C. Alexander, "A Construct of the Image and a Method of Measurement," *Journal of Communication* 21 (June 1971): 170-78.

28. Boulding, *The Image,* p. 7 (italics are Boulding's); see also Colin A. Hughes, *Images and Issues: The Queensland State Elections of 1963 and 1966* (Canberra: Australian National University Press, 1969), pp. 186-87.

29. Ibid., pp. 5-6.

30. Downing, "What Is a Brand Image?", p. 14.

31. Avery Leiserson, "Problems of Methodology in Political Research," in Heinz Eulau et al., *Political Behavior* (Glencoe, Ill.: The Free Press, 1956), p. 58.

32. We recognize that such methodological openness besets the reader with many problems in keeping up with the specific limitations to be placed on given sets of findings, but some perplexity is preferable over a too-simplified clarity whose only justification is rigid adherence to a single "brand" of research procedure; cf. Michael Haas and Theodore L. Becker, "A Multimethodological Plea," *Polity* 2 (Spring 1970): 267-94.

Chapter Two

Candidate Images and Voting Behavior

In recent decades social scientists have accumulated a great deal of information about the electoral process and why people vote as they do. In this chapter we will examine theories and interpretations based on that information. Our purpose is to determine the role assigned to candidate images in explaining voting. We first consider published studies of voting that treat candidate images as one of the many facets of elections. Then we take a preliminary look at published studies devoted almost exclusively to delineating the character of candidate images (we examine many of these studies in more detail in subsequent chapters). Finally we consider a key description of voting behavior that relates the images of candidates to those of political parties and issues. Along the way we shall raise several questions pertinent to understanding candidate images, which we hope to respond to in succeeding chapters by reporting previously unpublished information and analyses.

Candidate Images in Major Studies of Voting

Systematic voting research dates back to the 1920s, but the earliest studies have scarcely anything to say about candidate images.[1] Then beginning in the 1940s and extending into the present, social scientists conducted a number of significant studies based on surveys of samples of voting populations. These studies, frequently labelled "the voting studies," have become increasingly sophisticated, both in techniques of gathering and analyzing data and in building theories of the voting act.[2] Among the most informative were the studies conducted under the auspices of the Survey Research Center, University of

Michigan (SRC).[3] Those studies contribute to building a description of voting that incorporates voters' responses to candidates relative to other influences. Later in this chapter we will discuss the SRC description in detail. At this point we will focus on other voting studies that tell us something about candidate images. Although few of them employ the terminology of candidate images, they clearly deal with related matters: for example, voters' "orientations" to candidates, the candidate's "appeal" and "personality," and to candidates' "stands." Through such language they introduce four important questions relevant to anyone interested in the image of political candidates.

How Do Voters Perceive Candidates?

In Chapter 1 we said that a candidate's image consists of how voters perceive him or her; moreover, those perceptions are a function of both the subjective knowledge possessed by the voters (that is, the images of the voters) and the messages transmitted by the candidate. One of the first major voting studies, which examined the 1940 presidential campaign, addressed itself to the question of how voters perceive candidates. In May of 1940, interview teams visited every fourth house in Erie County, Ohio. From a total of 3000 persons sampled, researchers selected four groups of 600 individuals. Each group was, in effect, a miniature sample of the county. Three of those groups of 600 were interviewed once each—one in July, one in August, one in October; the fourth group was interviewed once every month from May to November. Thus, this voting study used a "panel design"—that is, a technique in which a sample of people is interviewed repeatedly over an extended period of time. In this instance the panel design was arranged to determine if sampled Erie County residents changed their voting intentions, perceptions of candidates, and other political behavior during the course of the presidential campaign.[4] This ground-breaking research effort resulted in significant findings in a variety of areas. What it indicated about how voters perceive candidates, however, interests us most at the moment. Put simply, the Erie County researchers found that voters give selective attention to the content of campaigns, exposing themselves predominantly to campaign messages that support their images of candidates: "People selected political material in accord with their own taste and bias, including even those who had not yet made a decision."[5]

Later voting studies seemed to confirm a "rule of selective perception."[6] A leading example was research conducted in Elmira, New York, in conjunction with the 1948 presidential campaign. This study also involved a panel design. From a panel of 1029 persons interviewed in June, reinterviews were conducted once in August, once in October, and once in November.[7] Regarding voters' perceptions of candidates, the study clearly indicated that the partisanship of the voter had a great deal to do with how he regarded the candidates. For one thing, partisans tended to perceive their preferred candidate's stand on issues as favorable to their own, perceiving their candidate's stand as similar to their own and the opponent's stand as dissimilar; further, they often failed to perceive any differences that they might have with their own candidate or any similarity of stands they might share with the opposition. Finally, voters who felt strongly about their choice of candidates were more likely to misperceive their candidate's stand as favorable to their own even though they held views that differed from their party's candidate.

Out of these early voting studies, then, emerged a conventional wisdom that said that voters perceive candidates in highly selective ways: they selectively pay attention to some candidates and not to others; they selectively perceive, or misperceive, the content of messages to conform to their predispositions (usually partisan); they selectively remember and forget messages and their content; and, in short, they selectively respond to candidates' appeals.

Assuming such a rule of selective perception exists, what are the implications for the images of political candidates? If a candidate's image consists of voters' perceptions (perceptions flowing from the voters' various images) of the candidate's attributes as conveyed by his messages, then we have a specific message-image relationship—that of candidate message-voter image. If we invoke selective perception as a rule, it biases that relationship at the outset since voters perceive only the messages of their preferred candidate; they do not perceive those of his rival. The voter's image of his desired candidate may change when confronted with the candidate's appeals simply because the voter is paying attention; but since the voter is not even likely to pay attention to the opposition, his image of the opposing candidate is fixed relatively early in the campaign. Put differently, with his chosen candidate the voter establishes an image-message relationship, the effects frequently being to reinforce the voter's image of the candidate; with the opposition candidate the voter makes no such linkage and his image of the candidate is unlikely to be reinforced or diminished (or, if he does reach such an image-message relationship with the opposition, the voter misperceives the rival's stand so as to protect or enhance his own voting image).

But critics have challenged the rule of selective perception in recent years. To begin with the evidence that people do, in fact, expose themselves chiefly to messages that are in accord with their predispositions is more equivocal than first reading of the early voting studies suggested. For example, David Sears and Jonathan Freedman reexamined the Erie County and Elmira studies to determine just how much de facto selective perception occurred. They found that in Erie County, selective exposure was considerably more common among Republicans than Democrats. In fact, approximately one-half of Democrats were exposed primarily to Republican propaganda. Even 30 percent of Republicans turned primarily to pro-Democratic propaganda. And in Elmira they found a similar pattern: the Republicans' exposure was 54 percent pro-Dewey and the Democrats' exposure was 57 percent pro-Truman, leaving large portions of both sets of partisans attending to the other party's cues.[8] Not only is there no clear-cut evidence that people selectively attend to campaign messages, according to Sears and Freedman there is also no conclusive evidence that people have a psychological preference for supportive information: "Under some circumstances people seem to prefer information that supports their opinions; under other circumstances, people seem to prefer information that contradicts their opinions."[9] A final and more recent indication that the verdict is still out on the question of selective perception tends to support earlier interpretations, however. In a study of the 1968 presidential election, in a sample of residents in San Mateo County, California, the researcher found that voters with a preferred candidate selectively perceived that candidate's position on issues to make it congruent with their own, thus lending support to the thesis propounded in the Elmira study of two decades earlier.[10]

In summary, we are left with a question about how voters perceive candidates. That is, are voters' perceptions of candidates so selective that the voter is solely responsible for "the candidate's image" or are voters' subjective appraisals sufficiently open-minded that they can respond to any candidate's message? We will attempt to answer this question in Chapter 4, when we consider whether the candidate's image is "perceiver determined" or "stimulus determined."

Do Campaigns Make a Difference?

Political candidates typically spend huge sums of money to make themselves known, gain support, and ultimately win votes. According to the Citizen's Research Foundation studies over two decades, the total costs of campaigns for all elective offices in the United States rose from $140 million in 1952 to an estimated $400 million in 1972.[11] With so much money being spent, we might assume that campaigns make a great deal of difference, but do they?

Among types of changes that can occur during election campaigns, we shall consider two, namely, changes in voter preferences and in images of political candidates. In relation to the former, voting studies indicate a rule similar to that of selective perception, which we might call the "rule of minimal effects." Again, the Erie County study established conventional wisdom. Researchers outlined three effects of the presidential campaign on vote intention—it *activated* the indifferent, *reinforced* partisans, and *converted* the doubtful. By and large, the major effect was to reinforce partisans in decisions they had made before the campaign; 53 percent of those in the Erie County panel were so reinforced. Another 17 percent either changed their vote intention from one candidate to another during the campaign, moved from decided to undecided, or from an undecided stance to a choice. Fourteen percent were persons who began with an indifference to the campaign but later became sufficiently interested to vote. The remaining 16 percent apparently were unaffected in any way.[12]

The first voting studies also concluded that a high percentage of voters reach their decisions very early in presidential campaigns, then seldom waver thereafter. Succeeding studies confirmed this view (see Table 2.1). In fact, studies of presidential voting from 1948 through 1972 indicate that, on the average, 40 percent of voters make up their minds before the presidential nominating conventions, about one-fourth do so during the conventions, and the remaining one-third arrive at their voting decision during the campaign.[13]

Table 2.1   Time of Voting Decision, 1948-1972 Presidential Elections

TIME OF DECISION	ELECTION						
	1948	1952	1956	1960	1964	1968	1972
Before the nominating conventions	42%	31%	60%	31%	42%	35%	46%
During the nominating conventions	30	36	18	32	25	24	18
During the campaign	28	33	22	37	33	41	36
Total	100	100	100	100	100	100	100
$n =$	357	1139	1213	1361	1077	957	1486

Source: Survey Research Center/Center for Political Studies, University of Michigan, and the Interuniversity Consortium for Political Research.

Yet one-third of an electorate is considerable and, if the presidential nominating conventions are considered closely linked to the campaign, as many as six voters in ten may be subject to some influence from campaign messages. Moreover, if reinforcement of one's position is considered an important effect of a campaign (as we think it is), then certainly campaigns make a difference.

The proportion of people not particularly affected by campaigns is probably higher in congressional than in presidential elections. The evidence is that in congressional races a considerably higher percentage of people know something about the incumbent, if there is one, than know anything about the challenger. Further, a lack of knowledge of the challenger seems to be associated with voting principally along party lines in congressional elections, thus suggesting that the campaign may have less effect than it does in presidential contests.[14]

Of course, we are speaking strictly of the general election campaigns for the presidency and for congressional seats. Primary elections to nominate party candidates generally awaken the interests of far fewer citizens; yet for those who take part, primaries alert voters to the identities of challengers, reinforce those voters already committed to candidates, and assist the undecided in reaching decisions. One effect of primaries is to permit voters with similar party loyalties to sort themselves out along issue and/or ideological lines. There is evidence, for example, that one result of the 1972 presidential primaries was they they afforded George McGovern the opportunity to capture the votes of liberal Democrats, while conservative Democrats divided among the several other contenders.[15]

As the voting studies suggested, election campaigns also make a difference regarding the images of political candidates—especially if the images of a candidate are ambiguous and ill-defined in the early campaign. Researchers in Erie County, for example, pointed out that if the campaign message (which, in part, helps define the candidate's image) has a variety of possible meanings, the candidate's appeal helps identify for a voter "the way of thinking and acting that he is half aware of wanting." Such was the case with

Wendell Wilkie, Republican presidential candidate in 1940. To the extent that people knew little more of him than that he was a "poor boy who made good," (1) poor people were able to believe that he would not forget them; (2) the rich could believe that he would serve their interests; and (3) the middle-class voters could perceive in his success the virtues of hard work and thrift: "Structured sterotypes are too well defined to be useful. The unstructured are catchalls into which each voter reads the meanings he desires."[16]

Few published voting studies consider directly the question of image change during campaigns. Two studies, however, dealt with changes in the images of presidential candidates Lyndon Johnson and Barry Goldwater in 1964. In one study, pollster Thomas Benham asked respondents in nationwide surveys before and after the campagin to select from a list of favorable and unfavorable qualities those most typical of each candidate. Listed were such qualities as "warm and friendly personality" and "shows good judgment" (favorable), as well as "too much of a politician" and "promises anything to get votes" (unfavorable). At the start of the campaign, voters were more inclined to ascribe favorable qualities to Johnson than to Goldwater. The effects of the campaign, according to Benham, were to highlight negative qualities in Johnson's image while diminishing some of the favorable; for Goldwater there was an improvement in favorable qualities, but negatives also received greater mention.[17]

Political scientist John Kessel also examined the 1964 campaign. Relying on the nationwide survey conducted by the Survey Research Center of the University of Michigan, Kessel compared what respondents said they liked and disliked about the candidates at three different time periods—respondents interviewed before October 4, between October 4 and 17, and between October 18 and November 2. He found that images were fairly stable during the campaign, but that the perceptions of both candidates' experience and ability declined. Respondents came to perceive Johnson as less trustworthy, and positive references to his character and background diminished. Voters viewed Goldwater as more likeable as the campaign progressed, but his stands on domestic policies and relations to various groups became more of a liability.[18]

In the 1968 presidential election, the American Institute for Political Communication (AIPC) studied changes in the images of the candidates. Using a panel design, the AIPC interviewed approximately 300 sampled persons in the Milwaukee metropolitan area (a locale that had been a bellwether for metropolitan areas in the nation) at five separate periods from March to early December. During each of the five interview periods, respondents rated leading contenders for the nomination (and later, the party nominees after their selection) on a five-point scale on each of five characteristics: leadership ability, political philosophy, speaking ability, intelligence, and honesty. Across the campaign Hubert Humphrey's rating improved in all five categories; Richard Nixon's changed less than Humphrey's, but generally the change was on the positive side (especially in the category of "intelligence") and occurred among respondents in the adjoining county rather than in the city or its suburbs; the ratings of George Wallace declined during the campaign.

Respondents also rated the vice-presidential candidates. The Democratic candidate, Edmund Muskie, improved his image in all categories during the campaign (indeed, by late October his ratings were better than those of his running mate), whereas little change occurred in Spiro Agnew's image. (Curtis LeMay was added as Wallace's vice-presidential candidate too late to receive early ratings. His late October rating was slightly unfavorable, but more positive by a slim margin over that of Wallace.)[19]

For purposes of cross-national comparison it is helpful to examine one remaining study, that of party leaders in the 1964 British parliamentary elections. In surveys in two Yorkshire constituencies, researchers asked respondents to rate the three party leaders—Conservative Alec Douglas-Home, Labourite Harold Wilson, and Liberal Jo Grimond—on a dozen qualities such as "inspiring," "strong," "persuasive," "sincere," and so forth. Surveys were conducted before, during, and following the campaign. Generally, aspects of the leaders' images dealing with strength changed. Specifically, Wilson's public reputation for strength increased, Home's decreased, and Grimond's profile improved. The researchers observed:

> Voters tend to perceive a campaign as presenting a challenge to a leader to show his mettle. If he is believed to be performing well, he will be upgraded for qualities of strength, confidence, and persuasiveness. If he is considered to be responding inadequately, his standing on those traits will suffer accordingly. Since a typical campaign rarely poses any *moral* tests of a leader's worth, ratings of politicians for such qualities as fairness and straightforwardness are less likely to be affected.[20]

Published voting studies, then, indicate that campaigns do make a difference, at least in marginal ways, by activating, reinforcing, or changing voting intentions and by effecting shifts in the perceptions of specific traits comprising a candidate's image. In Chapter 5 we will examine the relationship between candidate images and campaign styles, and in Chapter 6 we examine the effects of campaigns on candidate images in more detail, offering a compilation of unpublished data that we think assists us in interpreting those changes.

Do the Communications Media Make a Difference?

The earliest voting studies assumed at the outset that voting is essentially an individualistic act, something like buying a product in a store, in which an array of candidates confronts each voter. Thus, just as advertising acts on each buyer's predispositions, the mass media could influence individual voters in their final selection. The studies' findings forced a revision of this general view. For one thing, since relatively few people actually made their decision during the campaign it became apparent that most voters selectively scanned the media for reinforcing messages—especially messages reinforcing decisions made on the basis of deeply held party loyalties—rather than reaching a choice just prior to the elections, as would a buyer having less strongly felt brand loyalties just prior to the time of purchase.

A second revision pertained to the alleged individualistic character of the voting act. Both the Erie County and Elmira studies revealed the social nature of voting, discovering that patterns of political preference among members of social groups (such as the voter's contact with his family, friends, co-workers, and other peers) were more influential in governing voting decisions than was exposure to the mass media. The Erie County study offered the view that "personal influence," not media influence, was critical. Instead of being persuaded by campaign appeals in the media, voters go to "opinion leaders" for advice, people they trust and respect. These opinion leaders (21 percent of the panel group in the Erie County study) monitor the media for campaign messages on which to make judgments. Hence, there is a "two-step flow" of information: step one is from mass media to opinion leader; step two is from opinion leader to voter via personal contact and influence.[21]

Finally, the assumption that voters are even sufficiently interested in campaigns to pay attention to political information in the mass media required revision. The Elmira study indicated that the more exposure people had to the campaign in the mass media, the more correct was their information and their perception of the candidates' stands on issues. However, the more exposure of voters to mass communication, the less they changed their positions. Rather, they simply voted as they had probably decided prior to the campaign.[22] The earlier Erie County study noted that exposure itself is a function of interest in the campaign; that is, the more interest, the more exposure. Finally, later studies revealed that greater interest in the campaign was associated with strength of party loyalty, which was also associated with early voting decisions based on that loyalty. Hence, we have a paradox: the people most exposed to campaign information in the mass media are those least likely to change their minds because of it; those who might change their minds because of such exposure (the indifferents and independents) are the least likely to pay attention to the political media.[23]

In sum, either because people go elsewhere for information or simply pay no attention to the campaign or are unlikely to be influenced by the media if they do pay attention, early voting studies indicated that media exposure made relatively little difference in vote intentions, But, lest the role of the media be dismissed too lightly, consider a few qualifications to the above "rule of minimal media effects."[24]

First, although the data are sketchy, experimental studies suggest that exposure to communications among people with varying interest in the campaign may be a significant influence on how they perceive candidates and the decisions they ultimately reach. Such conclusions thus clash with interpretations that minimize media influence, based on surveys during presidential elections.[25]

Second, the voting studies indicate that media effects are minimal (other than reinforcing effects) so long as people have strong partisan loyalties and make their decisions accordingly. But if party loyalties diminish and a higher percentage of people either consider themselves independent of party or have only weak partisan loyalties, media exposure can take on a more important role. Such is the argument of a recently published book by Walter DeVries and Lance Tarrance. Using a number of surveys of Michigan voters conducted in the 1960s, they contend that voters are increasingly prone to split their vote rather than voting straight party tickets. These ticket-splitters constitute a new class of political independents. Unlike the independents of the 1950s, who had little interest in politics, these independents of the 1970s are better educated, of higher income and occupational status, and generally more politically active. And, what is of prime importance, the mass media play a key role in assisting the ticket-splitters to evaluate candidates independently of partisanship. Television newscasts, advertisements, documentaries, and special broadcasts along with newspaper stories and editorials are particularly vital to the decision making of this new class of independents.[26]

Tending to confirm the view that the mass media play a greater role in shaping voting decisions than was once thought is John P. Robinson's examination of the effect of newspaper editorials. In 1972, for instance, self-designated Independents voted very much in accordance with what candidate was endorsed by the newspaper they most often read. Independents reading pro-McGovern newspapers were nearly twice as likely to vote for McGovern (50 percent) as were Independent voters exposed to pro-Nixon papers (26 percent). Even controlling for the fact that many people first choose a candidate and then read newspapers supporting their choice, Robinson found that the McGovern vote among the Independents and Democrats who read pro-Nixon newspapers was 12 percent less than among the voters reading pro-McGovern papers. Looking at presidential elections in the last two decades, Robinson concluded that in close contests (those of 1960 and 1968, for example) newspapers definitely contributed to the shift of votes among Independents.[27]

Third, even in those cases where exposure to the mass media does not contribute directly to changes in voting intentions, it probably affects how people perceive candidates, what they know, feel, and propose to do (in other words, their images of political candidates). Robert D. McClure and Thomas E. Patterson demonstrated this in a study using pre- and postelection interviews in Syracuse, New York, in the 1972 presidential election. McClure and Patterson sought to examine what changes in beliefs about candidates result from exposure to televised news and political advertising rather than surveying merely how voters evaluated the candidates. Thus the cognitive, rather than the affective, dimension of the candidates' images was the focus of attention (see Chapter 1). Their findings indicate that televised network news had very little impact on changes in voters' beliefs, but exposure to televised political advertising had an immediate and direct influence on voters' beliefs about the candidates. Particularly striking was the discovery that even among voters having little other interest in the campaign, there was considerable change in perceptions as a result of exposure to televised political commercials.[28]

Fourth, there may be different effects on candidate images depending on what type of media voters are exposed to. For example, a study of the images of candidates in the 1952 presidential election suggested that television improved Dwight Eisenhower's image but did little for his opponent, Adlai

Stevenson's. Stevenson, on the other hand, benefited among persons who obtained their impressions of him through radio.[29] We shall examine the relationship between media exposure and candidate images in presidential elections from 1952–1972 in more detail in Chapter 6.

Finally, even if the mass media have relatively little effect on immediate campaign voting intentions and images (and we have suggested such a verdict may be premature), it is still possible that media exposure has long-term political effects. Between election campaigns, the mass media provide citizens with a secondhand view of reality, filtering or interpreting world events of which people have no firsthand knowledge. Popular expectations of rises or falls in the economy, whom to trust and whom not to trust, prospects of peace or war, feast or famine—all bear the mark of media influence. Certainly, who will or will not be a candidate owes much to media coverage of potential contenders. In short, the interelection influence of the media on citizens' images of the political environment leaves lasting effects not so easily discerned in studies focusing solely on the campaigns.[30]

Do Partisan Self-Images Make a Difference?

At each stage in this review of what published voting studies tell us about the images of political candidates, we have alluded to the significance of party loyalties. The voter's partisanship influences his selective perceptions of candidates, and the strength of that loyalty is associated with his interest in the campaign, exposure to media, and degree of commitment to party candidates. In addition, as we shall see later in this chapter, partisan loyalties play a major part in constructing a description of voting behavior that helps us relate candidate images to other factors that influence voting. Preliminary to that discussion, we want simply to state here what the notion of party loyalty implies and what voting studies suggest about the role of party loyalties in formulating the images of political candidates.

A recent voting study speaks not of party loyalties but of "partisan self-images."[31] This phrasing is congenial with our general concerns in this book. A *partisan self-image* is a person's image of himself as psychologically identified with a political party: an individual supports a given party in a lasting sense as he identifies his perceived intrinsic and enduring interests with it. For a person with a partisan self-image, the phrase "I am a Democrat" or "I am a Republican" takes on a deeply felt meaning that extends beyond legal membership in a party or merely voting for a party's candidates in a single primary or general election. Such a partisan self-image, or identification, is an affective tie that influences the citizen's outlook on politics:

> Identification with a party raises a perceptual screen through which the individual tends to see what is favorable to his partisan orientation. The stronger the party bond, the more exaggerated the process of selection and perceptual distortion will be. Without this psychological tie, or perhaps with commitments to symbols of another kind, the Independent is less likely to develop consistent partisan attitudes.[32]

For partisans, the appeals of their preferred party and its leaders are powerful cues conditioning how to view elections and how to vote. Partisan self-images may be strong, moderate, or weak; the stronger the self-image, the greater the likelihood that the voter will follow his party's cues. The potential effect on the image of a political candidate is clear: "Merely associating the party symbol with his name encourages those identifying with the party to develop a more favorable image of his record and experience, his abilities, and his other personal attributes. Likewise, this association encourages supporters of the opposite party to take a less favorable view of these same personal qualities."[33]

We usually find the expected relationship between partisan self-images and candidate images. In surveys conducted by the Survey Research Center during presidential campaigns since 1952, Americans have generally held positive images of the candidates regardless of their own partisan self-images. This may reflect a "positivity bias," the tendency of Americans to evaluate political figures in a positive way whether they know anything about them or not.[34] Both Democrats and Republicans perceived Dwight Eisenhower favorably in 1952 and 1956. When asked what they liked and disliked about the candidate, the ratio of favorable to unfavorable comments for Eisenhower was 2:1. Similarly, Adlai Stevenson in 1952 received a favorable rating, again a 2:1 ratio of positive over negative comments. But in 1956, Stevenson's ratio fell to 1:1. In 1960 neither Richard Nixon nor John Kennedy had much of an image advantage over the other, but Republicans perceived Nixon favorably, and Democrats did so for Kennedy. In 1964 Lyndon Johnson's positive image transcended partisanship, at least compared to Barry Goldwater's: Johnson's positive to negative ratio was 2.5:1, Goldwater's was 1:2. In 1968 Hubert Humphrey and Richard Nixon gained an advantage in positive perceptions only among their party loyalists; both had more positive images than George Wallace. Finally in 1972, Nixon's image tended to be more favorable, even among many Democrats, than that of his opponent, George McGovern. Overall the ratio of positive to negative comments about Nixon was 1.6:1, but for McGovern there were twice as many negative as positive remarks.[35]

In Chapter 3 we describe in detail the traits that voters associate with the images of presidential candidates. Here, however, it is useful to note that the partisan anchoring of the images of presidential candidates extends to such specific traits as are illustrated in Table 2.2. There the reader finds the percentages of positive references made to each of the major party's candidates in nationwide surveys conducted by the Survey Research Center for each presidential election from 1952 to 1972. Respondents stated what they liked and disliked about the candidates in each survey, and the responses were coded into the seven image traits designated in Table 2.2. The pattern is clear. With only rare and relatively minor exceptions, the percentage of positive comments about each candidate's traits is highest among members of that candidate's party and lowest among the opposition, with Independents being arrayed in between. The exceptions include Lyndon Johnson's receiving a slightly higher rating in relation to groups among Independents in 1964

Table 2.2 Image Traits of Presidential Candidates by Partisanship of Respondents, 1952–1972

PERCENTAGE POSITIVE COMMENTS

IMAGE TRAITS	1952			1956			1960			1964			1968			1972		
	D	I	R	D	I	R	D	I	R	D	I	R	D	I	R	D	I	R
For Democratic Candidates																		
Experience/ability	91	86	63	79	55	32	81	61	40	96	93	85	87	72	43	48	37	16
Background/character	72	67	45	69	41	34	65	51	33	69	45	30	67	40	33	11	10	09
Personal attraction	67	53	40	63	20	27	84	74	53	75	49	37	61	48	27	51	32	22
Party representative	77	36	16	78	28	15	87	49	11	89	47	13	66	15	07	41	11	04
Domestic policies	78	60	25	71	56	27	73	67	19	77	62	30	82	48	29	52	38	13
Foreign policies	86	24	24	42	21	16	71	42	11	82	52	33	64	41	29	50	36	17
Relation to groups	95	87	41	94	86	55	92	78	36	96	98	73	93	68	57	84	79	47
For Republican Candidates																		
Experience/ability	47	60	72	75	91	97	80	95	97	27	40	76	61	89	93	88	91	98
Background/character	83	85	97	41	79	97	80	95	97	22	41	70	31	50	73	39	59	63
Personal attraction	92	95	97	42	56	76	57	74	90	48	52	60	31	45	65	38	51	82
Party representative	10	38	84	07	46	91	54	58	82	06	14	55	13	61	89	12	67	71
Domestic policies	47	67	88	33	53	78	22	53	95	19	37	64	40	65	85	34	46	31
Foreign policies	69	86	91	87	94	94	35	33	82	17	38	53	55	71	92	60	70	78
Relation to groups	42	76	90	25	54	84	52	87	95	05	17	57	11	56	89	17	24	67

Source: Survey Research Center/Center for Political Studies, University of Michigan, and the Interuniversity Consortium for Political Research.

and Richard Nixon's higher rating on foreign policies among Independents in 1960.

Generally, a candidate's approval on each trait is well above a majority among members of his own party. Adlai Stevenson, however, fell short of majority approval on foreign policies among Democrats in 1956, as did George McGovern in 1972 on "experience and ability," "background and character," and "representative of the Democratic party." No Republican has failed to receive a majority of positive comments from his partisans on all traits, not even Barry Goldwater. But partisans also frequently refer to the opposition's candidate in positive ways. Thus, Democrats responded positively to certain of Dwight Eisenhower's traits in both 1952 and 1956, and to certain of Nixon's traits in his three bids for the presidency; Republicans similarly have found at least one trait of Democratic candidates to praise in each election, with the exception of McGovern in 1972. Not reported in Table 2.2 are references to George Wallace as a third-party candidate in 1968. Overall, positive references to Wallace were highest among self-styled Independents. He received highest marks for personal traits, lowest for experience and ability. Generally, then, partisan self-image has a direct influence on the images voters have of political candidates. We will explore the role of that and other influences in Chapter 4.

The Content of Candidate Images: Measurement and Findings

The studies we have reviewed have not generally been directly concerned with examining the content of candidate images; rather, we have had to tease out of published studies findings that are relevant to our interests. We now examine a series of studies that do explore candidate images. First we examine how they have done so; then we turn to the key findings from those investigations.

How Are Candidate Images Studied?

Underlying systematic research into any problem is a design, or plan, of the assumptions, rules, steps, and procedures required to conduct the inquiry. Studies of candidate images have used three general types of designs. The major voting studies, such as those of the Survey Research Center, employ *survey designs*, which are a means of gathering information, or data, about a large number of people by interviewing a few of them. (There are excellent introductions to the intricacies of survey research available.)[36] Simply put, an investigator using a survey design tries to define a "population," or group of people, that he wants to study—as, for example, all Americans of voting age. Then using one of a number of possible sampling plans—probability, random, quota, stratification, or weighting—the investigator seeks a "representative sample" of that population selected in such a way as to assure that the conclusions he draws from the sample (within calculable margins of error) apply to the entire population and not merely to those persons who happen to be sampled.

In most voting studies samples vary from as few as 400 to as many as 3000 persons. The accuracy of such samples—that is, the degree of assurance the investigator has that what he finds in the sample exists to the same degree in the population from which he drew it—depends on such things as how he obtained the sample, the degree of diversity in the sampled population, and the size of the sample. For example, a properly drawn representative sample of all Americans of voting age, containing 2500 persons, would provide a "tolerated error" of around 2 percent at the 95-percent "level of confidence." That is, the researcher could be assured that in 95 out of 100 samples like that drawn, the value for any given characteristic (say, the percentage of persons having a favorable image of a candidate) would be within plus or minus 2 percent of the value in his sample (for example, if he found that 40 percent had a positive image of a candidate, he could expect that within the population 38–42 percent would hold a positive image).

Sample surveys are extremely useful in estimating the distribution of voters' images of candidates in a given population. In subsequent chapters, we report a considerable body of information based on nationwide sample surveys conducted between 1952 and 1972 to examine such problems as the effects of campaigns and media exposure on the perceptions of candidates (Chapter 6) and the effects of candidate images on voting (Chapter 8). A variant of the sample survey, the *panel design,* helps scientists estimate the effects of selected influences on voters during a campaign, as in the Erie County, Elmira, and Milwaukee studies cited previously. The possibility that interviewing people repeatedly in a campaign may influence their images (since interviewers' questions are messages creating message-image relationships) necessitates control panels with but one interview with each person in the panel, as checks for contamination from the reinterviewing process.

Experimental designs also permit investigators to study changes in voter images over time. In the classical experimental design, researchers assign "subjects" randomly to "test" (or "treatment") and "control" groups. Each group's members express their perceptions in interviews or through questionnaires. Investigators then expose test groups to selected stimuli (for instance, the candidates' appeals) but do not expose the members of the control group. Finally, both groups are retested to measure the difference in perceptions of members of the two groups, thus providing indicators of any changes in the test group exposed to the selected stimuli. The simplicity of such a design is clear. There is a variety of ways, which we need not elaborate here, that experiments are sometimes contaminated, thus yielding misleading conclusions.[37] Nonetheless, several significant pieces of social research,[38] including insightful studies of the effect of campaign communications on candidate images,[39] have successfully employed experimental designs.

Whereas survey designs use large representative samples to make statements about populations, experimental designs typically employ a relatively small number of subjects. Thus, whereas researchers usually conduct a large-sample survey only once during an election campaign (because of costs, practical difficulties, and so on), a properly designed experiment using small numbers of subjects in each of successive experiments can be repeated several times. Hence, the experimental researcher builds confidence in his findings not by using large samples but by repetitive experiments under varying controlled conditions. He is less interested in describing the distribution of attributes in a population (such as voters' perceptions) than in answering the question, "Does a given stimulus, or treatment, under given conditions yield given predicted effects?"

A difficulty with experiments is that they frequently place people in artificial situations far removed from the everyday events that may surround substantive problems under investigation. As a result, it is possible to design an informative experiment to examine the effect of media exposure on voters' images of candidates but it may be difficult to translate the findings to "real-world" situations. In election campaigns, for instance, it is nearly impossible to shelter members of a control group from all exposure to presidential candidates via the mass media while systematically treating test-group members with the candidates' messages.

Researchers often modify the classical experimental design by conducting "quasi-experimental" research.[40] Instead of the classical pretest-posttest control group design, the experiment may continue without a control group. In studying the effects of electoral victory or defeat on the images of political figures (see Chapter 7), we tested voters' perceptions prior to and after the 1972 presidential election. It was scarcely feasible to use a control group of voters so oblivious to the election as not to know who won or lost. Also, it is sometimes impossible to obtain a representative sampling of subjects, then randomly assign them to test and control groups. Two researchers, for example, investigated the effects of nominating conventions on the voters' perceptions of political figures. Having neither the finances nor the facilities to employ randomly selected subjects, they tested students in introductory and upper-division psychology classes.[41] Or, it may be desirable to conduct a "posttest-only" experiment, in which investigators obtain subjects' perceptions after an unexpected event that allows no pretest. Studies of the impact of the assassination of President John F. Kennedy on public attitudes were necessarily of this nature and resulted in survey research of a quasi-experimental nature.[42] In sum, there is a variety of potential compromises with the classical experimental design. So long as both investigators and those who use the results understand the risks involved in interpreting specific research under such conditions, quasi-experimental designs provide useful information.

In the subsequent chapters we rely on information from small-sample and quasi-experimental designs as well as large-sample surveys to explore the character of candidate images. These studies employ "Q-methodology" and its associated techniques.[43] In Q-designs, relatively small samples (typically no more than 100) with known characteristics respond to various stimuli selected to represent a variety of potential perceptions of some object (an event, person, or idea). Subjects in Q-studies seldom represent populations in such known ways that a researcher can say that the distribution of images among his subjects parallels that in any sampled population. Yet Q-methodology is a powerful quasi-experimental tool that permits researchers to operationalize a definition of image (that is, have people perform specific operations revealing their images),[44] explore the variety of potential images of specific candidates, and gauge the effects of certain influences on those images. We urge readers to consult Appendix A to familiarize themselves with Q-methodology and our adaptation of it to the study of candidate images. Specifically, we will report Q-studies of candidate images in the 1970 campaign for the United States Senate in Missouri (see Chapters 3 and 4), in a hypothetical congressional campaign (see Chapter 5), in the preconvention period of the 1972 presidential election (Chapters 3 and 4), and in the 1972 presidential election (Chapter 7).

Techniques of Gathering Information

Regardless of their design, studies of the images of political candidates ultimately gather data from people regarding their perceptions. Three principal techniques are employed—free choice, rating, and ranking. Using *free-choice techniques*, researchers ask respondents or subjects a series of questions to get them to volunteer their views of candidates. A variation is to ask respondents to choose characteristics they perceive in a candidate from a list provided them by interviewers. Researchers analyze the content of the responses and construct a profile approximating the images of the candidates in the voters' minds. For example, in its quadrennial studies of presidential elections, the Survey Research Center asks the following questions of its large-size, representative sampling of Americans of voting age:

> "Now, I'd like to ask you about the good and bad points of the two candidates for president. Is there anything in particular about (Democrat) that might make you want to vote *for* him?" "Is there anything in particular about (Democrat) that might make you want to vote *against* him?" "Is there anything in particular about (Republican) that might make you want to vote *for* him?" "Is there anything in particular about (Republican) that might make you want to vote *against* him?"

Typically many respondents think of nothing good or bad about either candidate or about a single candidate. We have found that almost one-half (47 percent) of the respondents in SRC studies of presidential elections from 1952 to 1972 have volunteered no responses. The proportions of people articulating no response have ranged from a high of 51 percent in 1952 to a low of 43 percent in 1964. Doris Graber, finding that in 1968, 42 percent of the SRC sample in that presidential year made no comments about the candidates, argued that many people simply do not labor to assemble a presidential image.[45] If images consist of only what can be measured by this technique, that statement may well be true. But the technique itself places a premium on people's ability to articulate their feelings. Those unable to do so may still have perceptions of the candidates but simply be less facile than the articulates. Thus in considering such findings, we should keep in mind that they probably represent only the images of a self-selected sample of political artic- ulates rather than a cross-section of all voting-age adults.

Many of these articulate people name several things they like and dislike about candidates. SRC researchers accumulate the responses from individual interviews and classify them into a series of categories: separate positive and negative references to the candidate's experience and ability, character and background, personal attraction, domestic policy stands, foreign policy stands, relation to groups, and relation to his political party. In Chapters 6 and 8 we will see how such information, despite its possible limitations, contributes to our understanding of candidate images.

Rating techniques require a respondent/subject to rate the degree that, in his perception, a candidate possesses a particular attribute. For example, several studies dealing specifically with candidate images employ the "seman- tic differential."[46] Researchers provide respondents with sets of polar adjec- tives (hot-cold, strong-weak, good-bad); respondents then rate the candidate on a seven-point scale. If a respondent thinks Richard Nixon is "hot," for instance, he may rate him "1"; if cold, "7"; if lukewarm, "3," and so forth. Readers will recall that the Milwaukee study, referred to earlier in this chap- ter, had people rate the 1968 presidential and vice-presidential candidates on five-point scales pertaining to selected aspects of candidate images. The SRC investigators, in addition to the free-choice techniques, in recent years have employed a rating technique. Specifically, they have had respondents rate candidates on a "feeling thermometer"—a thermometer running from 0 to 100 degrees enabling a person to declare how warm or cool he feels toward the candidate.[47] Rating techniques are common in many of the studies of the content of candidate images discussed below. We will also apply a rating technique in Chapter 5 to discuss voter images of the "ideal" congressional candidate.

Ranking techniques provide interviewees with a series of attributes, characteristics, or statements about a candidate. The respondent ranks the attributes, usually from those he believes most characteristic of the candi- date to those he finds least characteristic. *Q*-sorting, the data-gathering tech- nique associated with *Q*-methodology, is a sophisticated ranking technique that permits a person to portray his set of perceptions about a candidate by sorting statements, perhaps several dozen, from most to least characteristic. Appropriate statistical means (see Appendix A) make it possible to compare the individual *Q*-sorts of all respondents and to define the discrete images of candidates held among a sample of persons. The analyses in Chapters 3, 4, 5, and 7 rely on such sortings.

Another instance of ranking procedures was employed by Leonard Gordon to explore the dimensions of candidate images. Gordon presented his respondents with an instrument devised to measure the importance people place on six types of values in interpersonal relationships—support, conform- ity, recognition, independence, benevolence, and leadership. This "Survey of Interpersonal Values" consists of thirty sets of three items per set. One set, for example, consists of the following three items: "to be free to do as I choose," "to have others agree with me," "to make friends with the unfortunate." Gordon asked each respondent to place himself in the position of a specified political candidate, then select the item in each set he thought the candidate would consider most important and the item of least importance to that candidate. With such procedures, Gordon concluded that voters prefer candidates they perceive as benevolent (kind, generous, and helpful) and as having a touch of humility, that is, as not being egotistical and not needing to be thought of as an important person. But he found that most of his subjects believed that candidates value power and influence, characteristics negatively related to the voters' preferences.[48]

Of course free-choice, rating, and ranking techniques each have merits and defects. Free choice permits an in-depth questioning about perceptions of candidates, but problems invariably arise regarding the classification of rambling, diffuse, and bizarre responses into content categories. Rating techniques are easily and quickly administered and the results are subject to facile analysis. Yet they raise the problem of whether voters really think of candidates in the ways researchers use to construct the various rating scales: do they in fact tap images of voters, or do they test instead the traits those who made up the scales find interesting as researchers? Ranking procedures engender practical difficulties in administration and, unless the items being ranked are statements voters typically make about candidates, they too may impose an image rather than discover it.

Of course, the study of candidate images also requires techniques for analyzing the fund of data obtained by various techniques adapted to survey, experimental, and quasi-experimental designs. Rather than introduce those here, we merely note that there are analytical techniques at varying levels of sophistication.

What Is the Content of Candidate Images?

In Chapter 1 we discussed the cognitive, affective, and conative (knowing, feeling, and action) components of images. Studies of the content of candidate images typically focus on how voters feel—that is, whether they like or dislike specific candidates—revealing that there are several dimensions to the affective component. For example, recall that in analyzing the content of voluntary responses to questions about what people liked or disliked about presidential candidates in the campaign of 1952-72, SRC investigators classified the distribution of positive and negative comments into such categories as background and experience, leadership ability, personal qualifications, and policy stands.[49] Researchers examining the images of party leaders in the 1968 Canadian federal elections, employing free-choice responses, were able to classify responses along several dimensions similar to the SRC's efforts.[50] Recall also the AIPC's fivefold classification of the dimensions of candidate images in the 1968 presidential campaign—leadership ability, political philosophy, speaking ability, intelligence, and honesty.[51]

Studies employing the semantic differential generally attempt to measure both belief and feeling components in candidate images by using three major scales. Two of these scales ask respondents to rate candidates on the basis of how active and strong they believe the candidates are. A third scale seeks respondents' evaluations of candidates, or the feeling component. Thus Blumler and McQuail studied the images of party leaders in the 1964 British parliamentary elections, finding the strength dimension of candidates' images more prone to change during the campaign than the activity or evaluative dimensions.[52]

Finally, the Langs argue that even the "television personality" of political candidates actually consists of three overlapping dimensions—television performance (an image of whether or not the candidate's performance is appropriate and effective on television), political role (whether the candidate seems to be capable and competent for a particular political task), and personal image (whether viewers see in the candidate human qualities, feelings, and emotions that they sympathetically share).[53]

Thus we encounter one of the major problems revealed by previous studies of candidate images—that is, although there is agreement that a candidate's image is multifaceted and composed of numerous discrete elements, there is no consensus on what dimensions make up the content of those images. To say that voters respond to a candidate's "image," however, suggests the desirability of identifying specific dimensions most salient to voters. As we shall see in Chapter 3, we find that different voters perceive quite different particulars in a given candidate. These differences lead to the possibility that *each candidate possesses various images,* rather than a single "candidate image."

Sources of Candidate Images

In addition to a concern with various dimensions of candidate images, major studies deal with the question of the sources of candidate images. The question frequently posed is, "Are the images of candidates defined primarily by attributes the candidate 'projects' to voters, or are candidate images the results of qualities voters 'project' on candidates?" The notion that the candidate's image stems from traits that he or she emphasizes and projects to voters (the "image theory") obtains confirmation in published research. But so does the alternative view that voters see in candidates what is favorable to themselves and distort or ignore what is unfavorable (the "perceptual balance theory").

Separate studies by the McGraths and Roberta Sigel illustrate the alternative perspectives. The McGraths asked each of eighty selected members of Young Democrats and thirty-nine Young Republicans in 1960 to rate himself, John Kennedy, and Richard Nixon on a six-point scale for each of fifty pairs of attributes (ambitious/not ambitious, extrovert/introvert, dynamic/relaxed). The presumption was that if the candidates' appeals were most important, there would be little difference in the ratings of the two partisan groups; if the perceptual balance theory held, however, each partisan group would rate the candidate of their party more favorably and the opposition's candidate less favorably. Members of the two parties had similar self-concepts. Then on twenty-nine of the test items they placed Kennedy toward one end and Nixon toward the other end of the opposed pairs. But, "Neither party group tended to differentiate between self and opponent more than between self and preferred candidate, as would have been expected from the perceptual balance hypothesis."[54]

Political scientist Roberta Sigel challenged the McGraths' conclusions. She argued that the items the respondents in the McGraths' study used to rate the candidates were not sufficiently politically relevant to trigger partisan-based perceptions. Using interviews with a random sample of 1350 Detroit voters, she had voters select from a list of ten traits, the three they thought important in a president (thus constituting their idealized image). She then had voters rank their preferred and nonpreferred 1960 presidential candidates on the same list of attributes to obtain respondents' perceptions of Kennedy and Nixon. The items included honesty, intelligence, independence, careful spender of public money, ideas, sympathy with the "little man," humility, friendliness, good speaker, and good television personality. She found Democrats more likely to assign traits to Kennedy corresponding to their image of the ideal, while Republicans did so for Nixon. She concluded, "Our data suggest that the different images of the two parties are deeply engraved on the perception of voters and influence their candidate perception more than do the candidates themselves."[55]

Related to this question of the sources of candidate images is Weisberg and Rusk's effort to explore dimensions of candidate evaluation. Relying on the "feeling thermometer" employed in recent SRC studies to measure the affective component of candidate images, they have delineated two "dimensional antecedents," or sources, of how people evaluate candidates. One is partisan: If candidates are unknown to voters or there are no issues that divide the electorate, the voter's partisan self-image provides a basic dimension along which to rate candidates. But there is also an issue dimension contributing to an ideological focus on candidates. When issues in a campaign have little relation to conventional party lines, voters evaluate candidates who take stands on those issues in other than partisan ways. If party divisions do not reflect, for instance, divisions on such issues as a war in Vietnam or domestic law and order, voters concerned with those issues evaluate candidates independently of party. Although both dimensions, partisan and issue, are significant anchors for candidate evaluation, any lessening of the strength of partisan self-images heralds a relative rise in the influence of the issue dimensions, perhaps contributing to the emergence of issue candidates—that is, candidates with issue messages as more fundamental than party affiliation to their images.[56]

These questions of the sources of candidate images and whether the dominant influence on candidate images consists of the candidates' appeals (are "simulus determined") or of voters' predispositions (are "perceiver determined") are not easily resolved, at least not as frequently posed. Several studies grapple with them,[57] but as in the case of the dimensions of candidate images, there is no consensus. We will reconsider the sources and influences in Chapter 4, focusing particularly on the degree that self-, idealized, and partisan images are related to voters' images of specific candidates.

Changes in Image Content

A final set of important questions dealt with by studies pertaining directly to examining candidate images consists of those related to changes in the content of images related to nominations, campaigns, victory, or defeat. Since we deal specifically with these matters in succeeding chapters (especially Chapters 6 and 7), at this point we will suggest only the major outlines of previously published research.

Unavoidably, questions of what changes in the content of images occur as a result of political events reintroduce the stimulus- versus perceiver-determined controversy. Take as an example what the effect of being nominated as a party's candidate has on perceptions of the nominee. Published findings support the hypothesis that partisans more positively evaluate a political figure once he becomes their party's candidate. Thus, a quasi-experimental study conducted in 1960 employed a semantic differential to obtain subject's ratings of prominent political figures before and after the major parties' nominating conventions. With respect to John Kennedy there was a polarizing effect; that is, Democrats became more positive toward him after his nomination, Republicans more negative. There was no polarizing effect, however, for the Republican nominee, Richard Nixon. Apparently Democrats and Republicans were already polarized with respect to Nixon prior to his expected nomination; but since Kennedy was less well known, his nomination affixed to him his party's symbol, which divided partisans as it had already done for the more widely known Nixon.[58] The study underscores the oft-stated belief that partisanship is a significant moderator of image change.

There are relatively few studies that measure the effects of specific campaign events on how voters perceive candidates. The presidential campaign of 1960 provided an exception, as several researchers weighed the effect on that election of televised debates between candidates John F. Kennedy and Richard M. Nixon. The research tended to emphasize which candidate won or lost votes as a result of the debates. Findings indicated that Kennedy profited among previously "undecided" voters (who were Independents who had been indifferent to the election's outcome) where his "youthful" image improved as a result of a comparison of his "vigor" with the older and more experienced Nixon.[59] In general, survey and experimental studies revealed a process of selective distortion among viewers of the television debates. Voters who preferred one of the candidates before the election perceived the debates in ways to reinforce prior images. Image-maintaining perceptual mechanisms functioned to preserve a favorable image of the candidate of the preferred party and an unfavorable image of the candidate of the opposition.[60]

Often ignored in the study of image changes in a campaign are the effects of one candidate on another, or more specifically, the extent that the actions and reactions of candidates to one another influence voters' images of them. When, for example, one candidate simply ignores (publicly, at least) the existence of his opponent—as did Richard Nixon ignore George McGovern in 1972—what effect does this have on how voters perceive both the candidate and his opponent? Or, what effects are there when one candidate hammers away at his challenger, levelling personal as well as partisan attacks? The *comparative* quality of candidate images, the extent to which voters reach comparative judgments of contenders, receives all too little examination in image research.[61]

Several studies have examined the effects of victory or defeat on the images of political leaders. Generally, regardless of the partisanship of the voters, the image of the winner improves substantially across a variety of dimensions, whereas voters perceive the loser as less politically potent, but do so with a warmer feeling toward him. Researchers suggest a "fait accompli effect" whereby voters opposed to election, sensing that he will win or has won, adjust their perceptions to relieve the conflict between what they know (the election of an opposing candidate) and feel (a preference for the loser). Voters who supported the winner no longer perceive the opposing candidate as a threat and thus evaluate him more positively.[62] We will examine this effect further in Chapter 7.

In summary, the relatively few studies directly concerned with the images of political candidates raise more questions than they answer. For example, What are the dimensions, sources, and changes in candidate images? Before exploring these questions in the remaining chapters of this work, let us attempt a preliminary statement of the importance of candidate images based on a widely used description of the voting act.

Candidate Images and the Voting Act

The studies of American elections and voting of the Survey Research Center span over a quarter of a century. With admirable persistence and sophistication, the SRC has undertaken nationwide surveys of the population of voting age during presidential and congressional elections. From the rich store of data and interpretations spawned by these efforts emerges a description of voting that permits a tentative estimate of the role of candidate images.

Basic to the SRC's description is the concept of the "normal vote."[63] Earlier in this chapter we discussed the importance of partisan self-images in guiding voters' perceptions of candidates. According to SRC studies, when asked whether they think of themselves as Republicans or Democrats, voters usually identify with one of the major parties. The affective ties implied when people think of themselves as partisans constitute partisan self-images. Upon probing, approximately three-fourths of Americans interviewed in SRC election studies indicate they have such self-images. Table 2.3 depicts the distribution of partisan self-images in recent years according to the strength of partisanship.

For the whole of the national electorate or for any subgroup within it (such as Catholics, union workers, Southerners, and so on) it is possible to estimate a *normal vote*—that is, the expected partisan division of the vote on

Table 2.3 The Distribution of Partisan Self-Images in the United States, 1952–1972

QUESTION: "Generally speaking, do you usually think of yourself as a Republican, a Democrat, an Independent, or what? (IF REPUBLICAN OR DEMO-CRAT) Would you call yourself a strong (R) (D) or a not very strong (R) (D)? (IF INDEPENDENT) Do you think of yourself as closer to the Republican or Democratic party?"

	OCT. 1952	OCT. 1954	OCT. 1956	OCT. 1958	OCT. 1960	NOV. 1962	OCT. 1964	NOV. 1966	NOV. 1968	NOV. 1970	NOV. 1972
Democrat											
Strong	22%	22%	21%	23%	21%	23%	26%	18%	20%	20%	15%
Weak	25	25	23	24	25	23	25	27	25	23	27
Independent											
Democratic	10	9	7	7	8	8	9	9	10	10	11
Independent	5	7	9	8	8	8	8	12	11	13	13
Republican	7	6	8	4	7	6	6	7	9	8	10
Republican											
Weak	14	14	14	16	13	16	13	15	14	15	13
Strong	13	13	15	13	14	12	11	10	10	10	10
Apolitical, Don't know	4	4	3	5	4	4	2	2	1	1	1
Total	100%	100%	100%	100%	100%	100%	100%	100%	100%	100%	100%
n =	1614	1139	1772	1269	3021	1289	1571	1291	1553	1802	2697

Source: Survey Research Center/Center for Political Studies, University of Michigan, and the Interuniversity Consortium for Political Research.

the basis of the distribution of partisan self-images. In the period since the SRC began its studies, there have been considerably more people of Democratic than Republican partisan self-images. That fact, adjusted to take into account how Independents without partisan self-images might vote and the differing voting rates among partisans, results in an estimated nationwide normal vote in past elections of Democrats obtaining approximately 54 percent of the two-party vote. Such a normal vote rarely emerges in presidential elections; in fact, the Democratic candidate has failed to win a majority of the two-party vote in four of the last six presidential elections, while in 1960 he achieved less than the expected normal vote and in 1964 substantially more. Although a normal vote occurs more frequently in congressional elections, these also vary considerably.

In the SRC's description of voting, partisan self-images are a long-term force transcending specific elections. If only that long-term force shaped the outcome of specific elections, there would be almost a one-to-one correspondence between the distribution of partisan self-images and the division of the two-party vote. But, specific to every election are short-term forces that influence (1) how many people vote and who they are, and (2) whether people vote in accordance with their partisan self-images. How interested people are in an election is one factor that helps determine if they will vote. Extensive news coverage of presidential elections, for example, is a short-term force contributing to greater interest and turnout than occur in less-publicized congressional races. As a rule Democrats vote at lower rates than Republicans, a variation explained in part by the greater proportions of members of lower socioeconomic classes who identify with the Democratic party yet who have fewer opportunities, resources, and motivations stimulating them to vote. Generally, the stronger the partisan self-images of voters, the less influence short-term forces governing turnout have on them. Put differently, strong partisans are less likely to respond to short-term forces stimulating interest in specific elections, but they vote regularly in all elections; weak partisans require considerable prodding (say, from a highly vigorous candidate's campaign) to stimulate their interest in voting.

But there are also short-term forces in specific elections contributing to persons' defecting from their partisan self-images, thus crossing party lines and voting for the opposition. Here, too, the stronger the partisanship of voters, the less likely that short-term forces will contribute to their defection from their party. In recent presidential and congressional elections, defection rates seem to have increased. What short-term forces are associated with the rise? Among the key forces are the images people possess of the two parties, issues that are important to voters, and the images of candidates.

Voters have mental pictures of each of the major political parties in the United States. They have certain beliefs about what each party stands for; they feel they like or dislike the parties; and they are inclined to support or oppose a given party. Voters' party images differ from their partisan self-images. A voter's partisan self-image consists of his psychological identification with a party, that is, whether he thinks and feels himself a Democrat or a Republican. In contrast his party image is the picture he holds of a particular party—what he believes about it, whether he likes it or not, and whether he would vote for it. Typically a voter has one partisan self-image (he is a Democrat, a Republican, or whatever) but holds a party image of each contending political party in an election. When voters develop negative images in a specific election of the party they identify with, they may defect and vote for the opposition's candidates. In 1964, for example, one short-term factor contributing to Lyndon Johnson's capturing more than 60 percent of the two-party vote (a considerable deviation from the expected normal vote of 54–56 percent for a Democrat) was the negative image held by many Republicans of their own party. Two of every ten Republicans defected to the Democrats in 1964, many because they believed their party had been taken over by "reactionaries" and was no longer in the "mainstream." The 1964 SRC study reflected the negative image of the Republican party. When people volunteered what they liked and disliked about each major party, negative reference outnumbered positive comments about the Republican party by 1.3 to 1.[64] Or, consider that both Democrats and Republicans alike defected to the candidacy of George Wallace's American Independent Party in part because they found some appeal in Wallace's negative assessment of the two major parties: There was "not a dime's worth of difference" between them. (We shall have considerably more to say about the role of party images in elections in Chapter 8.)

There is increasing evidence that voters defect from their long-term partisan self-images in specific elections because of issues they feel important and differences they detect between the parties and the candidates on those issues. This view is of relatively recent origin, for SRC studies of presidential elections in the 1950s indicated that voters were not much interested in issues and not particularly informed regarding them, at least the issues deemed relevant by investigators. To the extent that issues were important to voters, they were very vague—such as the "nature of the times"—an assessment voters might make of current conditions and then, if they were bad, hold the "ins" responsible by voting for the "outs."[65] For the most part, however, voters were more likely to be influenced by their party affiliation or by candidate images than by the issue positions of the candidates or parties.[66]

Issue voting became more frequent in the 1960s. For instance, comparing the policy positions of voters in the SRC's presidential election surveys on six issues from 1956 to 1968, Gerald Pomper found an increased policy distinctiveness of partisans. That is, by 1968 a voters' partisan self-image was a good predictor of his stand on such matters as federal aid to education, government provision of medical care, federal enforcement of school integration, and so forth. This had not been the case in the late 1950s. Moreover, he found that voters were increasingly aware of differences in the positions of the major parties in the same policy areas. In short, not only were partisan self-images of voters related to policies, so were their images of the parties themselves.[67] Other researchers found that when respondents freely selected the issues most important to them (in contrast to having researchers list a series of issues), voters were able to evaluate which party was most likely to take the action they desired.[68] Finally, students of voting behavior began to measure issue voting in increasingly sophisticated and sensitive ways. Those measures also suggested that issue voting is a significant short-term force stimulating defections from partisan loyalties and, hence, deviations from the normal vote. Instead of merely attempting to correlate voters' stands on issues with their vote, investigators asked voters to identify their stands on issues, then identify how they perceived each of the candidate's stands. That procedure revealed that the voter's proximity to candidates—that is, how close the voter perceives his stand to be to each of the candidates—is a useful predictor of the vote.[69]

The proximity of voters' issue positions to perceived stands of the candidates is, of course, one way the images of candidates act as short-term forces, for the perceived issue positions of candidates constitute one dimension of the candidate's image. This simply means that there is a close, perhaps inextricable, relationship between issue voting and candidate images as short-term forces. Indeed, there is a strong suggestion that candidates inject issue politics into the public domain, thus increasing the likelihood of candidate-issue voting.[70]

Studies using SRC data underscore the importance of candidate images as short-term forces in other ways. First, a negative feeling toward the candidate of one's preferred party may override moderate and weak partisan self-images and produce defections. Just as many Republicans defected to Lyndon Johnson in 1964 because they perceived their party in negative ways, so also did those Republicans with a negative image of Barry Goldwater defect. In 1972, 43 percent of self-proclaimed Democrats voted for Richard Nixon, at least some because of a negative perception of George McGovern.[71] Second, certain candidates have had particular attributes that bothered voters. In 1960, for instance, John F. Kennedy's Catholicism was a major factor contributing to a narrow victory in which he received much less of the normal vote than expected of a Democrat.[72] Third, just as candidate images are closely linked to issue voting as a short-term force, so also are they related to party images. Indeed, the images people have of the two parties are chiefly a function of what the electorate hears and responds to in the candidates. This is especially true of Independents and weak and moderate identifiers.[73]

In sum, the SRC's description of voting is that partisan self-images are a long-term force contributing to a normal vote division between the two major parties. Short-term forces specific to each election, acting on weakly identified partisans and on Independents, can produce defections from party loyalties and deviations from the normal vote. These party, issue, and candidate forces overlap. It is thus not likely that voters draw sharp distinctions among the three but rather have a vague sense of the proximity between their partisan self-images and issue positions and those of parties and candidates. To the extent that any of the three short-term forces stands out, *voters' images of the candidates* probably provide the major explanation of defection from partisan self-images and, in that context, the best single predictor of voting behavior.[74]

This description of voting and the impact of candidate images on voting rests on the notion that large portions of the population have partisan self-images. But, as we saw earlier, a few recent studies contend that there is a weakening of partisanship in this country. There is an apparent rise in the proportions of the voting-age population claiming independence from either party,[75] an increase in split-ticket voting,[76] and stepped-up rates of defections from party loyalties, even among strongly identified partisans, as seen in the fact that defection rates doubled from the 1950s to 1960s.[77] What implications does this hold for the role of candidate images? We suspect that candidate images will have an even greater impact on voting in the future. Perhaps, as some have argued, issue candidates will be more common than partisan candidates (that is, candidates articulating issues that may run counter to their partisan affiliations). Yet partisanship and issues are only two dimensions of candidate images. A decrease in voters' partisan self-images coupled with greater concern over issues simply implies different criteria voters can use to perceive candidates and formulate their images of those candidates. To the extent that party loyalties decline, voters may be even more prone to shift from one party to another in the face of the image-message relationship they establish with each candidate.

Conclusion

This review of numerous studies of voting and candidate images leaves us with conclusions as well as questions. Voters respond to candidates on the basis of the images they have of them and, hence, candidate images are a significant short-term force in elections. Sometimes attractive candidates win the hearts of a few of even the most stalwart members of the opposition party. But in what does the attractiveness lie? Clearly it lies in the expressions (messages) of the candidate coupled with the impressions of the voters, based on their images or dispositions to perceive and respond in various ways.

From a plethora of studies we know a little about these perceptual/response dispositions of the electorate. We know that they are anchored in the partisan self-images of voters, but partisanship alone does not account for all images voters have of candidates. We know that for many voters these perceptual/response dispositions form early in the campaign, but that a sufficiently large number are influenced by the campaign, particularly as portrayed in the media, to suggest that campaigns make at least some difference. We know that there are many dimensions to candidate images and several shadings of voters' beliefs, feelings, and proposals to act in relation to candidates. But what aspects of candidate images change in what ways and under what conditions? We do not know as yet. It is time, therefore, to look more closely at the candidate message–voter image relationship so that we can learn more about the content, sources, and changes in candidate images and better gauge the effects of those images on voting behavior.

Notes

1. For an example of one of the earliest studies, see Stuart Rice, *Quantitative Methods in Politics* (New York: Alfred A. Knopf, 1938).

2. There are several excellent reviews of the early voting studies. See, for instance, Peter H. Rossi, "Four Landmarks in Voting Research," in Eugene Burdick and Arthur J. Brodbeck, eds., *American Voting Behavior* (Glencoe, Ill.: The Free Press, 1959), pp. 5–54; Alvin Boskoff and Harmon Zeigler, *Voting Patterns in a Local Election* (New York: J. B. Lippincott, 1964), chap. 1; and David O. Sears, "Political Behavior," in Gardner Lindzey and Elliot Aronson, eds., *The Handbook of Social Psychology*, 2nd ed. (Reading, Mass.: Addison-Wesley, 1969), vol. 5, chap. 41. A critique of these studies is Walter Berns, "Voting Studies," in Herbert J. Storing, ed., *Essays on the Scientific Study of Politics* (New York: Holt, Rinehart and Winston, 1962), chap. 1.

3. In recent years, a reorganization of the Institute for Social Research at the University of Michigan resulted in the Survey Research Center's being succeeded by the Center for Political Studies (CPS). Most of the voting studies we shall refer to were products of the SRC, and we shall employ the designation SRC or SRC/CPS for the unit as appropriate throughout this book. The leading studies of the SRC consist of the following by Angus Campbell et al., *The Voter Decides* (Evanston, Ill.: Row, Peterson, 1954); *The American Voter* (New York: John Wiley, 1960); and *Elections and the Political Order* (New York: John Wiley, 1966). A major study in the same vein of British voting is David Butler and Donald Stokes, *Political Change in Britain* (New York: St. Martin's Press, 1969). An excellent review of these studies is Kenneth Prewitt and Norman Nie, "Election Studies of the Survey Research Center," *British Journal of Political Science* 1 (October 1971): 479–502.

4. Paul F. Lazarsfeld et al., *The People's Choice* (New York: Columbia University Press, 1944).

5. Ibid., p. 80.

6. For review of findings concerning selective exposure, attention, perception, and recall, see Joseph T. Klapper, *The Effects of Mass Communication* (Glencoe, Ill.: The Free Press, 1961).

7. Bernard R. Berelson et al., *Voting* (Chicago: University of Chicago Press, 1954).

8. David O. Sears and Jonathan L. Freedman, "Selective Exposure to Information: A Critical Review," *Public Opinion Quarterly* 31 (Summer 1967): 194–213.

9. Ibid., p. 212.

10. Drury R. Sherrod, "Selective Perception of Political Candidates," *Public Opinion Quarterly* 35 (Winter 1971–72): 554–62.

11. Congressional Quarterly Guide, *Current American Government* (Fall 1972): 49.

12. Lazarsfeld et. al., *The People's Choice.*

13. See William H. Flanagan, *Political Behavior of the American Electorate*, 2nd ed. (Boston: Allyn and Bacon, 1972), p. 110.

14. Donald Stokes and Warren Miller, "Party Government and the Saliency of Congress," *Public Opinion Quarterly* 26 (Winter 1962): 531–46.

15. For details on the partisan direction of decisions, see Arthur H. Miller et al., "A Majority Party in Disarray: Policy Polarization in the 1972 Election" (paper delivered at the Annual Meeting of the American Political Science Association, New Orleans, Louisiana, September 1973). On presidential primaries in general, see James W. Davis, *Presidential Primaries: Road to the White House* (New York: Thomas Y. Crowell, 1967).

16. Lazarsfeld et al., *The People's Choice*, p. 85.

17. Thomas W. Benham, "Polling for a Presidential Candidate: Some Observations on the 1964 Campaign," *Public Opinion Quarterly* 29 (Summer 1965): 185–99.

18. John H. Kessel, *The Goldwater Coalition* (Indianapolis: Bobbs-Merrill, 1968), pp. 274–76.

19. The American Institute for Political Communication, *Anatomy of a Crucial Election* (Washington, D.C.: American Institute for Political Communication, 1970).

20. Jay G. Blumler and Denis McQuail, *Television in Politics* (Chicago: University of Chicago Press, 1969), pp. 238–39.

21. Lazarsfeld et al., *The People's Choice*.

22. Berelson et al., *Voting*, chap. 11.

23. Philip E. Converse, "Information Flow and the Stability of Partisan Attitudes," *Public Opinion Quarterly* 26 (Winter 1962): 578–99; Edward C. Dreyer, "Media Use and Electoral Choices: Some Political Consequences of Information Exposure," *Public Opinion Quarterly* 35 (Winter 1971–72): 544–53.

24. Klapper, *The Effects of Mass Communication*; Percy H. Tanenbaum and Bradley S. Greenberg, "Mass Communication," in Paul R. Farnsworth, ed., *The Annual Review of Psychology* 19 (1968): 351–86; Otto Larsen, "Social Effects of Mass Communication," in Robert E. L. Faris, ed., *Handbook of Modern Sociology* (Chicago: Rand McNally, 1964), pp. 349–81; Walter Weiss, "Effects of the Mass Media of Communications," in Lindzey and Aronson, eds., *Handbook of Social Psychology*, pp. 77–195.

25. Charles N. Brownstein, "Communication Strategies and the Electoral Decision-Making Process: Some Results from Experimentation," *Experimental Study of Politics* 1 (July 1971): 37–50; James W. Dyson and Frank P. Scioli, Jr., "Communication and Candidate Selection: Relationships of Information and Personal Characteristics to Vote Choice" (paper delivered at the Annual Meeting of the Southern Political Science Association, Atlanta, Georgia, November 1972). That experimental studies find exposure to communications influential in changing candidate images whereas survey research does not is in part a function of the designs themselves; see Carl I. Hovland, "Results from Studies of Attitude Change," *American Psychologist* 14 (1959): 8–17.

26. Walter DeVries and V. Lance Tarrance, *The Ticket-Splitter* (Grand Rapids, Mich.: William B. Eerdmans, 1972). See also Walter Dean Burnham, *Critical Elections and the Mainsprings of American Politics* (New York: W. W. Norton, 1970).

27. Robinson's study results are reported in "Newspapers Play Important Role in Recent Elections: Help Nixon in '68 and '72," *Newsletter*, Institute for Social Research, The University of Michigan (Winter 1974): 3–4; and in John P. Robinson, "The Press as King-Maker: What Surveys from Last Five Campaigns Show," *Journalism Quarterly* 51 (Winter 1974): 587–94, 606.

28. Robert D. McClure and Thomas E. Patterson, "Television News and Political Advertising," *Communication Research* 1 (January 1974): 3–31.

29. Ithiel de Sola Pool, "Television: A New Dimension in Politics," in Burdick and Brodbeck, *American Voting Behavior*, pp. 236–61.

30. Kurt Lang and Gladys Engel Lang, "The Mass Media and Voting," in Burdick and Brodbeck, *American Voting Behavior*, pp. 217–35.

31. Butler and Stokes, *Political Change in Britain*, p. 37.

32. Campbell et al., *The American Voter*, p. 133.

33. Ibid., p. 128.

34. David O. Sears and Richard E. Whitney, *Political Persuasion* (Morristown, N.J.: General Learning Press, 1973).

35. For an example of such studies, see Angus Campbell, "Interpreting the Presidential Victory," in Milton C. Cummings, Jr., ed., *The National Election of 1964* (Washington, D.C.: The Brookings Institution, 1966).

36. A useful review of survey works is Herbert McClosky, "Survey Research in Political Science," in Charles Y. Glock, ed., *Survey Research in the Social Sciences* (New York: Russell Sage Foundation, 1967), pp. 63–143. The intricacies of sampling are explained in Frederick J. Stephan and Philip J. McCarthy, *Sampling Opinions* (New York: John Wiley, 1963); and Charles H. Backstrom and Gerald D. Hursh, *Survey Research* (Evanston, Ill.: Northwestern University Press, 1963); see particularly Leslie Kish, *Survey Sampling* (New York: John Wiley, 1965).

37. The best primer on experimental research is Donald T. Campbell and Julian C. Stanley, *Experimental and Quasi-Experimental Design for Research* (Chicago: Rand McNally, 1963).

38. See Carl I. Hovland et al., *Experiments on Mass Communication* (Princeton, N.J.: Princeton University Press, 1949), as one of many examples of such research.

39. Consult Brownstein, "Communication Strategies," as an example.

40. Campbell and Stanley, *Experimental and Quasi-Experimental Designs for Research.*

41. Bertram H. Raven and Philip S. Gallo, "The Effects of Nominating Conventions, Elections, and Reference Group Identification upon the Perception of Political Figures," *Human Relations* 18 (August 1965): 217–29.

42. See, for example, Paul B. Sheatsley and Jacob J. Feldman, "The Assassination of President Kennedy: Public Reactions and Behavior," *Public Opinion Quarterly* 28 (Summer 1964): 189–215.

43. The basic work on Q-methodology is William Stephenson, *The Study of Behavior: Q-Technique and Its Methodology* (Chicago: University of Chicago Press, 1953).

44. This relationship of Q-methodology to the study of images is spelled out in William Stephenson, *The Play Theory of Mass Communication* (Chicago: University of Chicago Press, 1967), chap. 3.

45. Doris A. Graber, "Personal Qualities in Presidential Images: The Contribution of the Press," *Midwest Journal of Political Science* 16 (February 1972): 46–76.

46. The semantic differential is explained in Charles E. Osgood et al., *The Measurement of Meaning* (Urbana, Ill.: University of Illinois Press, 1957).

47. See Herbert F. Weisberg and Jerrold G. Rusk, "Dimensions of Candidate Evaluation," *American Political Science Review* 64 (December 1970): 1167–85; and Jerrold G. Rusk and Herbert F. Weisberg, "Perceptions of Presidential Candidates: Implications of Electoral Change," *Midwest Journal of Political Science* 16 (August 1972): 388–410, for examples of the use of the "feeling thermometer" in studying candidate images.

48. Leonard V. Gordon, "The Image of Political Candidates: Values and Voter Preference," *Journal of Applied Psychology* 56 (1972): 382–87.

49. Campbell, "Interpreting the Presidential Victory."

50. Gilbert R. Winham and Robert B. Cunningham, "Party Leader Images in the 1968 Federal Election," *Canadian Journal of Political Science* 3 (March 1970): 37–55.

51. American Institute for Political Communication, *Anatomy of a Crucial Election.*

52. Blumler and McQuail, *Television in Politics,* chap. 12.

53. Kurt Lang and Gladys Engel Lang, *Politics and Television* (Chicago: Quadrangle Books, 1968), chap. 5.

54. Joseph E. McGrath and Marion F. McGrath, "Effects of Partisanship on Perceptions of Political Figures," *Public Opinion Quarterly* 26 (Summer 1962): 241.

55. Roberta S. Sigel, "Effect of Partisanship on the Perception of Political Candidates," *Public Opinion Quarterly* 28 (Fall 1964): 493.

56. Weisberg and Rusk, "Dimensions of Candidate Evaluation."

57. See Colin A. Hughes and John S. Western, *The Prime Minister's Policy Speech: A Case Study in Televised Politics* (Canberra: Australian National University Press, 1966); Don Byrne et al., "Response to Political Candidates as a Function of Attitude Similarity-Dissimilarity," *Human Relations* 22 (June 1969): 251–62; Blumler and McQuail, *Television in Politics*; Ira S. Rohter, "An Attribution Theory Approach to the Formation of Candidate Images: Theory and Research Paradigm" (paper delivered at the Annual Meeting of the American Political Science Association, Chicago, Illinois, September 1971).

58. Raven and Gallo, "The Effects of Nominating Conventions."

59. Lang and Lang, *Television and Politics,* chap. 6.

60. Hans Sebald, "Limitations of Communication: Mechanisms of Image Maintenance in Form of Selective Perception, Selective Memory and Selective Distortion," *Journal of Communication* 12 (June 1962): 142–49.

61. An exception is Robert O. Anderson, *A Rhetoric of Image Communication,* (unpublished Ph.D. dissertation, University of Missouri-Columbia, Columbia, Missouri, December 1971).

62. See Lynn R. Anderson and Alan R. Bass, "Some Effects of Victory or Defeat upon Perception of Political Candidates," *Journal of Social Psychology* 73 (1967): 227–40; Raven and Gallo, "The Effects of Nominating Conventions"; I. H. Paul, "Impressions of Personality, Authoritarianism, and the *fait accompli* Effect," *Journal of Abnormal and Social Psychology* 53 (1956): 338–44; G. Striker, "The Operation of Cognitive Dissonance on Pre- and Post-election Attitudes," *Journal of Social Psychology* 63 (1964): 111–19; and J. Korchin, "Reconstruction of Attitude following a National Election: The *fait accompli* Effect," *American Psychologist* 3 (1948): 272 (abstract).

63. See Philip E. Converse, "The Concept of a Normal Vote," in Campbell et al., *Elections and the Political Order,* chap. 2.

64. Campbell, "Interpreting the Presidential Victory," p. 266.

65. See V. O. Key, Jr., *The Responsible Electorate* (Cambridge, Mass.: Belknap Press, 1966); and Philip E. Converse et al., "Continuity and Change in American Politics: Parties and Issues in the 1968 Election," *American Political Science Review* 63 (December 1969): 1083–1105.

66. Sidney Verba and Norman H. Nie, *Participation in America* (New York: Harper and Row, 1972), p. 104.

67. Gerald M. Pomper, "From Confusion to Clarity: Issues and American Voters, 1956–1968," *American Political Science Review* 66 (June 1972): 415–28.

68. David E. Repass, "Issue Salience and Party Choice," *American Political Science Review* 65 (June 1971): 389–400.

69. See Miller et al., "A Majority Party in Disarray: Policy Polarization in the 1972 Election"; David M. Kovenock, "Status, Party, Ideology, Issues, and

Campaign Choice: A Preliminary, Theory-Relevant Analysis of the 1968 American Presidential Election" (paper delivered at the Eighth World Congress of the International Political Science Association, Munich, Germany, August–September 1970); and Roy E. Miller and John S. Jackson, III, "Split Ticket Voting: An Empirical Approach Derived from Psychological Field Theory" (paper delivered at the Annual Meeting of the Southern Political Science Association, Atlanta, Georgia, November 1973).

70. John O. Field and Ronald E. Anderson, "Ideology and the Public's Conceptualization of the 1964 Election," *Public Opinion Quarterly* 33 (Fall 1969): 380–98.

71. Miller et al., "A Majority Party in Disarray."

72. See Philip E. Converse et al., "Stability and Change in 1960: A Reinstating Election," in Campbell et al., *Elections and the Political Order,* chap. 5; cf. Ithiel de Sola Pool et al., *Candidates, Issues and Strategies* (Cambridge, Mass.: MIT Press, 1964), chap. 3.

73. Donald E. Stokes, "Some Dynamic Elements of Contests for the Presidency," *American Political Science Review* 60 (March 1966): 19–28.

74. Richard W. Boyd, "Presidential Elections: An Explanation of Voting Defection," *American Political Science Review* 63 (June 1969): 498–514; Peter B. Natchez and Irvin C. Bupp, "Candidates, Issues, and Voters," *Public Policy* (1968): 409–37.

75. *The Gallup Opinion Index* (October 1973): 19–23; *The Harris Survey* (November 19, 1973).

76. DeVries and Tarrance, *The Ticket-Splitter.*

77. Miller et al., "A Majority Party in Disarray," p. 88.

Chapter Three

What People See in Political Figures: The Content of Candidate Images

What are the major criteria voters use in perceiving political candidates? Are there regularities in the ways they think about candidates? Do they typically organize the attributes that they perceive in similar ways? These questions are seldom addressed in a systematic fashion in research on voting behavior. Although there is a large body of data about popular perceptions of candidates, as we saw in Chapter 2, such information rarely provides direct answers to the types of questions posed here. Instead these data have been directed toward other problems, especially the extent to which voters' evaluations of candidates coincide with other components of the voting decision such as partisan self-images,[1] and the effect of campaign events on how voters change their views of selected attributes of candidates.[2] These are important problems, of course, but the content of candidate images is a significant and intriguing problem in its own right. The contemporary attitude toward "merchandising" candidates and the crises of credibility and confidence in recent presidential administrations are two important instances in which more complete knowledge of the "stuff" of which political leaders' images are made would contribute to a better understanding of the political order. In this chapter we seek to contribute to that knowledge by relating candidate images to the more general phenomenon of interpersonal perception, exploring the principal traits people see in candidates, and advancing the notion of "image types" as a way of helping us understand the content of candidate images.

45

Candidate Images: A Problem of Interpersonal Perception

Prior research presents a confusing array of suggestions about the content of candidate images, and the current effort to synthesize those findings is exploratory at best. Yet research into more general problems of interpersonal perceptions points toward the path this synthesis should take.[3] That research suggests that three major things help shape people's perceptions of other people. To begin with, the perceived person is in some kind of *situation*. For example, the political candidate seeks public office; he purports to be, or aspires to be, a public official. In that context there are relatively well-defined expectations about seeking and holding public office that structure any candidate's relationship with potential voters. Next, the candidate is a specific person and thereby an *object-of-perception*. He displays attributes that reflect his conformity to situationally and culturally defined expectations and reflect his own unique personality. Voters must determine what about him is relevant to their electoral decision by examining both his qualities as a fellow human and the extent to which these qualities fit public prescriptions for the office he seeks. It is thus the potential voter who determines what is really relevant about this "person-in-situation" (candidate) who confronts him. As a result, the *perceiver* is the third major element in the interpersonal relationship. The perceiver brings to this relationship his own characteristic way of orienting to others in general and to political candidates in particular.

These three elements of perception present in interpersonal relationships require further specification if they are to help us understand the content of candidate images. As an object-of-perception, the candidate appears before voters in many guises, or roles. These roles constitute the various relationships of the candidate with the voters. Although there is a wide variety, we may classify such diversified roles broadly as indicative of two basic components underlying the image projected by the candidate: (1) his political role and (2) his stylistic role. *Political role* pertains to the acts of any candidate that are particularly germane to his position as, and aspirations to be, a community leader. Relevant here are his acts and qualifications as a public official—past, present, and future—and as a politician representing partisan interests within the community. *Stylistic role* (or, the candidate's political style), on the other hand, refers both to those things a candidate does that are not directly political (as, for example, the Kennedy family's well-known penchant for touch football or Richard Nixon's alleged preference for John Wayne movies), the manner in which he performs in order to impress voters with his capabilities (whether through the mass media or in personal contacts), and the distinctly personal qualities he exhibits as a human being (physical appearance, bearing, honesty, integrity, compassion). Thus the candidate as an object-of-perception exhibits at least four faces to the public: leader, partisan, dramatic performer, and person.[4] The contender for elective office is an actor playing a political role which voters perceive as leader and/or as politician, and each has a characteristic style that reflects the candidate's performance and/or personal qualities.

Now, what of a second element in the interpersonal relationship to which we liken candidate imagery—that is, the voter—who in this case is the perceiver? We must first admit that we know relatively little about the ways voters orient themselves to receive messages about candidates. However, research on perceptions of other kinds of objects, including perceptions of advertising, concepts, and communication sources, suggests the likelihood of a multidimensional response to candidates.[5] These studies typically use the semantic differential (see Chapter 2) to explore a perceiver's reactions to a perceived object's actions, implying a subjective appraisal of the action context by the perceiver. The inference from these studies is that voters probably respond to what a candidate says or does by taking his message and reading various types of meaning into it. A large body of research using semantic differential scales provides a clue to the principal types of meanings voters might use. For many kinds of objects, three independent types of semantic differentiation mediate between a perceiver and the object of his perception. These are the "evaluative," "potency," and "activity" components of meaning.[6] Put in simple terms, voters may respond to a candidate by evaluating him as good or bad, and by assessing his strength and his capacity for action. Other types of meaning might come into play, but for the moment this tripartite semantic differentiation is sufficient to indicate the voters' principal modes of perceptual orientation.

Finally, as stated above, the candidate's perceived attributes and the voter's perceptual "screen," or blinders, are framed within a situational context. The varying situations of electoral contests do not lend themselves very well to systematic analysis across several elections. However, one aspect of electoral situations that we find has broad applicability from one election to another consists of public expectations regarding the attributes of desirable political leaders. These expectations define a set of attributes for the "ideal public official," preferred traits that act as a standard voters use in assessing the perceived attributes of actual candidates for public office.[7]

In summary, studies of interpersonal perception—that is, how people see one another—suggest three elements in the relationship between candidate and voter perceptions: (1) the candidate is an object-of-perception portraying political and stylistic roles in relation to voters; (2) the voters are the key perceivers responding to candidates in varying degrees as good or bad, strong or weak, active or passive; and (3) candidate images, occurring in a situational context, are compared by voters to their mental images of the ideal public official, which they carry from election to election. We explore each of these elements contributing to the content of candidate images in the remaining pages of this chapter, drawing on the many published studies of candidate images, but more especially on our own Q-studies.

Our exploration looks first at the attributes of candidates on which voters' judgments vary as to the value, relevance, and applicability of any given attribute for a specific candidate. But these attributes are only fragments of images, and the analysis that follows assuredly does not deal with these fragments in all their particularities. Instead we attempt to reduce the very large number of ascribed attributes to a much smaller number of *image traits*, or groupings of similar attributes.[8] The components of candidate role (leader, politician, performer, person) and voter orientation (evaluative, potency, activity) are, hypothetically, image traits. Yet taking only this small number of traits into consideration, voters are still left with numerous ways of subjectively ordering the traits into whole images. The fact that voters tend to organize the traits they see in a candidate into only a few comprehensive political pictures, or patternings,[9] suggests that not all possible arrays of image attributes for a given candidate ever appear. In other words, voters conform more or less to a small number of *image types* in reaching a conclusion about what a political candidate is like. The following analysis of the content of candidate images begins with that hardiest of perennial queries about such images, the importance of stylistic versus political role considerations in popular perceptions of electoral contenders.

The Candidate as Actor: Political Role versus Stylistic Role

Candidates often present themselves to the electorate primarily as political actors by emphasizing such things as their past records and experiences as public officials, their qualifications and abilities to perform tasks associated with the office they seek, their stand on political issues, or their partisan and group affiliations. In contrast, others depict themselves stylistically as men of exceptional character or as dynamic showmen skilled in the arts of rhetoric. Still others work to publicize an attractive package combining political and stylistic qualities. Certainly, candidates prefer to throw the best light possible on all their attributes. However, studies of candidate imagery suggest that whether we look at candidates' images as represented in the communication media or at the attributes of candidates that voters recall, there are differences in the candidates' projection and the voters' reception of these varying traits.

Doris Graber has analyzed the content of newspaper items published about the presidential and vice-presidential contenders in 1968.[10] She found that personal traits of candidates received more newspaper attention than any others. "Style" (image projection) and professional capacities received the next most emphasis. Whereas stylistic roles captured the most press attention, political role attributes were reduced largely to comments about general leadership traits, with only slight coverage devoted to the contenders' abilities to handle problems in crucial policy areas—for example, Richard Nixon's experience and interest in making foreign policy. The press largely ignored the candidates' political philosophies and their qualifications to handle specific tasks.

Image Traits of Presidential Candidates: 1952-1972

Survey Research indicates that the public does seem to be constrained by the range of image attributes presented in the press. The quadrennial SRC/CPS studies of the American electorate (as described in Chapter 2), on the average, classify about 55 percent of the volunteered statements about presidential candidates as "personal attributes."[11] However, these personal attributes include the candidates' "experience and ability" and their "background and character," as well as statements about their "personal attraction" (see Table 3.1a). If personal attributes are limited to "personal attraction," only for Dwight D. Eisenhower and John F. Kennedy did these perceptions clearly dominate the public image of presidential candidates.[12] More typically, voters attempt to locate salient cues about the candidates' prospects as public officials, but they are limited by the raw material provided by the candidates and other campaign sources, including the mass media. Beyond the usual presumption of intelligence in candidates, voter responses vary greatly from candidate to candidate.[13]

In 1952, for example, Eisenhower provoked favorable comments on his military record, his integrity, and his pleasant personality. Adlai Stevenson, on the other hand, drew favorable responses for his political experience, his record as a public servant, his education, and his ability as a speaker. These image attributes were grounded to a large extent in the actual performances of these two men and no doubt were reinforced by campaign messages. But by 1956, Eisenhower's record and experience as a military leader had been eclipsed by his incumbency. Accompanying this change was increasing favorable attention to his integrity, and most especially, his pleasant personality. Positive qualities of Stevenson as voters saw them did not change noticeably, but his negative qualities received more emphasis: lack of experience (he had been out of public office for some years and confronted an incumbent president, after all) and lack of credibility (an egghead image, flippant humor).[14] These aggregated image profiles acted as caricatures, suggesting that the public perceived the facets of candidate projections largely in high relief rather than precise detail.

Political imagery is dynamic, changing with time and circumstance. This dynamic character is especially notable in the case of Richard Nixon. He has been around so long as a presidential candidate and, moreover, the events transpiring in the aftermath of Watergate have done so much to tarnish his image that the positive perception voters held of him in his first effort to win the office in 1960 are easily forgotten. Almost three-fourths of references to Nixon in 1960 were positive, an even better performance than that achieved by Eisenhower in his campaigns. Nixon's strong traits were in voters' perceptions of his experience, ability, background, character, personal attraction, and foreign policy. (Remember, this was the Nixon who had helped Jim Hagerty and Sherman Adams govern the country after Eisenhower's heart attack, who had won the Kitchen Debate with Nikita Khrushchev, and had remained cool under fire in Latin America.) But Nixon's image even then suffered in his political affiliations, both as party representative and in his relation to groups (his sympathy toward business, the well-to-do). His opponent in 1960, John F. Kennedy, likewise had an overall positive image. References to Kennedy's experience, ability, background, and character were positive, although not to the same degree as Nixon's. But for personal attractiveness, as representative

Table 3.1a Image Traits of Presidential Candidates, 1952–1968

YEAR AND CANDIDATE	EXPERIENCE/ABILITY		BACKGROUND/CHARACTER		PERSONAL ATTRACTION		PARTY REPRESENTATIVE		ISSUES		DOMESTIC POLICY		FOREIGN POLICY		RELATION TO GROUPS		TOTAL	
	n[a]	%+[b]	n	%+	n	%+	n	%+	n	%+	n	%+	n	%+	n	%+	n	%+
1952																		
Stevenson	722	85	688	64	292	55	838	50	110	66	197	64	62	52	169	86	3078	65
Eisenhower	1597	57	1039	89	355	94	409	34	81	47	231	67	552	82	112	63	4357	69
1956																		
Stevenson	545	59	776	52	370	39	627	51	63	41	210	59	196	25	181	91	2968	52
Eisenhower	1117	88	1105	81	830	55	341	31	93	60	305	53	645	70	249	56	4745	70
1960																		
Kennedy	396	64	1033	52	453	73	314	61	63	73	197	58	108	47	109	83	2673	60
Nixon	557	91	529	73	255	65	357	49	79	66	135	54	321	79	93	38	2326	71
1964																		
Johnson	803	92	1002	51	299	61	341	62	159	60	598	66	210	61	261	94	3673	69
Goldwater	213	50	864	43	224	52	360	23	400	33	793	34	408	32	153	14	3415	36
1968																		
Humphrey	578	77	912	49	463	48	546	42	286	36	495	63	301	53	267	87	3848	56
Nixon	595	81	1110	50	359	46	369	44	196	52	449	61	428	74	180	31	3686	57
Wallace	352	16	1126	22	297	62	221	26	256	56	1029	46	147	51	288	22	3716	35

IMAGE TRAITS

[a] n stands for the number of statements conforming to the category in response to open-end questions asking respondents for favorable and unfavorable comments about the candidates.

[b] %+ indicates the percentage of comments that are favorable.

Source: Survey Research Center/Center for Political Studies, University of Michigan, and the Interuniversity Consortium for Political Research.

of the Democratic party, in his relation to groups, and in domestic policy, respondents more often found Kennedy appealing.[15]

The presidential race of 1964 between Johnson and Goldwater provides the first of two instances in the 1952–1972 period when one of the two major party candidates had a decidedly negative image. (The other occurs in 1972; also, George Wallace's image in 1968 was negative.) On every image trait Lyndon Johnson received a substantial percentage of positive responses in 1964; only on attributes relating to background and character were positive remarks limited to a bare majority. In contrast, aside from personal attraction and experience/ability, comments about Barry Goldwater were preponderantly negative.[16]

The second instance of a major party candidate with a negative image was in the McGovern–Nixon campaign of 1972. Were the images of the two candidates in 1972 simply mirror images of 1964? Certainly, the overall image of McGovern, using the SRC/CPS measure, was clearly as negative as Goldwater's had been, and the image of Nixon in 1972 was generally positive (see Table 3.1b). However, there are noteworthy differences between 1964 and 1972, largely accountable to the built-in advantage in image accruing to even an unpopular Democratic candidate by reason that he is a Democrat. References to McGovern's relations to groups, hence, were very largely positive (76 percent), whereas Goldwater had been seen in the most negative light for this image trait of any presidential candidate in the two-decade period (only 14 percent positive references). Moreover, whereas Johnson's image had been positive on every trait in 1964, Nixon's image had the traditional flaws

Table 3.1b   Image Traits of Presidential Candidates, 1972

IMAGE TRAITS	McGOVERN		NIXON	
	n^a	$\%+^b$	n	$\%+$
Experience/ability	106	35	429	92
Leadership qualities	288	10	133	56
Personal attributes	518	38	474	55
Party representative	252	23	86	32
Management of government	33	27	133	59
Philosophy of government	278	29	81	57
Domestic policy	378	37	430	47
Foreign policy	417	38	774	68
Relation to groups	235	76	183	25
Miscellaneous/other	147	33	225	62
Total	2652	35	2948	61

[a]n stands for the number of statements conforming to the category in response to open-end questions asking respondents for favorable and unfavorable comments about the candidates.

[b]$\%+$ indicates the percentage of comments that are favorable.

Source: Survey Research Center/Center for Political Studies, University of Michigan, and the Interuniversity Consortium for Political Research.

associated with Republican candidacies, generally negative remarks for his partisan affiliation and his relations to groups. Finally, the personal attributes of Goldwater such as experience/ability and personal attraction were his assets; for McGovern in 1972, however, respondents seemed to perceive this personal side as a serious defect, especially as 90 percent of references to his leadership qualities were negative.

For a number of reasons including the entry of George Wallace as a third-party candidate, 1968 provides an atypical election for the study of candidate imagery. Both major party candidates had overall positive images, but Hubert Humphrey and Richard Nixon were scarcely as appealing as most former candidates of their respective parties. Both candidates did well in assessments of their experience/ability; neither was particularly admired for background/character or personal attractiveness. The bulk of positive comments about Humphrey were associated with his relations to groups; regarding Nixon they were related to his foreign policies. In sum, aside from these two traits, little distinguished the images of the two major party candidates.[17] As Table 3.1a indicates, the overall assessment of George Wallace's traits was roughly equivalent to Goldwater's 1964 level (and what was to become George McGovern's 1972 level). Wallace's strongest trait was personal attractiveness, but aspersions of his ability and reputation flawed the overall image.

This review of image traits of presidential candidates from 1952 to 1972, as uncovered and aggregated by SRC/CPS national surveys, can be summarized as follows. With the exceptions of Goldwater, Wallace, and McGovern, the images of presidential candidates are generally positive, perhaps exemplifying the tendency that Sears has found among Americans to evaluate political figures in a positive light, whether the respondent knows them or not.[18] However, it appears that persons nominated as candidates of each party acquire certain image strengths and weaknesses associated with their respective parties. If there has been a source of strength for the images of Republican candidates (with the exception of Goldwater), it tends to lie in the area of foreign policy, an area where Democratic candidates have their ups and downs. Yet Democrats generally pick up an image bonus because they are perceived as being related to preferred, or at least broader, social interests and, importantly, they are Democrats; Republican candidates are not perceived as related to broad, approved social groups and are, unfortunately for their images, Republicans.

When we move from presidential to other elections, we obtain a slightly different picture of the traits voters emphasize most in candidates. In 1968, for example, in races for the U.S. Senate and/or for governor in thirty-one states, voters emphasized the "political record" as a positive attribute of incumbents, with general references to a candidate's party as being among the most frequent characteristics voters like or dislike. In such subpresidential contests, leadership (that is, experience) and partisanship (that is, references to the candidate's party) were generally more strongly correlated with how people voted than they were in the presidential election, where voters placed a higher premium on perceived issue stands of the contenders. Yet in 1968, across all presidential, senatorial, and gubernatorial contests the perceived personal qualifications of the various candidates was the image trait most strongly related to voters' decisions.[19]

How Voters Rate Image Traits

The survey studies of political campaigns typically use open-end questions for which responses are subsequently coded for analysis in the researchers' own categories. Many respondents are unwilling or unable to articulate their impressions of candidates, and in jotting down the remarks of the articulate who do respond, interviewers may miss the subtleties of meaning that comprise the rich, full picture that voters have of candidates. Although the SRC studies provide a store of invaluable information about the public assessment of leading American political figures since the early 1950s, their data are not very conducive to systematic, rigorous analysis of perceptions of candidates as political actors playing multifaceted roles upon the political stage. Similarly, survey studies using attribute checklists seldom exhibit a concern for systematic sampling or analysis of the varied dimensions of candidate role attributes. There are studies, however, that help us understand how people respond to candidates' political and stylistic role attributes. Let us examine them briefly.

The American Institute for Political Communication study of the 1968 presidential campaign surveyed a general population sample and a companion group of "influentials" in the Milwaukee metropolitan area in a four-stage panel design.[20] As noted in our description of the AIPC project in Chapter 2, the AIPC included an item that asked respondents to rate candidates on each of five characteristics: leadership ability, political philosophy, speaking ability, intelligence, and honesty. If the latter two attributes are taken as representative of the candidate's image as a person, then the question provides a reasonably straightforward comparison of the four role traits discussed above—leadership, partisanship, dramatic performance, and personal qualities. The AIPC reports respondents' ratings for various contenders from the March and June surveys—ten and twelve candidates, respectively—and for the three final candidates in September and in October. These ratings are averaged for four population groups, including Milwaukee City, the suburbs, Waukesha County, and the Milwaukee influentials. On the average, respondents tended to rate the candidates highest for intelligence. Honesty was typically scored next highest, followed closely by speaking ability. Leadership ability generally averaged fourth, and political philosophy scored lowest. Moreover, these relative rankings of the five attributes held across the four groups of respondents.

Ratings of individual candidates exhibited the same pattern with a few notable, but not surprising, exceptions. Ronald Reagan was consistently rated highest for his speaking ability. George Wallace also tended to receive the highest rating for his speaking ability, especially from influentials and Milwaukee City residents. Harold Stassen (a perennial candidate for public office, who was still trying in 1968)—and to a lesser extent, Eugene McCarthy and George Romney—received highest ratings for honesty. Stassen and McCarthy gained their lowest ratings for leadership ability rather than political philosophy. In general, Milwaukee voters did not rate the leadership abilities and political philosophies of 1968 presidential contenders poorly; rather they simply responded more eagerly to stylistic role traits.

Another positive step in the direction of measuring the role traits that people see in candidates appears in a study of the 1959 British parliamentary elections.[21] Trenaman and McQuail had respondents rate Harold Macmillan, the Conservative party leader, and Hugh Gaitskell, the Labourite leader, on each of thirty attribute scales. To uncover the basic dimensions underlying these attributes, they then divided the respondents into Conservative and Labour supporters and factor analyzed each group's responses for each of the party leaders. Labour supporters' ratings of Gaitskell exhibited three basic components: "the complete leader," "personal trust," and "unsuitability." The first represents a conglomeration of attributes that mark what people believe to be a statesman. It exemplifies a halo effect—that is, one of party loyalists' generalizing socially desirable features to their party's leader yet deemphasizing perceptions of any partisan traits of that leader. This "complete leader" trait includes both political and stylistic role elements. The "personal trust" factor is clearly a stylistic component. For example, respondents see Gaitskell as kindly, humorous, straightforward, fair-minded, and honest.

The "unsuitability" component is of special interest. First, it indicates that even among Labourites, not all partisans support their party's standard bearer. Second, it has both a positive and negative side, for it indicates that the cognitive content of the candidate's image (see Chapter 1) is that Gaitskell is either an impotent but benevolent politician or a competent but self-seeking leader. In other words, those people who perceived Gaitskell as a bumbling but kindly leader rejected the image of successful opportunist; others reversed these images.

Trenaman and McQuail found that Conservative supporters perceived a different set of traits in their party's leader. Again three basic dimensions emerged, but Macmillan's traits were "strong leadership," "culture," and "unsuitability." However, Macmillan's "strong leadership" differs from Gaitskell's "complete leadership" only in degree of emphasis. Among his supporters, this first component in Macmillan's image is also a generalized "statesmanship" factor. The "culture" factor is largely stylistic, pointing to a witty urbaneness that transcends immediate political events. In fact, this analysis implies that some respondents perceived Macmillan's urbaneness and sophistication (a much-admired quality in a few Conservative circles) as a flaw in his character, shielding him from harsh political realities and thus rendering him less decisive than they might like. Hence, "culture" has important negative implications for political role considerations. The "unsuitability" factor is again two-sided, matching strong but unpleasant and ineffectual leadership against a polished but weak-willed personality.

Finally, analyses of partisans' images of the opposition party leaders are very similar. Two dimensions emerge in both cases, which may be appropriately labelled "impotency" and "personal benevolence." The first emphasizes political roles, but in a negative fashion. Positive attributes, on the other hand, comprise the more stylistic "personal benevolence" trait.

Q-Studies of Image Traits

In general, then, the factor-analytic study by Trenaman and McQuail makes clear that there is some regularity in the traits that compose candidate images, yet there are also important variations that depend on the candidates themselves. More particularly, the analyses show that political and stylistic roles are independent of one another in some cases, but in other instances merge to depict quite different conceptions of candidates as political actors. Such image traits—statesmanship, personal benevolence, impotent leadership—have been further substantiated in studies of American political campaigns, using a different method of data collection, the Q-sort.[22] These studies varied considerably in campaign context. The first was an investigation of the Stuart Symington–John Danforth contest for the United States Senate in Missouri (1970), the second focused on Democratic party presidential contenders (1971), and the last involved the 1972 presidential campaign. (See Appendixes B, C, and D for details.)

Through the use of a statistical technique called *factor analysis*,[23] it is possible to uncover the clusters of attributes that people performing Q-sorts (see Chapter 2) rank in similar ways when ordering their likes and dislikes about specific candidates. These clusters, or factors, constitute image traits. These factors may be interpreted in many ways, but the concept of role provides a consistent analytical approach that cuts across the three electoral situations in which respondents ranked three different sets of statements depicting attributes of candidates. Consequently we shall refer to these image components as role traits.

In each study we selected statements under the general rubric of the four segments of the candidate's political and stylistic roles discussed above—leader, partisan politician, dramatic performer, and person. But the traits that we were able to identify from the data analyses were more numerous and varied than we would have expected. Table 3.2 presents these role traits along with an indication of which candidates respondents assessed according to the thirty-three different image components. The labels we have applied to each of the thirty-three traits in Table 3.2 summarize (we think, not too arbitrarily) the kinds of statements used in our Q-sorts that factor analyses indicate make up each trait. Although the labels are relatively self-explanatory, we invite readers to consult the appendixes as a means of checking, in doubtful cases. Listing a particular trait for a given candidate simply means respondents judged him on the basis of it—either finding the trait present or lacking in the candidate. These image traits thus represent judgmental categories that respondents may or may not use in evaluating candidates, varying with the respondents themselves (perceivers), the candidates under consideration (objects-of-perception), the particular campaign situation, and the particular set of attribute statements being sorted. We offer the role traits discovered in these Q-sort studies as a first approximation of the universe of relevant role traits found in candidate images.

Note that despite the considerable dispersion of traits within and across the studies, several traits appear with great regularity. These include the distinctly political roles of the partisan and the decisive leader; and the stylistic

Table 3.2 Role Traits of Political Candidates Found in Three *Q*-Sort Studies

ROLE TRAIT	SENATORIAL CAMPAIGN (1970)		DEMOCRATIC PRESIDENTIAL CONTENDERS (1971)				PRESIDENTIAL CAMPAIGN (1972)	
	SYMINGTON	DANFORTH	MUSKIE	HUMPHREY	KENNEDY	McGOVERN	McGOVERN	NIXON
Political Roles								
1. Liberal Politico	X	X		X	X	X		
2. Conservative Politico	X	X		X	X	X		
3. Partisan		X		X	X	X		X
4. Professional Politico			X				X	X
5. Maverick	X	X	X	X	X	X		
6. Partisan Leader							X	X
7. Secular Leader	X	X						X
8. Experienced Leader			X					
9. New Face	X		X		X			
10. Decisive Leader	X	X			X	X	X	X
11. Representative							X	
12. Conciliator							X	
13. Agitator		X					X	X
Role-Style Combined								
14. Statesman	X	X	X	X	X	X	X	X
15. Titan	X	X	X		X	X		X
16. Activist	X	X	X		X	X		X
17. Traditionalist			X					X
18. Arrogant Incompetent			X					
19. Malevolent Manipulator	X	X		X	X	X	X	
20. Credible Communicator								X

Table 3.2 (continued)

ROLE TRAIT	SENATORIAL CAMPAIGN (1970)		DEMOCRATIC PRESIDENTIAL CONTENDERS (1971)				PRESIDENTIAL CAMPAIGN (1972)	
	SYMINGTON	DANFORTH	MUSKIE	HUMPHREY	KENNEDY	McGOVERN	McGOVERN	NIXON
Political Styles								
21. Man of Integrity, Conviction	X	X	X	X	X		X	X
22. Good Joe; Empathic Person	X			X	X	X	X	X
23. Pragmatist	X			X			X	X
24. Moral Exemplar		X			X		X	X
25. Mature Person				X				
26. Stable Person				X				
27. Honest Person								X
28. Oily Person			X	X				
29. Insensitive Person; Disagreeable Person	X	X	X	X	X			
30. Man of Culture; Sophisticate	X	X	X	X	X			X
31. Dynamic Communicator	X	X	X	X		X	X	X
32. Dull Performer		X						
33. Uninformative Communicator			X					

roles of the man of integrity and/or conviction, the empathic person, the sophisticate, and the dynamic communicator. And certain roles combine political and stylistic attributes: the statesman, the titan, and the malevolent manipulator. Other roles may be universal in their applicability but do not emerge in all three studies, perhaps due to changes in the sets of statements used in each instance. These would include the liberal politico, the conservative politico, the maverick, and the activist.

Findings

Our findings indicate several things about the content of candidate images. In the first place, the partisan, conflictual character of political candidacy emerges in several role alternatives. Respondents assess candidates as a representative of a political party, but they also judge candidates for party regularity or conformity to party standards. Both mavericks and partisan leaders illustrate the tendency. The maverick is one who challenges party standards. The partisan leader, present in the images of the 1972 presidential candidates, is not absolutely bound by his partisan affiliation; instead, the partisan leader shapes those standards through his articulation and dissemination of the positions and achievements of the party. Beyond his party affiliation, the candidate may project an ideological stance that further delineates his actions in the arena of political conflict.

Second, the candidate exhibits certain characteristics as a working leader apart from his political affiliations. The trait that seems to interest voters most in this regard is his decisiveness—Is he firm, bold, and prompt in deciding among alternatives? Other working traits of the leader are more likely to be campaign- or candidate-specific.

Third, respondents assess the candidate's political style most often in relation to his character as a person, recognizing three distinct dimensions (roles) of this personal visage. Thus voters may ask, "Is he a man with integrity and conviction?" This means not only that he can be trusted to do as he says, but indeed that he has a sense of direction and purpose to guide what he says. However, the man of integrity is not necessarily one who *cares* for his followers' well-being. Hence, voters may be concerned with the candidate's capacity for empathy. Or, if his capacity as an empathic person is unknown or lacking, is he at least affable, a "Good Joe"? Somewhat less certain as to its desirability is the role of the sophisticate—Is the candidate educated, witty, polished?

Fourth, the candidate's political style includes his acts not only as a person but also as a dramatic performer. However, voters seem largely to respond to this performance (when considered in conjunction with other role traits) *unidimensionally*, lumping his expressive attributes into the singular role of the dynamic communicator. As such, he is informative, warm, organized, and so on, or he is not.

Finally, the candidate may be perceived as playing certain roles that combine elements of both political role and political style. Probably the most important of these is the role of the statesman, constituted largely of leadership and personal attributes high in positive evaluative tone, but sometimes encompassing also attributes of the politician and the performer. And the statesman role is decidedly *not* partisan. The candidate may at the same time be measured for his conformance to the role of the malevolent manipulator—a role marked by ambition, cunning, and deceit. The roles of statesman and of malevolent manipulator are not necessarily polar opposites, but they are both marked by a decidedly evaluative character. Two other roles having a similar semantic basis are the titan and the activist. The titan is marked by his power and mastery whatever he may be doing, and the activist likewise is distinctive for his activity and zest.

The other roles presented in Table 3.2 seem to be either campaign-specific or candidate-specific. However, role traits that appear to be campaign-specific may actually be more universal and were not found in some instances because relevant attribute statements were not included in Q-samples. The partisan leader role seems almost certainly a casualty of statement selection in the 1970 and 1971 studies.[24] Other campaign-specific roles that may be of more universal applicability are the representative, the conciliator, and the moral exemplar.[25] The sixteen remaining roles are more or less candidate-specific, and in most cases they are truncated or alternate versions of the more universal roles. For example, the secular leader is the statesman shorn of his fine qualities as a person. In-depth case studies beyond the scope of this book would be required to uncover the reasons for the emergence of these particularized roles.

It appears, then, that voters basically dichotomize the role behaviors of candidates according to political role and political style. Some evidence suggests that voters typically find the political style of candidates more palatable, but this does not necessarily imply a lack of awareness or appreciation of the political roles of those candidates. If, as Graber's findings imply, campaigns do not produce adequate dissemination of information about the political roles of candidates, in contrast to abundant publicity of their political styles, voters nevertheless respond to both dimensions of candidates' performances as political actors.[26] At the same time, certain role categories—for example, the statesman and the titan—go beyond mere role behaviors to suggest a broader context of meaning that is applied to perceptions of candidates.

The Action Context of the Candidate's Image

As an object-of-perception, the candidate is normally available to voters only indirectly. Few voters meet candidates face-to-face or even see contestants for public office at a distance at campaign rallies. Instead the candidate's visage and message are transmitted to voters by campaign workers and the communications media. Indeed the candidate is more of a *sign* to voters of a concrete person seeking a position of public trust than a living fellow human. It is up to the voter to interpret that sign for its social significance to him. As a sign, the candidate *becomes* meaningful to voters through the union of two processes. One consists of the candidate's efforts (and those of other political actors, whether his supporters or detractors) to project qualities he wants voters to see in him; the other consists of the tendencies of voters to project on the candidate qualities they are predisposed to see in him regardless of what attributes he publicizes (the process of selective perception suggested by the early voting studies we reviewed in Chapter 2). The meaning of a candidate's image thus arises from what politicans and voters jointly project in campaigns. These projections reflect the campaign situation, the attributes of the object-of-perception, and the predispositions of the perceiver.

Having examined candidate attributes in the last section, let us now consider a second element in the processes of interpersonal perception that gives meaning to candidate images—that is, the voters' attributions of meaning. We refer to this element as the *action context* of the candidate's image to emphasize that it involves actions of the voter on his own volition to relate himself to the candidate and to answer such questions as, "Is the candidate good or bad for me?" and "Is he stronger or weaker than his opponent?"

The action context of the candidate's image is thus voter centered. The kinds of meaningful relations individual voters establish with candidates vary considerably. In measuring these variations, researchers have relied primarily on the semantic differential, a technique designed specifically to measure the meaning of an object to an individual. As applied to the study of candidate images, the semantic differential has been used to (1) identify the principal dimensions voters use in attributing meaning to candidate images,[27] (2) measure changes in candidate images,[28] and (3) locate the images of candidates in "semantic space," that is, to portray graphically how people rate candidates as compared to how they respond to other political concepts and objects such as issues, groups, and political parties.[29] Our present concern is with studies of the first variety, which tell us how people assign meanings to candidate images by associating the political figures with certain commonly used words—the process we call *semantic differentiation*.

The early studies of semantic differentiation by Osgood and his associates were directed toward uncovering basic dimensions people use in interpreting all kinds of signs. As we noted in Chapter 2, subjects simply rated selected concepts (such as "Republican," "wife," "me as I am," or "me as I would like to be") on a series of seven-point, bipolar rating scales. They discovered three such dimensions—*evaluative* (designated "E" and constituting the individual's evaluation of the object or concept being rated and thus representing a favorable-unfavorable dimension), *potency* (designated "P" and referring to the individual's perception of the power or strength of the object or concept), and *activity* (designated "A" and reflecting the person's impression of the dynamic qualities of the object-of-perception).[30]

George Suci investigated semantic differentiation of candidate images among a random sample of voters in the 1952 presidential campaign.[31] Setting a pattern for future candidate image research, he used a questionnaire with specific scales selected for their representativeness of the E, P, and A factors. His findings were important and had considerable subsequent validation. In the first place, the E and P factors were highly correlated in the representation of political figures; that is, the good political leader was also the strong political leader.[32] Three different groups of partisans—Taft Republicans, Eisenhower Republicans, and Stevenson Democrats—all exhibited a single major dimension (moving from a pole of "fair-strong-active" toward another at "unfair-weak-passive") in their differentiations of twenty political concepts.[33] This pattern contributed to the discovery of three bipolar "characteristic attributes," or traits, that shape political judgments. The first trait Suci named "benevolent dynamism versus malevolent insipidness." The second trait he called "malevolent dynamism versus benevolent insipidness." The third trait polarized weak-active against strong-passive with no evaluative content.

The three partisan groups differed in their ratings of five political leaders—MacArthur, Churchill, Eisenhower, Stevenson, and Kefauver—only in the degree of benevolent dynamism they attributed to each.[34] In those instances where partisans did not attribute benevolent dynamism to the political figures, a partisan group's ratings were ambiguous rather than indicative of a definite alternative to the most flattering trait. For example, Stevenson Democrats did not agree on whether Robert Taft was a good guy or a bad guy, but they could agree that he was dynamic. Republicans exhibited the same ambivalence toward Franklin Roosevelt. The images of Harry Truman were perhaps the most intriguing in the summer of 1952. Democrats perceived him as benevolently dynamic; Eisenhower Republicans characterized him as weak but active; and Taft Republicans agreed only that he was impotent, disagreeing on his activity and fairness. In conclusion, Suci pointed to the halo effect whereby partisans attribute the most desirable traits to their own candidates. He also revealed a broader tendency of Americans to describe political leaders generally in socially desirable terms. Finally, he demonstrated that semantic differentiation, if not always successful in distinguishing the qualitative components of candidate images, was adequate for measuring candidate images holistically.[35]

More recent studies employing the semantic differential generally reflect two basic dimensions common to candidate images. Anderson and Bass, in their study of presidential, gubernatorial, and senatorial contests in Michigan (1964), derived a general "evaluation" factor and a "political assertiveness" factor that were common to all six candidates both before and after the election.[36] Similarly, a study of five presidential contenders in the 1968 Oregon primary found varying image components specific to the candidates, but two factors, "genuineness" and "leadership," were common to all.[37] And, a study of audience response to a live speech given by Edmund Muskie uncovered a similar pair of factors in semantic ratings of Muskie before and after his speech, traits labelled "trustworthiness" and "demeanor."[38] Examination of the particular attribute scales constituting the various dimensions revealed in these three recent studies suggests a correspondence with Suci's revelaton of a strong correlation between the evaluative and potency factors and a more independent role for the activity factor. The "trustworthiness" trait, especially, seems to be a combination of evaluative and potency attributes, and the "leadership" and "demeanor" traits are largely composed of attributes relating to activity. Still, semantic dimensions clearly shift from study to study, usually with explicit acknowledgment of candidate-specific image components.

As with the Q-sort studies discussed previously, variations in the findings of semantic differential studies pertaining to image traits are also a product of different sets of attribute statements (hot-cold, strong-weak, fast-slow). Unlike the Q-sort studies we have reported, the semantic differential analyses are products of various researchers' using different statistical options and criteria, a problem especially associated with factor analysis.[39] Jack Douglas explored this problem of artifactual differences in semantic dimensions in a complex, but rewarding, design. He performed two different factor analyses on ratings of four candidates by the same respondents using two entirely different sets of adjective pairs (seventy-four pairs in each set).[40] He found that he obtained different results depending on which type of factor analysis and which set of adjective pairs he used.

Yet Douglas demonstrated that it is still possible to discover regularities in how people attribute meanings to political images by looking for clusters of attributes common to all of his analyses. The most general cluster he found, effect-achievement, resembles the evaluative dimension of Osgood and his colleagues. A second cluster, force-dominance-drive-dynamism, is equivalent to the potency dimension with some admixture of activity traits. A third cluster, trust-abrasive credibility, is a familiar dimension in communication research, typically labelled "trustworthiness." Refinement-cultivation-blandness as a fourth cluster is a familiar "culture" or "sophistication" trait. Finally, a contact-attraction-failure-import cluster, similar to the first cluster in its evaluative tone, is even more suggestive of the political context of the rated concepts. Generally, Douglas's work supports the prominent role given to semantic differentiation in studying candidate imagery but suggests the necessity for more exploration and caution in the construction and use of semantic categories.

Still another proviso in accepting semantic categories as basic components of candidates' images is evident in findings from our Q-sort study of Democratic presidential candidates conducted in 1971. In that study a balanced, interactive design incorporated semantic categories with role orientation in the selection of statements.[41] The two experimental categories, together with their interaction, explained on the average about 15 percent of the variance in each of the candidates' images. Role orientation accounted for almost half of this explained variation, about 7 percent; and another 5 percent was due to interaction effects. Thus only about 3 percent of the variance in the candidates' images was due to semantic categorization alone, a marginal effect indeed. Still that margin may have an influential interpretive import despite the lack of quantitative explanation.

One indication of this importance of semantic categorization in candidate images is the part played by the potency factor. In the first place, the "titan" trait described earlier as a combination of political and stylistic role attributes is primarily a cluster of attributes linked by their emphasis on strength: for example, "confident person," "serious speaker," and "brave politician," as found in some respondents' images of John Danforth. The "titan" trait, although varying in role emphases and evaluative tone, may have universal applicability. Moreover, assessment by voters of the potency of candidates is subject to changes of fortune through campaign events, and especially to that climactic point of victory or defeat, a point we shall return to in Chapter 7.[42] It is enough to say that semantic differentiation pervades how voters represent candidates' actions to themselves, even if it is difficult to measure. In this respect the action contexts that shape perceptions of candidates are similar to still another major variable shaping candidate images—idealized conceptions of the public positions that candidates seek, by which voters may gauge their merits.

The Ideal Public Official as Political Hero

The preeminent concern of political science with power and its exercise extends to trying to understand what people expect of those who seek and hold power. What qualities do people want in a public official, their ideal wielder of power? In most candidate image research, popular conceptions of the ideal official are often assumed but rarely investigated. Yet voters do hold images of their ideal and such expectations, and they invoke at least some of these expectations in assessing candidates.[43]

Trenaman and McQuail, for example, argue that loyalists of various parties see comparable traits in their respective party leaders and they offer this as evidence that voters share certain judgmental categories of what all leaders ought to be like. These expectations, organized into a hierarchical complex of the role traits, constitute an image type. Voters first expect a leader to exhibit strength. Next voters judge the candidate as a person, particularly for such attributes as humility and integrity. Finally, they assess the candidate's disposition toward voters, demanding an abiding sense of compassion, empathy, and concern for all sectors of the community. A strong leader, a man of integrity, an empathic person—these all distinguish the ideal governmental official. Still, these assertions derive from assessments of actual candidates, whatever the validity of the authors' interpretation of these data as idealizations of leaders by their followers.[44] In addition, the expectations of Britons may not be universally applicable to other societies.

A recent study by Leonard Gordon provides both confirmation and refutation of the varous Trenaman and McQuail conclusions. Adopting a view similar to the one we have argued—namely, that candidate images are a product of interpersonal perceptions—Gordon asked respondents to complete a Survey of Interpersonal Values (SIV, described in Chapter 2). Recall that this instrument measures how much people value support (being treated with understanding by others), conformity (doing what is socially accepted), recognition (being looked up to), independence (having the right to do as one wants), benevolence (helping others), and leadership (being in charge of others). One week prior to the 1970 congressional elections, Gordon had 210 randomly selected respondents rank five political figures in order of preference—Spiro Agnew, Hubert Humphrey, John Lindsay, Edmund Muskie, and Richard Nixon—and asked the same respondents to complete the SIV as they believed each political figure might. By comparing preferences for the candidates with respondents' perceptions of the interpersonal values of those candidates, Gordon drew conclusions about what aspects of candidates' images are most strongly associated with voter preference. The trait voters wanted most in their preferred candidate was benevolence, that is, a kind, generous, helping individual. What they did not want was a candidate who values power (leadership) and seeks recognition for himself. Thus there is partial confirmation of the Trenaman and McQuail emphasis on empathy and compassion (benevolence) and on humility (downgrading of recognition), but contrary findings regarding allegedly ideal traits such as strength and power.[45]

Actually, American scholars often note a popular emphasis on personal qualities in images of political figures rather than preferences for a strong leader.[46] But in probably the most ambitious effort to reveal Americans' expectations of the ideal public official—the president, in this instance—Sigel hypothesized that a sample of Detroit citizens desired a strong president above all, who would know how to lead and make his will prevail.[47] Although confirming this hypothesis, her data were less substantial in relation to the proposition that citizens ignore "virtues" and other personal qualities in favor of attributes related to power. For example, she asked respondents to select from a list of ten attributes the three most important qualities to be found in a president. More than three-quarters of the sample selected honesty. Intelligence was a distant second, followed by independence and careful spending. Given such data, the argument that attributes most often selected as desirable in the ideal president are more "job relevant" than "personalized" is not persuasive.

The rankings of the ten attributes of the ideal president in the Sigel study, as well as those deriving from a similar question put to British voters (but only those voters having television sets) in the 1964 general election,[48] appear in Table 3.3. To be sure, the attributes listed in the two surveys are altogether different, but the rankings certainly suggest a similar composite image held by most Americans and Britons of the ideal public official. He is Sir Lancelot (before Guinevere), a shining knight battle-tested and clothed in the code of chivalry. Americans perhaps more easily visualize Gary Cooper, ever ready to make the walk at high noon.[49] Unfortunately, the technique used in these studies, the attribute checklist (and severely limited at that by the restriction to three choices), actually allows no more than gross distributional assertions about the attributes most preferred in political leaders. Do voters visualize a shining knight, a Western hero, or some other cultural hero as a distinctly desirable social type to exercise political leadership?[50] No direct answer to such a question has been made, perhaps because it is unanswerable. But the question is important inasmuch as it suggests that a conception of the

Table 3.3 Attributes of the Ideal Leader, Selected as Most Characteristic in Surveys of American and British Voters

TRAITS IN THE AMERICAN STUDY BY SIGEL (n = 1342)[a]	%	TRAITS IN THE BRITISH STUDY BY BLUMLER AND McQUAIL (n = 660)	%
Honesty	78	Straightforward	43
Intelligence	55	Hard-working	41
Independent, a man who is not run by others	45	Sincere	40
		Strong	40
Careful spender of public money	36	Confident	36
A man with a lot of ideas of his own on how to solve problems	24	Fair-minded	33
Sympathy with the lot of the little man	20	Inspiring	22
		Able	17
Humility	10	Persuasive	12
Friendly	8	Likeable	11
Good speaker	8	Kindly	4
Good television personality	2	Unassuming	2

[a] The n is the number of respondents.

Source: The data from the American study are reprinted from Roberta S. Sigel, "Image of the American Presidency—Part II of an Exploration into Popular Views of Presidential Power," *Midwest Journal of Political Science* 10, no. 1 (February 1966): 130, by permission of the Wayne State University Press. The data from the British study are reprinted (in percentagized form) from Jay G. Blumler and Denis McQuail, *Television in Politics: Its Uses and Influence* (1969), p. 115, by permission of the University of Chicago Press and Faber and Faber Ltd.

ideal public official as a component of the images of actual candidates is no mere list of desirable traits to be ticked off in a mechanical fashion in assessing candidates; instead, people acquire an overall vision of the political hero and use it as a gauge to determine the acceptability of actual candidates.

Our own *Q*-sort studies of the Democratic presidential contenders in 1971 and the Nixon-McGovern campaign in 1972 provide some evidence of such a culturally validated political hero. (Again, readers should consult the appendixes for details.) In both instances respondents exhibited a much greater consensus in their rankings of attributes of the ideal president than in their rankings of the same attributes for actual candidates. There were fifty-eight "consensus" attributes in the 1971 study for the ideal president, five for the self-image, and nine, ten, twenty-one, and eight, respectively, for Muskie, Humphrey, McGovern, and Kennedy.[51] In the 1972 study the numbers of consensus items for the ideal president, Nixon, and McGovern were thirty-two, six, and eight, respectively. Tables 3.4 and 3.5 array the items scored highest and lowest in these consensual images of the ideal president. Contrary to the argument by Trenaman and McQuail, there is not a hierarchical array of image traits but rather a holistic pattern intermingling several such traits on the positive and negative poles of these modal *Q*-sort descriptions of the ideal president.

The image of a strong, cultured leader emerges in the high positive rankings of such statements as practical person, masterful person, productive politician, skillful speaker, informative speaker, and trained leader, and the negative rankings of cowardly politician, feeble leader, indecisive leader, inexperienced leader, timid leader, and unintelligible speaker in the 1971 study (see Table 3.4). But these attributes did not receive the greatest stress, for respondents placed greatest emphasis on the integrity and conviction of the president as a leader and a person. However, they assigned much less weight to empathic qualities although they did not ignore such traits; for example, the ideal president is a fair person and not disagreeable, insensitive, or stingy. Still, these Democratic partisans from the American Midwest do not differ too greatly in their conception of the ideal political leader from that presented in the British study. The major exception, or addition, is the

Table 3.4   A Consensus Image of the Ideal President across Three Types of Democratic Partisans (1971)

ATTRIBUTE	AVERAGE z-SCORE
Mature Person	1.6
Fair Person	1.5
Hard-working Leader	1.4
Active Leader	1.4
Calm Person	1.3
Stable Person	1.3
Clean Politician	1.2
Practical Person	1.2
Masterful Person	1.2
Productive Politician	1.1
Skillful Speaker	1.0
Serious Speaker	0.9
Informative Speaker	0.9
Trained Leader	0.9
Stingy Person	−0.9
Unintelligible Speaker	−0.9
Insensitive Person	−0.9
Timid Leader	−1.0
Disagreeable Person	−1.0
Inexperienced Leader	−1.1
Oily Person	−1.1
Small Person	−1.2
Aimless Politician	−1.2
Indecisive Leader	−1.2
Feeble Leader	−1.3
Shallow Person	−1.4
Cowardly Politician	−1.4
Shady Politician	−1.6
Deceitful Leader	−1.8

Note: A z-score is a transformed value equal to the number of standard deviations from the mean. The mean is equal to zero in the present case.

emphasis Americans place on activity; the ideal president is very definitely an active and hard-working leader.[52]

The consensual image of the ideal president emerging from a study of college students in 1972 (see Table 3.5) conforms better to the hierarchical pattern suggested by Trenaman and McQuail, especially in the positively ranked statements. Again the ideal president emerges as a strong, decisive leader endowed with conviction and integrity, who cares for the needs of his followers. Note, however, that the appearance of pandering to the public, actively seeking public popularity, is objectionable.

Yet even if the political hero is a social type worthy of investigation, this social typing as a component of candidate images remains to be substantiated. The only systematic evidence that can presently be offered is the slight but definite tendency for respondents to evaluate candidates, at least via Q-sorts, separately against a standard of the ideal official, rather than to compare candidates directly against one another in the absence of some idealized conception. Thus in the 1970 study of the senatorial campaign in Missouri, the average interpersonal correlation of ideal senator sorts with Symington sorts was .55, and with Danforth sorts, .59; the average correlation of Danforth

Table 3.5   A Consensus Image of the Ideal President across Supporters of McGovern and of Nixon from Ten College Campuses, 1972

ATTRIBUTES	AVERAGE z-SCORE
He arrives at decisions through careful consideration and analysis of all available information.	1.8
He is a statesman and a leader who explains to the people as much as possible the reasons behind his actions or proposals.	1.5
He is a good administrator.	1.5
He has the highest degree of honesty, integrity, and intelligence.	1.5
He takes a firm stand on pertinent issues but does not disregard the views of others.	1.4
He is concerned with the public as a whole, not a collection of minority and majority groups.	1.3
He is calm, analytical, and cautious yet bold and decisive in carrying out his plans.	1.3
He attempts to bring people together in common goals.	1.0
He can unite people in support of his policies.	0.9
He earnestly wants to be liked and respected.	−0.9
He exhibits warmth and personal appeal on television.	−1.2
His personal magnetism and physical attractiveness are positive assets.	−1.3
He is a middle-of-the-roader.	−1.3
He should be elected as a result of his party allegiance because talk is cheap and all candidates promise great things.	−1.6
His voice, speech patterns, expressions, and cool appearance are more important than the mere words of his speech.	−1.8
He is proof that Madison Avenue advertising techniques make television appearances more effective.	−2.1

and Symington sorts was .41.[53] In the 1972 study of the Nixon–McGovern campaign, the correlations were lower but the pattern was similar: Ideal President–McGovern, .31; Ideal President–Nixon, .07; McGovern–Nixon, −.13.[54] The average intrapersonal correlations in the more complex 1971 study of Democratic presidential contenders are presented in Table 3.6. Only Humphrey Q-sorts deviate from the pattern, with their stronger correlations with Muskie sorts. These results suggest that the image of the ideal official provides a general model against which actual candidates may be compared.

Table 3.6 Average Intrapersonal Correlations among Q-Sorts Describing the Ideal President and Four Democratic Presidential Contenders

	IDEAL PRESIDENT	EDMUND MUSKIE	HUBERT HUMPHREY	GEORGE McGOVERN	EDWARD KENNEDY
Ideal President	1.00	0.60	0.45	0.58	0.56
Edmund Muskie		1.00	0.53	0.47	0.50
Hubert Humphrey			1.00	0.41	0.46
George McGovern				1.00	0.51
Edward Kennedy					1.00

If the ideal official functions as such a model, why, then, aren't images of the candidates more similar than they are? One reason is that although images of the ideal official are culturally validated, there are slight but important deviations from that cultural ideal; that is, there are more or less distinct types of images of the ideal official. Ideal image types found among college students in our 1972 study are an example. Table 3.7 presents the statements that are most discriminating of the four major types of images of the ideal president, held by students.

The first type stresses the decision-making activities of the president and is less concerned about his empathic qualities and his role as a moral exemplar. A Type II individual, in contrast, is more concerned that the president be an empathic person, one who not only cares for people but is also flexible and not tied to partisan or group affiliations. The man of integrity and conviction is more appealing to the Type III person; the ideal president is a man of honor and sincerity, a moral exemplar, and a man of reason who articulates his convictions. The final type stresses the partisan advocate who is honest but not necessarily a highly empathic person. Moreover, the Type IV person may reject the role of the moral exemplar as a requirement for the ideal president.

Given such variations in popular images of the ideal public official (albeit variations within the context of a high degree of cultural consensus),[55] what regularities are there in the ways voters compare political candidates to the ideal, beyond the dimensionalities apparent in candidate images, that is, their traits? Are there only two images, one positive and one negative, each

Table 3.7 Discriminating Items in Four Ideal President Image Types[a]

Q-SORT STATEMENT	I	II	III	IV
His words, actions, and manners always reflect the dignity and honor of the office.	1	0	3	0
He sticks to his decisions once they are made.	2	-2	-1	-2
He articulates what the party stands for and always tries to show how each action or proposal is moving toward that goal.	-1	-4	0	3
He is capable of maintaining party unity on major issues.	0	-2	1	1
He carries the image and platform of his party to the people.	0	-3	-1	1
He has a record of good, honest service for his party.	-1	-2	1	0
He does not mirror the policies of any one party.	2	1	-2	-1
He makes deals without compromising his principles.	1	1	-3	0
He listens to other advisors' opinions first and then feels free to do what he thinks is best for everyone.	2	0	0	-1
When he is wrong, he admits it.	1	3	0	3
He has a faith in God and is not afraid to express it.	-3	-2	2	-3
He is a good family man.	-4	-3	-1	-4
He is of high moral character.	1	2	2	-2
He is imaginative, experimental, and hip.	-2	-1	-4	-2
He is natural and sincere and does not appear to be trying to impress people.	-1	1	2	0
He is ambitious.	0	-1	0	-3
He is a person capable of deep emotion and warmth.	-2	1	-1	-2
The central quality which gives depth and substance to all the others is his quality of caring.	-1	3	1	0
He expresses himself intelligently and clearly so that the educated and uneducated alike understand what is said.	0	2	3	3
He appeals to reason rather than people's emotions and prejudices.	-1	3	2	2

[a] Scores for the various statements are the modal, or most frequently given, scores for a statement by persons constituting a given type—I, II, III, or IV. Scores range from a +4 indicating a statement that respondents found most characteristic of their ideal to -4 for traits least desirable in their ideal.

varying only to the extent that it is suffused with perceived idiosyncratic attributes of a given candidate? Or are there no regularities, and as a consequence, the traits of each candidate appear in unique fashion for each and every voter?

From Candidate Image Traits to Candidate Image Types

Selected attributes of specific candidates pervade the popular imagination, frequently because the communications media emphasize them. The Eisenhower smile and the Kennedy charm served as cases in point. But although voters respond to such single attributes, the unique qualities they see in any given candidate seldom stand in isolation from other characteristics they see in the candidate. Voters combine the numerous traits they perceive, emphasizing some and minimizing others, into an overall picture of what the candidate is like. They then compare this combination of their cognitions, affects, and conations toward the candidate (the tripartite aspects of any image, which we described in Chapter 1) with their image of the ideal. Of course, different voters come up with different combinations of perceived traits of actual candidates, just as they imagine that they want different things in their ideal. These various combinations, or overall pictures, whether of the ideal public official or of "real" political figures, we call *image types* to differentiate them from the *image traits* that comprise them. To the extent that for a given candidate there are several image types, we can say there is no single candidate image but many candidate images.

The AIPC study of the responses of Milwaukee voters to the 1968 presidential contenders, discussed above, suggests that voters organize image traits into definite patterns with considerable consistency.[56] Voters tended to rate the personal qualities of most candidates most highly, followed in order by speaking ability, leadership ability, and political philosophy. Since the candidates were generally scored fairly high for all attributes, on the average, voters see all candidates as having the appropriate attributes to some degree but not necessarily in the most preferred order. Unfortunately the AIPC averaged the perceived image traits across many people in constructing a single view, and thereby probably obscured the variety of possible image types. In addition, the AIPC study employed a very small number of attributes, greatly reducing the possibilities of alternative image types.

A technique that permits a more rigorous, comparative assessment of such "whole" images of political candidates is the *Q*-factoring of *Q*-sorts.[57] As the purpose here is simply to explicate the notion of candidate image types, we present only the image typologies emerging from the study of the 1972 Nixon-McGovern contest. In the Nixon-McGovern confrontation, four distinct types of images of each candidate can be delineated from a sample of 100 college students. Each image type (as explained more fully in Appendix A) is comprised of all those persons sharing the same picture of a candidate. We shall describe both what picture each type represents and the characteristics of those respondents holding that image type.

Most of these college students perceived Richard Nixon as a "partisan campaigner" (Factor I) or as a "great statesman" (Factor II). The items that most sharply delineate these and the other types appear in Table 3.8. The "partisan campaigner" image of Nixon accents his role as a party leader who articulates his own party's interests with little concern for the interests of others. In classifying him as such, voters perceived Nixon as a man lacking in empathy and integrity. Moreover, they saw him as a leader who pursued his middle-of-the-road convictions with a large degree of uncomely stubbornness. Not surprisingly, the typical respondent perceiving Nixon in this unflattering fashion was a Democrat of liberal political persuasion with a stated intention of voting for McGovern.

The "great statesman," on the other hand, is a very positive image. Indeed it is a stereotyped implantation of the image of the American political hero on the person of Richard Nixon. As the "great statesman," he was perceived as a strong, decisive leader who stood above the partisan arena of

Table 3.8 Modal Scores of Discriminating Items in Four Types of Images of Richard Nixon, Found among 100 College Students

Q-SORT STATEMENT	I	II	III	IV
He is a good admininstrator.	1	4	4	1
He can unite people in support of his policies.	1	2	1	−3
He is not fearful of criticism.	−1	2	2	−1
He takes a firm stand on pertinent issues but does not disregard the views of others.	−1	4	−1	−1
He is concerned with the public as a whole, not a collection of minority and majority groups.	−2	3	2	−2
He is a statesman and a leader who explains to the people as much as possible the reasons behind his actions or proposals.	−1	4	1	−4
He is a middle-of-the-roader.	3	−1	3	3
He makes only those promises he has the ability to keep.	−4	0	−3	−3
He articulates what the party stands for and always tries to show how each action or proposal is moving toward that goal.	3	0	−1	0
He carries the image and platform of his party to the people.	4	0	−1	3
He represents the major policy stands of his party but he is flexible as the situation and public mood changes.	1	1	−4	0
He has a record of good, honest service for his party.	2	2	−4	1
He listens to others advisors' opinions and then feels free to do what he thinks is best for everyone.	2	1	−3	2
He has the highest degree of honesty, integrity, and intelligence.	−4	2	−2	−1
When he is wrong, he admits it.	−4	0	−4	−4
He is a good family man.	−3	−1	4	4
He is of high moral character.	−3	1	1	2
He is ambitious.	4	1	4	4
He is cool, calm, and collected in front of an audience.	2	−2	2	2
He exhibits warmth and personal appeal on television.	−3	−3	0	−3
His personal magnetism and physical attractiveness are positive assets.	−3	−4	0	−4
He is able to hold his audience's interest.	0	−2	4	−3
He is proof that Madison Avenue advertising techniques make television appearances more effective.	4	−4	−3	2

campaign politics in order to reach out to the wider interests of the entire community. The "great statesman" is a man of conviction and integrity and an exemplar of public morality, but these roles are less important in the image of Nixon as the "great statesman." Finally, persons having this image of Nixon found his media performance either poor or irrelevant, more likely the latter. The typical respondent holding this image was a Republican who described himself/herself as middle-of-the-road, and whose voting intention favored Nixon.

The third factor, or image type, has positive and negative sides. Persons scoring positively on this factor perceived Nixon as a "stubborn but ineffectual leader." They characterized Nixon as an empathic person, but his stubborn disregard of his party is more central to the image. Persons scoring negatively, in contrast, saw him as a "strong partisan leader." Peculiarly, in both instances Republicans are more likely to hold such images.

Finally, some respondents perceived Nixon as the "aloof stylist," a cold, calculating media performer. Yet he is of exemplary moral character and committed to his partisan convictions. This image type is largely found among liberal Democratic supporters of McGovern. The "aloof stylist" image of Nixon differs from that of the "partisan campaigner" primarily because the latter image type casts suspicion on his moral character.

Certain traits tended to be ascribed to Nixon by persons of all image types. His decisiveness as a leader was widely accepted, as was his relative inability to project human warmth as a media performer. Both of these traits received much attention in the mass media over the long span of his political career, and the message appeared to have gained wide acceptance.

The most prevalent image of George McGovern among these college students was quite unlike any of the Nixon image types (see Table 3.9).

Table 3.9   Modal Scores of Discriminating Items in Four Types of Images of George McGovern, Found among 100 College Students

Q-SORT STATEMENTS	I	II	III	IV
He is a good administrator.	1	−1	4	−1
He can unite people in support of his policies.	−3	1	2	0
He is not fearful of criticism.	2	−1	2	3
He is concerned with the public as a whole, not a collection of minority and majority groups.	3	−2	4	4
He arrives at decisions through careful consideration and analysis of all available information.	0	−3	3	−1
He articulates what the party stands for and always tries to show how each action or proposal is moving toward that goal.	−2	0	3	−2
He is capable of maintaining party unity on major issues.	−4	−3	0	−4
He does not mirror the policies of any one party.	−1	−2	0	4
He should be elected as a result of his party allegiance because talk is cheap and all candidates promise great things.	−4	1	−4	−1
He has the highest degree of honesty, integrity, and intelligence.	4	−2	2	0
When he is wrong, he admits it.	4	−3	1	3
He has a faith in God and is not afraid to express it.	1	1	−2	−3
He has a sense of humor.	0	2	−3	2
He is ambitious.	−1	4	0	0
The central quality which gives depth and substance to all the others is his quality of caring.	4	1	0	0
He is cool, calm, and collected in front of an audience.	−1	2	−1	2
He does not read his speeches; he delivers them!	−1	2	−2	3
His personal magnetism and physical attractiveness are positive assets.	−2	3	−3	−3
He is expressive but not overdramatic.	0	−1	−2	3
His perserverance, firmness, coolness, and aggressiveness clearly project a take-charge image.	−2	1	−1	−4
He is able to hold his audience's interest.	−1	3	−1	0
His voice, speech patterns, expressions, and cool appearance are more important than the mere words of his speech.	−4	0	−4	1
He is proof that Madison Avenue advertising techniques make television appearances more effective.	−4	1	−4	−2
He appeals to reason rather than people's emotions and prejudices.	3	−4	0	−3

Students—usually liberal Democrats favoring McGovern's candidacy—saw McGovern as "Mr. Nice Guy" (Factor I), a benevolent person with integrity and conviction. However, this image type also regarded him as ineffective as both a governmental and a partisan leader. Types II and III McGovern images are similar to the first two Nixon images in the arrays of attribute statements, although very dissimilar with regard to who holds the images. McGovern as the "partisan campaigner," however, was more empathic and less successful as a party leader. And as the "great statesman," McGovern was also seen as somewhat more empathic and perhaps more successful in articulating his positions. The typical respondent perceiving McGovern as the "partisan campaigner" was a middle-of-the road Republican supporter of Nixon, and the typical respondent visualizing McGovern as the "great statesman" was a loyalist liberal Democrat. The final type of McGovern image clusters respondents who characterized him as the "ineffectual campaigner." He is concerned with the wider public interest and a brave spokesman for that cause, but with all his sincerity and enthusiasm, he remains a bumbling, ineffective media performer. Nixon supporters were more likely to envision McGovern in this fashion. Generally, respondents tended to agree regarding McGovern's capacity for empathy and his lack of decisiveness. And very definitely, whatever the image type, McGovern was not perceived as a middle-of-the-roader.

These image types do not exhaust all the possibilities for patterning the image traits and/or attributes of candidates, nor would any of them necessarily appear for every political candidate. The "great statesman" and "partisan campaigner" image types seem to have general applicability in partisan contests, whereas other types are more dependent on the candidate or the campaign situation. In any case, the analysis demonstrates that to understand the images of political candidates requires more than simply determining the traits that voters use to describe candidates and then measuring the strength and favorability of those descriptions for specific groups of voters. How voters *organize* those traits into whole images of the candidates and the subsequent use they make of those images are also important topics for research. The research from the Nixon-McGovern campaign is only one more small step in exploring the dimensions and organization of candidate images.

Conclusion

Earlier in this book we proposed that candidate images are multifaceted phenomena of voters' beliefs, feelings, and tendencies in relation to several key dimensions or traits, combining the candidates' political and stylistic role attributes with voters' attributions of meaning in the areas of evaluation, potency, and activity. Later we suggested that each candidate typically possesses various images rather than a single candidate image. Taken together these perspectives point to the complementarity of image traits and image types.

Regularities occur in the groupings of attributes that voters assign to candidates. These traits of candidate images are relatively few, the actual number varying to some extent as a result of research techniques. Although Table 3.2 above lists thirty-three image traits, for example, only about half of these seem likely to emerge frequently in campaigns. And three role traits, perhaps best labelled as "strong leader," "man of integrity," and "empathic person," seemingly account for much of the variation among candidate images in a wide variety of electoral situations using different techniques of data gathering. These image traits are brought together in varying ways by voters as they view a given candidate. Regularities also appear in these patternings such that a large portion (40 percent and more) of the variation in voters' images of the candidate can be explained by a small number of basic types (usually three to five). This typal phenomenon emerges despite the large number of statements used in Q-sorts that might be arrayed in many different ways.

Notwithstanding these regularities, however, much of the variation in candidate images remains unexplained. There are doubtless several reasons for this. After all, whether through their own efforts or those of others, candidates project various facets of themselves (roles). In turn, voters learn of these candidate projections through diverse media at different times and with different expectations (for example, images of the ideal) and perspectives (for example, semantic orientations). Thus there is likely to be a large idiosyncratic element in each candidate-voter transaction. We turn to this basic transaction in more detail in the next chapter, with a view toward both describing the process and also looking at those constraints on candidate images that may help reduce the unexplained component in candidate images.

But at least three other factors affect the content of candidate images, which will require much more research in the future. One of these is the artifactual impact produced by the inadequacy of measurement techniques. No research instrument is fully adequate to the task of comprehensively measuring human images. Even the sophisticated Q-technique requires limits in design and scope, resulting in important facets of given image transactions going undetected. The most that researchers can do at present is to keep chipping away at the problem by using all the available resources.

Moreover the candidate may be perceived as a candidate, but he or she comes to that status with a history, having played and continuing to play other social roles. Certainly this is the lesson to be learned from the various studies that have brought into play the roles of leader, politician, dramatic performer, and person.[58] Perhaps voters view a given candidate in a segmental fashion, making a judgment for each socially recognized role and then merging several images from a number of role contexts to arrive at a single image when it comes time to vote. More research into social typing would be especially appropriate in this regard to determine whether some role images may be more or less easily converted into images approximating the political hero.

We also need to know more about this social type, the political hero, or the ideal public official. Some headway has been made here but with little extensive verification. Specific questions need answers. For example, we have seen evidence that there are variations in types of images of the ideal public official, but just how much deviation from the cultural norm is there, and how much deviation from that norm is permitted actual candidates? To seek the answers to such questions implies that simply predicting preferences, as most voting behavior research does, is not sufficient. We must know more about what voters find desirable in a leader and how they deal with the satisfactions and dissatisfactions of expressing their desires in making concrete choices among actual candidates. Our findings and those of others serve largely only to whet the appetites of image researchers.

Finally, what things influence the voters' images of political candidates? Why does one voter perceive one set of traits while another voter sees something quite different? What political, social, and ideological characteristics of voters are associated with their grouping themselves into differing image types? Chapter 4 proposes some tentative answers to these questions.

Notes

1. See, for example, Richard W. Boyd, "Presidential Elections: An Explanation of Voting Defection," *American Political Science Review* 63 (June 1969): 498-514.

2. See, for example, Sidney Kraus, ed., *The Great Debates: Background—Perspective—Effects* (Bloomington, Ind.: Indiana University Press, 1962), chaps. 11-18.

3. The following analysis draws heavily from Renato Tagiuri, "Introduction," *Person Perception and Interpersonal Behavior*, Renato Tagiuri and Luigi Petrullo, eds. (Stanford, Calif.: Stanford University Press, 1958), pp. ix-xvii. See also Kenneth E. Boulding, *The Image: Knowledge in Life and Society* (Ann Arbor, Mich.: University of Michigan Press, 1956), pp. 103-10.

4. Cf. Roberta S. Sigel, "Effect of Partisanship on the Perception of Political Candidates," *Public Opinion Quarterly* 28 (Fall 1964): 485, for a similar categorization. See also the description of a politician's "television personality" in Kurt Lang and Gladys Engel Lang, *Politics and Television* (Chicago: Quadrangle Books, 1968), pp. 186-203.

5. Cf., for example, David K. Berlo et al., "Dimensions for Evaluating the Acceptability of Message Sources," *Public Opinion Quarterly* 33 (Winter 1969-70): 563-73; and Arnold M. Barban and Werner F. Gruenbaum, "A Factor Analytical Study of Negro and White Responses to Advertising Stimuli," *Journal of Applied Psychology* 49 (August 1965): 274-79.

6. See Charles E. Osgood et al., *The Measurement of Meaning* (Urbana, Ill.: University of Illinois Press, 1957); and James G. Snider and Charles E. Osgood, eds., *Semantic Differential Techniques: A Sourcebook* (Chicago: Aldine, 1969).

7. Studies that lend support to this thesis include works by Roberta S. Sigel, "Effect of Partisanship on the Perception of Political Candidates"; and "Image of the American Presidency—Part II of an Exploration into Popular Views of Presidential Power," *Midwest Journal of Political Science* 10 (1966): 123-37; and by Dan Nimmo and Robert L. Savage, "Political Images and Political Perceptions," *Experimental Study of Politics* 1 (July 1971): 1-36; and "The Amateur Democrat Revisited," *Polity* 5 (Winter 1972): 268-76.

8. See Walter Mischel, *Personality and Assessment* (New York: John Wiley, 1968), pp. 41-72, for a detailed and cogent review of the trait approach to psychological analysis.

9. Cf. the discussion of party images in Joseph Trenaman and Denis McQuail, *Television and the Political Image: A Study of the Impact of Television on the 1959 General Election* (London: Methuen, 1961), pp. 204–6.

10. See Doris Graber, "The Press as Opinion Resource during the 1968 Presidential Campaign," *Public Opinion Quarterly* 35 (Summer 1971): 168–82; "Personal Qualities in Presidential Images: The Contribution of the Press," *Midwest Journal of Political Science* 16 (February 1972): 46–76; and "Press Coverage Patterns of Campaign News: The 1968 Presidential Race," *Journalism Quarterly* 48 (Autumn 1971): 502–12.

11. See Angus Campbell et al., *The American Voter* (New York: John Wiley, 1960), pp. 54–59; and Angus Campbell, "Interpreting the Presidential Victory," *The National Election of 1964*, Milton C. Cummings, Jr., ed. (Washington: The Brookings Institution, 1966), pp. 261–62. Similar "personalizing" has been found in studies of British and Australian party leaders: see David Butler and Donald Stokes, *Political Change in Britain: Forces Shaping Electoral Change* (New York: St. Martin's Press, 1969), pp. 373–81; and Colin A. Hughes and John S. Western, *The Prime Minister's Policy Speech: A Case Study in Televised Politics* (Canberra: Australian National University Press, 1966), pp. 32–37.

12. Campbell et al., *The American Voter*. However, Charles de Gaulle of France and, to a lesser extent, Pierre Trudeau of Canada exhibited similar personal charisma in survey results: see Philip E. Converse and Georges Depeux, "De Gaulle and Eisenhower: The Public Image of the Victorious General," *Elections and the Political Order*, Angus Campbell et al., eds. (New York: John Wiley, 1966), pp. 297–99; and Gilbert R. Winham and Robert B. Cunningham, "Party Leader Images in the 1968 Federal Election," *Canadian Journal of Political Science* 3 (March 1970): 40.

13. The presumption of intelligence as an attribute of political candidates is reflected in the ratings of Dwight Eisenhower and Adlai Stevenson in both 1952 and 1956, and also in the contenders in the 1968 presidential campaign (pre- and postconvention). See Elmo Roper, *You and Your Leaders: Their Actions and Your Reactions, 1936–1956* (New York: William Morrow, 1957), pp. 229–30, 273–74; and American Institute for Political Communication, *Anatomy of a Crucial Election: The 1968 Campaign* (Washington: The American Institute for Political Communication, 1970), pp. 19–24. On the other hand, candidates may find it necessary to shroud their intellectual capabilities in a cloak of humility; see Elliott White, "The Humility Complex in American Politics: Was Lincoln Created Equal?" *Publius* 1 (Spring 1971): 89–114.

14. Campbell et al., *The American Voter*. Roper, *You and Your Leaders*, using a trait checklist rather than an open-end question, reports different results. In the Roper poll, Eisenhower was most often described as intelligent with an attractive personality in 1952; to these attributes he gained considerable positive mention for his job experience and inspiration of confidence in 1956. Stevenson was most favorably assessed for his intelligence and experience in 1952, and for his intelligence and campaign abilities in 1956. Cf. also Department of Marketing, Miami University, *The Influence of Television on the Election of 1952* (Oxford, Ohio: Oxford Research Associates, 1954); Herbert H. Hyman and Paul B. Sheatsley, "The Political Appeal of President Eisenhower," *Public Opinion Quarterly* 19 (Winter 1955–61): 26–39; and Ithiel de Sola Pool, "TV: A New Dimension in Politics," *American Voting Behavior*, Eugene Burdick and Arthur J. Brodbeck, eds. (New York: The Free Press, 1959), pp. 236–61.

15. Cf. Philip E. Converse, "Religion and Politics: The 1960 Election," *Elections and the Political Order*, Angus Campbell et al., eds. (New York:

John Wiley, 1966), pp. 96–124; Campbell, "Interpreting the Presidential Victory"; Kraus, *The Great Debates*; and Kurt Lang and Gladys Engel Lang, "Ordeal by Debate: Viewer Reactions," *Public Opinion Quarterly* 25 (Summer 1961): 277–88.

16. Cf. Campbell, "Interpreting the Presidential Victory"; Peter B. Natchez and Irwin C. Bupp, "Candidates, Issues, and Voters," *Public Policy* 17 (1968): 409–37; and Benham, "Polling for a Presidential Candidate." Note, however, that Benham, using an attribute checklist, found that as many as a quarter of his preelection sample agreed on only two attributes for Goldwater: "speaks his own mind" and "has strong conviction," suggesting that personal integrity was the only positive trait that Goldwater projected with any success.

17. Cf. Herbert F. Weisberg and Jerrold G. Rusk, "Dimensions of Candidate Evaluation," *American Political Science Review* 64 (December 1970): 1167–85; and Jerrold G. Rusk and Herbert F. Weisberg, "Perceptions of Presidential Candidates: Implications for Electoral Change," *Midwest Journal of Political Science* 16 (August 1972): 388–410.

18. See David O. Sears and Richard E. Whitney, "Political Persuasion," *Handbook of Communication*, Ithiel de Sola Pool et al., eds. (Chicago: Rand McNally, 1973), pp. 253–89.

19. Barbara Hinckley et al., "Information and the Vote: A Comparative Election Study," *American Politics Quarterly* 2 (April 1974): 131–59. A number of other studies using either open-end questions or attribute checklists have been conducted in electoral contexts other than United States presidential elections: see John F. Becker and Eugene E. Heaton, Jr., "The Election of Senator Edward W. Brooke," *Public Opinion Quarterly* 31 (Fall 1967): 346–58; Walter DeVries, "Taking the Voter's Pulse," *The Political Image Merchants: Strategies in the New Politics*, Ray Hiebert et al., eds. (Washington: Acropolis Brooks, 1971), pp. 62–81; Colin A. Hughes, *Images and Issues: The Queensland State Elections of 1963 and 1966* (Canberra: Australian National University Press, 1969); David Butler and Donald Stokes, *Political Change in Britain: Forces Shaping Electoral Choice* (New York: St. Martin's Press, 1969), pp. 373–88; Peter Regenstreif, *The Diefenbaker Interlude: Parties and Voting in Canada–An Interpretation* (Toronto: Longmans Canada, 1965), pp. 56–83; and Winham and Cunningham, "Party Leader Images in the 1968 Federal Election."

20. AIPC, *Anatomy of a Crucial Election.*

21. Trenaman and McQuail, *Television and the Political Image*, pp. 48–51.

22. The *Q*-sort studies are described in detail in Appendixes B, C, and D.

23. Factor analysis is a widely used set of techniques for studying candidate images and voting behavior. Technical questions pertaining to factor analysis are discussed in detail in Chapters 5, 7, and 8, and with respect to *Q*-methodology in the appendixes.

24. Actually the "partisan" trait of the earlier studies may reflect both the narrower party worker and the broader party leader components of Nixon's image in 1972.

25. On the broader applicability of the moral exemplar role, see Sigel, "Image of the American Presidency," p. 131.

26. Cf. Roberta S. Sigel, "Image of a President: Some Insights into the Political Views of School Children," *American Political Science Review* 62 (March 1968): 216–26.

27. Osgood et al., *The Measurement of Meaning.*

28. See Charles N. Brownstein, "Communication Strategies and the Electoral Decision-Making Process: Some Results from Experimentation," *Experimental Study of Politics* 1 (July 1971): 37–50; Bertram H. Raven and Philip

S. Gallo, "The Effects of Nominating Conventions, Elections, and Reference Group Identification upon the Perception of Political Figures," *Human Relations* 18 (1965): 217–30; George Stricker, "The Operation of Cognitive Dissonance on Pre- and Postelection Attitudes," *Journal of Social Psychology* 63 (1964): 111–19; Stricker, "The Use of Semantic Differential to Predict Voting Behavior," *Journal of Social Psychology* 59 (1963): 159–67; and Percy H. Tannenbaum et al., "Candidate Images," in Kraus, ed., *The Great Debates*, pp. 271–88.

29. See Osgood et al., *The Measurement of Meaning*, pp. 104–24; Sidney Kraus and Raymond G. Smith, "Issues and Images," in Kraus, ed., *The Great Debates*, pp. 289–312; and Leonard L. Rosenbaum and Elliott McGinniss, "A Semantic Differential Analysis of Concepts Associated with the 1964 Presidential Election," *Journal of Social Psychology* 78 (August 1969): 227–35.

30. Osgood et al., *The Measurement of Meaning*, pp. 31–75.

31. Ibid., pp. 104–24.

32. Sigel, "Image of the American Presidency," using different techniques, found a similar correlation in characterizations of the ideal president.

33. Only the fair-unfair, strong-weak, and active-passive scales were used in analysis due to their high correlations with the other E, P, and A scales.

34. Cf. Pool, "TV: A New Dimension in Politics," pp. 251–52, who comments on the striking "parallelism between the images of Eisenhower and Stevenson" in 1952. Pool used an attribute checklist but his findings tend to corroborate the strong positive correlation of the E and P factors.

35. Cf. Stricker, "The Operation of Cognitive Dissonance on Pre- and Post-election Attitutdes," p. 117, who argues that the P and A factors add little variance in measuring candidate images as political predispositions are strongly loaded on the evaluative dimension.

36. Lynn R. Anderson and Alan R. Bass, "Some Effects of Victory or Defeat upon Perception of Political Candidates," *Journal of Social Psychology* 73 (December 1967): 227–40.

37. Bill O. Kjeldahl et al., "Factors in a Presidential Candidate's Image," *Speech Monographs* 38 (June 1971): 129–31.

38. Lynda Lee Kaid and Robert O. Hirsch, "Selective Exposure and Candidate Image over Time," *Central States Speech Journal* 24 (Spring 1973): 48–51.

39. See R. J. Rummel, "Understanding Factor Analysis," *Journal of Conflict Resolution* 11 (December 1967): 444–80, for a general overview. For more detailed information, see Rummel's *Applied Factor Analysis* (Evanston, Ill.: Northwestern University Press, 1970).

40. Jack Douglas, "The Verbal Image: Student Perceptions of Political Figures," *Speech Monographs* 39 (March 1972): 1–15.

41. See Appendix C. Note that "role" as used here refers specifically to the hypothesized traits of leader, politician, person, and speaker, and that the semantic categories are the E, P, and A factors.

42. Cf. Richard F. Carter, "Some Effects of the Debates," in Kraus, ed., *The Great Debates*, pp. 253–70; Jay G. Blumler and Denis McQuail, *Television in Politics: Its Uses and Influence* (Chicago: University of Chicago Press, 1969), pp. 228–39; and Dan Nimmo et al., "Effects of Victory or Defeat upon the Images of Political Candidates," *Experimental Study of Politics* 3 (February 1974): 1–30. Note also the findings in Tannenbaum et al., "Candidate Images," pp. 276–77, that while semantic differential ratings of Kennedy tended to remain static across the 1960 campaign, the ratings of Nixon, particularly the potency scales, declined.

43. On the relative stability of images of the ideal president, for example, as compared to the more fluid images of actual candidates, see Tannenbaum et al., "Candidate Images," p. 278; and Nimmo et al., "Effects of Victory or Defeat," pp. 6–7.

44. Trenaman and McQuail, *Television and the Political Image*, pp. 163–64. Cf., however, Sigel, "Image of a President," p. 224, who found that children's images (idealizations?) of John Kennedy tended to stress either personal benevolence (the empathic person role) or courageous efficiency (the strong leader role). The "man of integrity" image trait apparently did not emerge, further suggesting a generalized conception of the president, as children presumably are less likely to question motivations: for example, "Is he sincere?" See also Fred I. Greenstein, "The Benevolent Leader: Children's Images of Political Authority," *American Political Science Review* 54 (December 1960): 934–43.

45. Leonard V. Gordon, "The Image of Political Candidates: Values and Voter Preference," *Journal of Applied Psychology* 56 (1972): 382–87.

46. See, for example, Fred I. Greenstein, "Popular Images of the President," *American Journal of Psychiatry* 122 (November 1965): 523–29.

47. Sigel, "Image of the American Presidency." She further demonstrated that respondents also feared political power and would support institutionalized limitations on that power, such as the two-term limit of the Twenty-second Amendment.

48. Blumler and McQuail, *Television in Politics*, p. 115. In this survey respondents were asked to select three characteristics of politicians "most important for running the country" from a list of twelve attributes.

49. Cf. also the semantic differential analysis of the ideal president by Tannenbaum et al., "Candidate Images," p. 276. Here the ideal president as described by the average voter is very wise, fair, experienced, strong, active, deep, and calm. He would also be virile and colorful as well as slightly liberal, young, and warm.

50. On cultural heroes as social types, see the following works by Orrin E. Klapp: "The Folk Hero," *Journal of American Folklore* 62 (January–March 1949): 17–25; *Heroes, Villains, and Fools: The Changing American Character* (Englewood Cliffs, N.J.: Prentice-Hall, 1962); "Mexican Social Types," *American Journal of Sociology* 69 (January 1964): 404–14; and *Symbolic Leaders: Public Dramas and Public Men* (Chicago: Aldine, 1964). See also William Burlie Brown, *The People's Choice* (Baton Rouge, La.: Louisiana State University Press, 1969), for a suggestive account of what delineates the culture hero of American politics, at least as projected in campaign biographies.

51. The QUANAL computer program, devised by Norman van Tubergen of New York University and used for most of the factor analyses in these studies, provides the average standardized scores for cases (*Q*-statements in the *Q*-factor analysis), which are scored within one standard deviation across the basic types or factors. This higher degree of consensus is also indicated by the smaller number of factors that usually must be extracted in *Q*-factor analyses of ideal president sorts than for candidate sorts, to explain the same amount of variance.

52. Cf. Sigel, "Image of the American Presidency," p. 132, who reports that 81 percent of the Detroit residents believe that it would be good for a president "to have had to work as a young boy," suggesting that the work ethic pervades even the prevailing image of the ideal public official in the United States.

53. The correlations reported herein are Pearsonian correlations. For those readers not familiar with the use of correlation measures (on which we rely

frequently in this text), this correlation coefficient is a statistic ranging from +1.00 through 0 to -1.00. It depicts the interrelationship of two (or more) variables and indicates that the increase or decrease in the magnitude of one of the variables is associated with an increase or decrease in the magnitude of the other. A coefficient with a positive sign (+) indicates a direct association between the variables; one with a negative sign (-) indicates an inverse relationship. Although there are pitfalls in such oversimplifications, a useful guide to the importance of a coefficient is as follows: less than .20 is a slight correlation reflecting a negligible association; .20 to .40 is a low correlation marking a small association; .40 to .70 is a moderate correlation indicating a substantial relationship; .70 to .90 is a high correlation and marked relationship; and .90 to 1.00 is a very high correlation suggesting a dependable relationship. On these and other statistical matters, the reader should consult the appropriate appendixes and the technical portions of Chapter 5 and 8.

54. The lower correlations are in large part the result of a tendency toward stressing certain role aspects of the candidates' images—political roles for Nixon and stylistic roles for McGovern—than is the case in the more heterogeneous mixing of attributes in the image of the ideal president; see Nimmo et al., "Effects of Victory or Defeat upon the Images of Political Candidates."

55. Cf. White, "The Humility Complex in American Politics"; Brown, *The People's Choice*; Wayne C. Minick, "Politics and the Ideal Man," *Southern Speech Journal* 26 (Fall 1960): 16–21; and Charles G. Russell, "Diffusion Differences in State and National Elections" (paper presented at the Annual Meeting of the International Communication Association, Montreal, April 1973).

56. AIPC, *Anatomy of a Crucial Election*, pp. 17–23.

57. See Appendix A for a full elaboration of Q-technique in measuring candidate images.

58. See Tannenbaum et al., "Candidate Images," in which the images of Richard Nixon and John Kennedy were measured as "president" and as "TV performer."

Chapter Four

Beauty and the Eye of the Beholder:
Complexities of Image Transactions

Although people who study candidate imagery have generally avoided sharp debate over the *content* of those images, there have been clear, deep disagreements surrounding the *sources* of candidate images. The conflict divides researchers who support an "image thesis" (a view that images are determined primarily by the object-of-perception confronting a perceiver) from those who advance a "perceptual balance principle" (a view that stresses the primacy of the subjective predilections of a person attending and responding to stimuli). This chapter reviews that controversy by considering arguments of both views, examining the available evidence from earlier research on the subject, and exploring the implications for our understanding of the images of political candidates. At this point we introduce still another view of the image process, that is, the image as a transaction occurring between political stimuli and the responding citizen. There are various constraints on this transaction between candidates and constituents that warrant discussion as well—partisan and ideological anchoring, the images of relevant political figures other than the candidate (the ideal official, the self, and opposition candidates), group affiliations, communications, vote commitments, and attributive processes. We conclude by exploring the linkages between these constraints and the final product, the candidate image.

The Stimulus-Perceiver Controversy

The stimulus-perceiver controversy initially contrasts alternative conceptualizations of an image as a social phenomenon. Proponents of the image, or stimulus, theory define an image as all traits projected by a candidate to influence an audience.[1] For example, in Erving Goffman's description of the larger behavioral context, such traits are *expressions* a person gives off as a way of managing the *impressions* others have of him.[2] In comparison, advocates of the perceptual balance theory view images as held by audience members and simply attributed to political figures. In this theory, a candidate's "image" is thus little more than our personal values, hopes, frustrations, fears, and envies as we think they are reflected in the person of a politician. An otherwise ambiguously defined receptacle,[3] the candidate serves as an object of popular adulation or ridicule. Neither of these alternative views of what constitutes an image ignores the coupling of a stimulus with a response disposition of the perceiver, but each view does assert the primacy of one over the other.

The question of whether beauty is "only skin deep" or "in the eye of the beholder" is important, for it raises substantive disagreements over what factors determine voters' perceptions of political figures. The image thesis holds that perceptions are stimulus determined—that is, popular perceptions of a candidate are a function of the traits projected by that candidate. In contemporary America, the image thesis assigns an especially important role to the mass media in shaping voter perceptions. The perceptual balance principle, on the other hand, argues that perceptions are perceiver determined—that is, images are introjected into a candidate by voters in accordance with their own political values. The perceiver-determined thesis coincides with the voting studies reviewed in Chapter 2, which typically emphasize the enduring partisan loyalties held by voters as the principal shapers of perceptions of political leaders.

Image Projection or Perceptual Balance: The Argument

The image thesis as an explanation of political perception has arisen largely since the advent of television. However, the essential notion harks back to a continuing tradition in psychological research implicit in "connectionist" (contiguity, or conditioning) theories of learning.[4] Connectionist theories use a basic model typically labelled "S-R," which predicts response (R) from empirically determined rules related to stimulus (S) properties. These properties include the amount ("primacy") and location ("recency") of a given stimulus. Proponents of the image thesis in voting behavior research, however, have generally ignored S-R learning theories in developing measures and hypotheses regarding the influence of campaign stimuli on candidate images.[5] Instead they have operated with a narrow definition of image by measuring the extent of popular agreement that a specific attribute, or trait, emanates from a given candidate in relation to other candidates. The research problem typically has been to establish that campaign stimuli (that is, a candidate's projected qualities) actually influence voting instead of trying to determine how such stimuli combine in different ways to affect candidate images. Indeed, the image thesis is really more of a working assumption that a candidate's projected attributes influence votes than a theory explaining how people perceive candidates.

Although the image thesis contributes to our understanding of candidate images by calling our attention to the potential of election campaigns for developing a context within which citizens make voting judgments, the thesis goes too far. It implies that a candidate can assume virtually any image he wishes, project it at will, and have voters interpret it as he intends they should—without reference to the voters' personal predilections. To be sure, various campaign actors (campaign managers and advisers, journalists, and the candidates themselves) help create public images of the candidates. These creations are partially contrived, partially a product of chance factors in the campaign, and partially accurate reflections of the candidate's qualities. According to the image thesis, whatever the created and projected image, it is there for all voters to see and respond to in similar ways. Presumably the candidate projecting the more popular image attracts more votes and wins the election. But what makes one image more popular than another? And why do different voters respond to a given candidate's projected image in different ways? Among the critical flaws in the image thesis is that it supplies no answers to such questions.[6]

The principle of perceptual balance, in contrast, argues that perceivers' more or less deep-seated values or attitudes shape their perceptions of candidates. Party affiliations and, to a lesser extent, issue positions are key qualities of candidates that strike harmonious or dissonant chords within voters of varying predispositions. At first glance, this principle does not seem to differ from the image thesis, inasmuch as traits of the candidates themselves are essential elements in perception. However, the perceptual balance principle argues that one or another personal value is so important to voters that they seek out candidates having similar values. For a voter to maintain a homeostatic, or balanced, set of attitudes and to assure himself that his preferred candidate thinks as he does, he frequently unconsciously distorts certain information regarding the favored candidate and the opponent. This tendency yields the processes of selective exposure, perception, and retention revealed by early voting studies (see Chapter 2). Thus, once the voter makes a judgment regarding the candidates, he is likely to reinforce that decision by distorting the words and deeds of both candidates to assign most or all highly desirable attributes to the favored candidate and undesirable attributes to the opposition candidate. In this sense candidate images are perceiver determined.[7]

The perceptual balance principle implies not only that voters engage in selective distortion in exposure, perception, and/or retention but also that this distortion takes place in accordance with the principle of "assumed similarity." Extensive research by social psychologists suggests a definite tendency for individuals to perceive people they like as essentially similar to themselves.[8] Why not political candidates as well? It would follow that when a voter and a favored candidate share at least one attribute of central importance to the voter (say, being of the same political party), it is presumably easy for the voter to assign other attributes he sees in himself to his favored candidate.

The perceptual balance principle, then, is important for understanding how people acquire images of political candidates, for it sensitizes the researcher to the role of subjective categories (such as attitudes, beliefs, and values) in shaping political perceptions. Voters do not come to each campaign with a blank, open mind, a tabula rasa. A variety of prior experiences, values, and habits shape the voters' responses in attending to campaign stimuli. At the same time, however, the perceptual balance theory has certain fundamental weaknesses inherent in its application to research on candidate imagery. The assumed-similarity principle is an essential derivative of perceptual balance, yet most research demonstrating the empirical validity of the assumed-similarity principle occurred in small groups in which social distance between leaders and followers was relatively small. The presumption that the results of such research also apply to the relationship between candidate and voter, in which the social distance is much greater, is questionable. Furthermore, just how valid is the principle in reverse? In other words, do voters really assume the opposition candidate to be more dissimilar simply because he differs from them in one attribute, however important? Research has thus far provided no firm answer to this question.

As a result of the function of partisanship in American politics, the perceptual balance principle probably helps more to explain the images Republicans and Democrats have of candidates than it does to explain how strict Independents perceive electoral contenders.[9] When partisans have positive views of their party's candidate and negative ones of the opposition, it is easy (probably too easy) to evoke the perceptual balance principle as an explanation. The image thesis is then left with the task of pinpointing the sources of the images of candidates held by independent voters. Thus, returning to the argot of the general description of voting behavior as outlined in Chapter 2, the perceptual balance principle relies heavily on long-term factors such as partisan self-images to explain how voters formulate images of candidates; the image thesis takes the view that short-term forces peculiar to the election (especially the projected qualities of opposing candidates) shape the electorate's perceptions. In Chapter 8 we explore in more detail the relative effects of long-term and short-term images on voting. We mention it here, however, to indicate that the stimulus- versus perceiver-determined controversy extends to these key categories of analysis in the model of the voting act, as described in Chapter 2.

Image Projection or Perceptual Balance: The Evidence

On the basis of what we have said thus far, it is impossible to judge whether the image thesis or the perceptual balance principle helps more to explain what factors enter into the voters' formulation of candidate images. Unfortunately, a review of the studies that have most directly delved into the stimulus- versus perceiver-determined controversy renders it even more difficult to reach such a judgment. In what can be considered the seminal (but not definitive) study of the question, the McGraths admit that there is evidence for embracing both theories.[10] Using a fifty-item questionnaire similar to the semantic differential, they found that in 1960 Democrats and Republicans alike saw John Kennedy rather than Richard Nixon as the more ambitious, aggressive, striving, active, dynamic, and rebellious candidate. Thus, on this "activity" dimension, the McGraths found evidence of partisan agreement and thereby support for the stimulus-determined thesis. Overall, the two parties agreed in assigning Kennedy more of an attribute than Nixon on twenty-nine of the fifty items. Yet there were distinct partisan differences in rating the two candidates on the remaining twenty-one items, especially those measuring potency traits. Thus, members of each party saw their preferred candidate as more complex, hard, tall, deep, unchanging, and stable.

Such evidence on behalf of a perceiver-determined theory, however, did not deter the McGraths from concluding that their research generally supported the image thesis! They argued this on two counts. First, they had biased their data gathering in favor of the perceptual balance principle by selecting strong partisans (members of the Young Democrats and Young Republicans) as respondents. Failure to find clear partisan differences on a majority of traits among such respondents, they reasoned, was evidence enough that candidates were projecting distinctive images transcending partisan distortions. Second, they believed that they had found a "politically sensible pattern" for the 1960 presidential campaign in the fact that partisans agreed in rating Kennedy higher on the "activity" dimension while differing over which candidate rated higher on "potency."[11] However, they offered no evidence that Democrats and Republicans share a common admiration for "activity." If, in fact, Democrats in 1960 preferred an active leader, the trait they perceived in Kennedy was positive; if Republicans wanted a less active leader (say, in the tradition of Eisenhower), the "activity" they perceived in Kennedy could have been a negative quality. Hence, on the cognitive level partisans might have been concurring in their belief that Kennedy was more "active" (evidence of stimulus determination), but on the affective level they might well have been disagreeing over the appropriateness of that trait (evidence for perceiver determination).

In a later study designed to examine how people perceive political figures, Roberta Sigel offered another criticism of the McGraths' efforts. Sigel argued that, far from being biased in favor of the perceptual balance principle, the McGraths' research found scant evidence for the perceiver-determined view precisely because the types of questionnaire items they employed were unlikely to produce "stress" in politically relevant areas. To rate a candidate as "sensitive" or "rugged," for instance, is not likely to produce political differences. If researchers use measures tapping perceptions of politically relevant traits, however, partisan differences might emerge more consistently and, consequently, provide support for the perceptual balance theory.[12]

In reaching a conclusion that "partisan determination"[13] is a fruitful explanation of how people perceive political candidates, Sigel employed data from a random sample of the Detroit electorate in the 1960 presidential election. The questionnaire contained a list of ten traits—representing personality, political stands, abilities, and statesmanship—considered important in a president (traits tapping respondents' idealized images). Respondents also ranked their preferred and nonpreferred candidates on such traits. The basic evidence

Sigel offered to support a partisan determination consists of the strongly positive rank-order correlation between a given partisan group's (Democratic or Republican) image of the ideal president and that group's own candidate and the weak correlation between the ideal image and the group's image of the opposition candidate. No such discernible pattern could be found among independent voters.

Although these findings are impressive, one feature of the analysis detracts considerably from their conclusiveness. Sigel derived average rankings for each of the ten attributes for each partisan group—Democrats, Republicans, and Independents. Thus, "honesty" was the most important trait selected by Democrats (selected by over 75 percent), "intelligence" was second (selected by 51 percent), and so on down to tenth-ranked "good television personality" (chosen by only 2 percent). Yet, apparently each respondent did not actually rank each candidate on all ten attributes; rather, a given respondent ranked only the three attributes he believed best described each political figure.[14] In short, the rank orderings of the ten qualities were derived from having respondents select the "three most important" traits instead of using the preferable (and more reliable) procedure of having respondents rank all ten, then calculating average rankings on the basis of the complete set of responses.

Sigel also uses a second set of data on candidate images from the same Detroit survey, but here the results are considerably less conclusive. The data consist of responses to open-end questions regarding reasons voters gave as to why and why not they might vote for each of the candidates. Analysis revealed that the candidates' images differed considerably among voters willing to volunteer responses, but it revealed also that there was sufficient agreement on the candidates' images across Democrats, Republicans, and Independents to indicate that image differences were not merely reflections of partisan biases. Taken together these findings tend to favor the image thesis. Yet Sigel concludes the opposite on grounds that the candidates' images were "almost identical" to the images of their respective political parties.[15] Her tabulated findings show that Democrats, Republicans, and Independents agree in mentioning Nixon's greater experience as an asset by about two to one over the next most-often-mentioned attribute.[16] Was greater "experience" really a commonly perceived attribute of the Republican party in 1960? Perhaps it was, but Sigel provides no corroborating evidence, only an implicit assertion. And the case of Kennedy, while not so dramatic, shows that voters tended most often to cite his greater capability. Again, was this also true of the Democratic party image? Certainly the data show that candidate images are not independent of party affiliation but rather that those images are simply mirrored reflections of party images.

Sigel's conclusions, then, are in part an artifact of her data-gathering procedures. Her ranking procedure may have produced a partisan difference due to restricting choices to a ranking of only three attributes.[17] And the lack of partisan differences in the open-end responses is not especially surprising since these data largely reflect a self-selected sample of articulates (likely to be more involved and partisan), not the general population of Detroit voters. The conclusion here can only be that neither Sigel nor the McGraths adequately test either the image thesis or the perceptual balance principle.

There is one instance of potential agreement between the findings of the McGraths and Sigel. Sigel concludes that candidates' political attributes are more likely to be perceiver determined and that personal attributes are more likely to be stimulus determined. This parallels the McGraths' identification of potency attributes, which they concluded to be more perceiver determined, with political role attributes. However, the reader need only refer back to Chapter 3 to see that this identification of potency with political role attributes is not always clear-cut. Moreover, earlier studies have shown that partisans largely see the preferred candidate as stronger, but they do not necessarily see the opposition candidate as weak.[18] Such findings do not belie the role of perceiver-determined candidate images, although they do raise questions regarding the principle of perceptual balance.

A recent study of the effect of a speech by Sir Robert Menzies, the Australian prime minister, during the 1963 election campaign provides further evidence favoring perceiver-determined explanations.[19] This study was one of the first to incorporate a longitudinal design to measure changes resulting from a campaign event, a direct stimulus. Inasmuch as ratings of Sir Robert on several attributes increased favorably among his own Liberal party supporters in the immediate postspeech period and became more unfavorable among Labour partisans, the authors concluded that these image changes were perceiver determined. But measurements taken later in the campaign showed little permanent net change in either partisan group. Thus, once again, we see the central notion of perceptual balance theory cast in doubt as party members, in the long run, failed to perceive differences between preferred and opposition candidates.

Another longitudinal study, however, uncovered definite net changes during an election campaign. Blumler and McQuail, using semantic differential ratings of the party leaders' images in the 1964 British general election, obtained results generally, although not conclusively, favoring the image thesis.[20] While ratings of candidate attributes tended to change in the same direction as shifts in party attitudes, voters regardless of partisanship revised their assessments of the party leaders in similar ways, especially with regard to the perceived potency of the party leaders. Thus even Conservative party supporters of Sir Alec Douglas-Home agreed with Labour partisans in downgrading his "strength" and further agreed with Labourites in revising upward their estimate of the potency of Harold Wilson, the Labour leader. Since, as Blumler and McQuail argue, potency is a vital element that most voters expect to find in any political leader, it can hardly be concluded that partisan-based perceiver determination operated in this instance. If perceiver determination had an effect, it occurred through partisan differences in assessments of the moral character of the two leaders—that is, the evaluative dimension of images, discussed in Chapter 3.

Blumler and McQuail ask how Conservative voters could reconcile their declining regard for their leader and yet demonstrate strong support for their party. If perceptual balance, or a strain to attitudinal consistency, is a vital element in human perception, then Conservative voters had to adjust to a rather unsettling circumstance. One of many plausible explanations is that Sir Alec served as a scapegoat in explaining the Conservative party's failure in 1964. Voters downgraded his potency, condemning him for lack of vigor in leading the party to its "proper" fortune. Or, as Blumler and McQuail also suggest, perhaps there may have been no real stress for Conservative voters. If the party is the more salient political object to them, then it should not create perceptual stress to acknowledge the limitations of the flesh-and-blood leader as distinct from the party itself. Certainly, without further supporting data this latter explanation is more parsimonious than those derived from the perceptual balance principle.

A more recent study by Brownstein employed rigorous laboratory experimentation procedures to confront stimulus- and perceiver-determined views.[21] He exposed student respondents to campaign messages presenting hypothetical conservative and liberal candidates. The design examined the effect of message style, media, and interpersonal influence. To test the effect of message style, the students received one of the following sets of communications: (1) unambiguous, straightforward statements by each candidate speaking directly to the issues, (2) ambiguous, New Politics statements by each candidate, or (3) a combination of the two types. The media were television, radio, and print. Finally, some subjects had group discussions following media exposure while others did not. Of the communication variables, only message style had a significant systematic influence. Indeed, message style was a better predictor of vote choice then shared candidate-to-voter attitudes on the issues. More importantly, however, Brownstein found that when he combined message style and ideological orientation, prediction of voter choice was almost perfect. He concluded that the image thesis and the perceptual balance principle are complementary perspectives on the process of electoral decision making rather than competitive or exclusive.[22]

We can conclude only that the verdict has not been reached regarding the superiority of either the image or perceptual balance thesis. In support of the image thesis, voters gain a reasonably large amount of information about candidates, at least in high-interest campaigns. Yet this information tends to be highly selective, and voters seem indifferent or reluctant to articulate what they know. The primary objection to the image thesis is that it provides little hint about what kinds of campaign stimuli voters are most likely to be receptive to or, once exposed, how they fashion a response to those stimuli. Underpinning the perceptual balance principle is evidence showing that partisans typically have more favorable images of their party's candidate.

The perceptual balance principle goes beyond the image thesis in suggesting certain psychological processes that shape the response to political stimuli, but the principle is not always persuasive in this respect. In the first place, attempts to demonstrate a "strain to consistency" have not been overly successful. Such attempts are generally based on the principle of assumed similarity. Voters, however, not only perceive, but expect, some distance between themselves and the candidates. Further, voters frequently receive ambiguous stimuli that make efforts to develop consistency in their image difficult. In addition, the corollary assumption, that voters increase the displacement positively and negatively between the opposing candidates, is also questionable. But the evidence is that most voters rarely formulate distinctly negative images of a candidate. Rather they are inclined to look warmly on certain officials regardless of party.[23] Finally, the perceptual balance principle is also strongly associated with the notion of selective distortion of campaign stimuli, but research on such distortion is inconclusive at best.[24]

The Image as a Stimulus-Perceiver Transaction

If we examine the behavior of voters in the full context of an election campaign, we discover a set of transactions between leaders and followers. Those transactions suggest that a candidate's image is a function of both the characteristics people project on him and the qualities he tries to project to them. In a campaign voters are *exposed* to candidates, including their stands on relevant issues and their affiliations with relevant groups; the attitudes of the voters are *engaged* (whether through activation, reinforcement, or alteration); and voters reach some *intention* to support one candidate over the other, or to support neither. But the campaign is a continuous process through time—indeed, an American truism is that the "politicking" never stops—and to isolate exposures, attitudes, and voting intentions from one another as if they proceeded in a chronological sequence is a simplification.[25]

Although there is much that is worthwhile in both the image thesis and the perceptual balance principle, neither alone is a sufficient explanation of candidate imagery when one conceives of campaigns as continual transactions between office seekers and the electorate. Candidate images are products neither of projections emanating from campaign stimuli nor of response dispositions held independently by voters. Rather these images are reciprocal relationships reflecting continuing exchanges between leaders and followers, in which the former not only project selected attributes but must also imagine how followers perceive them; the latter perceive leaders and imagine how leaders perceive them. For either candidate or voter, the other is both an object with image traits to perceive and also an image-object of self-perception.[26]

Thus we find neither the image thesis nor the perceptual balance principle an adequate description of a very complex set of social transactions. The former fails by overlooking the multifaceted character of what the candidate projects and the voter receives; the latter does not take into account the voters' tendency to judge candidates along a variety of dimensions. Neither view fully recognizes the reciprocity that characterizes the transaction between the perceiver and the social object of that perception. Moreover, neither view formulates an operational definition of "image" in terms of this act of social perception, although both recognize, at least implicitly, the relationship between stimulus and response disposition.

Candidate imagery is in part an interplay, the continual engagement, of campaign stimuli with the response dispositions of voters.[27] But a more operationally focused conception views an image as a human, subjective construct imposed on an array of attributes projected by the candidate. Taken together, these two notions of imagery and image recognize the reciprocal, transactional character of political images. The latter, especially, stresses the holistic character of political perception. In other words, it recognizes the "imaginative" process involved in political perception, whereby a voter formulates a total, comprehensive picture of a candidate, an image fabricated from bits and pieces of information about the candidate and from the voter's predispositions.

As an example, the religious factor in the 1960 presidential campaign affected not only popular images of John Kennedy, the Catholic, but also those of Richard Nixon, the Protestant.[28] Religion became an important campaign stimulus with the advent of Kennedy as a serious contender for the Democratic nomination. Normally religion would be no more than a nominal factor in an American presidential campaign, but Kennedy's candidacy activated generally dormant emotional biases. Voters used their religious feelings in constructing images of the candidates. In 1960, then, religion as a component of candidate images was both stimulus (information that Kennedy was Catholic) and response disposition (the religious leanings of both Catholics and Protestants).[29]

Constraints on the Process of Image Formation

What factors influence the formation of images of office seekers? Among those that might be labelled perceiver determined, research most often cites partisanship and ideology. To these it is important to add voters' idealized images of the desirable public official, their various self-images beyond partisan and ideological self-images, and attitudes derived from their affiliations with reference groups.[30] In the category of stimulus-determined influences, there are the communications about politics that reach voters and the stimuli resulting from friends, neighbors, fellow workers, and other persons, which result in overall social judgments among people as they compare candidates. Related to perceiver- and stimulus-determined forces, combining elements of both, are the processes by which voters commit themselves to a candidate and then rationalize their choice by attributing certain qualities to their selection and his opponents. The interplay of these perceiver-determined, stimulus-determined, and attributive influences constrain the kinds of portraits voters paint of candidates. Some bear stronger relationships than do others to the types of images that result. To obtain a better understanding of the key factors, let us examine them briefly. Then it may be possible to speculate regarding which (if any) factor is the single best predictor of the types of images voters are likely to formulate.

Partisan and Ideological Anchoring

Party identification and ideological orientation are the most widely recognized political frames of reference that influence candidate images. A frame of reference is a person's internalized values, social norms, and beliefs resulting from his exposure to the political world.[31] Such a frame of reference orients an individual cognitively, affectively, and conatively toward candidates. However, most research on the influence of party identification and ideology on candidate images examines the affective and conative aspects of those images. Such studies have asked, for example, whether people who share the same party affiliation and/or ideological persuasion with a candidate feel favorably toward him and intend to vote for him.[32] This research has generally demonstrated that affective (favorability) and conative (vote intention) responses toward the candidate of one's own party are strongly biased on behalf of that candidate (as discussed in Chapter 2). Moreover, the stronger the identification with the party, the more favorable that image.[33] This does not mean, however, that members of a given political party necessarily hold negative images of the opposition party's candidate.[34]

Party Identification

The results of a study of the 1970 senatorial campaign in Missouri illustrate the relation of partisan self-images to candidate images (see Appendix B for details). With equal numbers of self-designated Democrats, Republicans, and Independents comparing Senator Stuart Symington (D) and Missouri Attorney General John Danforth (R) with the "Ideal Senator" via Q-sorts, the authors found that each set of partisans viewed their party's candidate as closer to their conceptions of the ideal senator than they did the opposition's nominee. Yet as Table 4.1 shows, images of the opposition party's candidate were still moderately, but positively, related to images of the ideal senator. Independents tended to visualize both candidates as less like their ideal senator than did partisans even of the opposition candidate. Partisanship apparently constrains candidate images but does not necessarily do so in accordance with the expectation derived from the perceptual balance principle—that is, selective distortion of the opposition candidate in a distinctly negative direction.

In a study of response to the Kennedy-Nixon television debates, using the semantic differential, Carter found similar results and labelled this phenomenon "image stereotypy," that is, giving the candidate virtually the same rating on every characteristic.[35] Among partisans stereotyping was strong for their own candidate but also appeared in their ratings of the opposition candidate. Image stereotypy is a form of selective distortion. However, rather than following the principle of assumed similarity and its corollary, assumed

Table 4.1   Average Intrapersonal Correlations across Partisan Groupings, among Q-Sorts Describing the Ideal Senator and Two Actual Candidates

Q-SORTS CORRELATED	DEMOCRATS	INDEPENDENTS	REPUBLICANS	ALL
Ideal Senator–Symington	0.66	0.46	0.52	0.55
Ideal Senator–Danforth	0.55	0.48	0.73	0.59
Symington–Danforth	0.44	0.37	0.41	0.41

dissimilarity, stereotypes derive largely from a tendency to perceive socially desirable traits in other persons, at least until available information counters such suppositions.[36] A notable example was the vast popular appeal that Dwight Eisenhower enjoyed prior to 1952; this popularity cut across all partisan groups.[37] With his acceptance of the Republican nomination in that year, however, his popularity among Democrats slipped noticeably (see Chapter 6). Still Eisenhower's personal appeal remained relatively high with Democrats, if not as stereotyped as before.[38]

There is some evidence that rather than changing the affective content of the image of a political figure associated with the opposition party, the partisan cognitively restructures his image to emphasize certain traits and deemphasize others. The evidence is indirect, however, since available studies measured candidate images at a point long after the political leaders were already definitely associated with political parties. There is a tendency in any case for partisan identifiers to perceive their own party's candidate in political role (job-related) terms and his opponent in stylistic (personal) ways.[39] More specifically, partisan followers typically perceive their leader as a "great statesman" and his opponent as a "partisan campaigner" (recall our discussion in Chapter 3). This cognitive response is more likely to occur in high-stimulus campaigns in which the candidates are relatively unable to avoid or obscure their partisan attachments. Thus in the Missouri senatorial campaign of 1970, John Danforth deemphasized his Republican party ties and was relatively successful in avoiding the image of a partisan campaigner among Democrats.[40] In the 1972 presidential campaign (see Chapter 7), however, neither Richard Nixon nor George McGovern seemed to project very successfully to the opposition an alternative to the partisan campaigner.

In sum, the voter's subjective party identification is a powerful influence on the ways he organizes into images the perceived traits of candidates. As with party identification, ideology acts as a political frame of reference, a subjective construct, shaping perceptions of political figures.

Ideological Orientation

As with the study of partisan self-images, investigations into the relationship of ideological anchorages have focused on the affective and conative aspects of candidate images.[41] Where voters have such ideological frames of reference and where candidates project clear-cut ideological appeals, voters make positive evaluations and definite vote intentions on behalf of candidates with whom they share ideological orientations.[42] In American politics, however, these conditions are less frequent than are those conducive to partisan responses to candidates. American voters' attitudes traditionally have lacked the recognizable consistency generally associated with ideological thinking about politics.[43] And the lack of issue content and ideological appeals in American political campaigns is proverbial. Although the conditions are seldom present for "real-world" studies of the relationship of ideology to political perception, appropriately controlled laboratory experiments creating those conditions are possible. In such studies the impact of ideology is apparent. And in those cases where actual candidates make ideological appeals, the effect also has been strong.[44]

We know much less about the influence of ideological anchoring on the cognitive component of political perception than we know of its effect on voting preferences. Byrne and his colleagues, however, found that although college students disagreed in evaluating fictional candidates issuing clear ideological appeals, the students could agree in their descriptions (including physical appearance) of the candidates. The major difference between conservative and liberal students was the tendency of the latter to describe the liberal candidate as more "flexible" and "humanitarian."[45]

Evidence of a linkage of ideology, party, and candidate has also been uncovered in studies of both British and American voting behavior.[46] For instance, using the thermometer scale ratings devised for SRC/CPS election surveys, Rusk and Weisberg mapped the dimensions of respondents' assessments of several political leaders, groups, and issues.[47] Two basic dimensions emerged—partisanship and ideology. Moreover, the ideological dimension appeared not as the conventional liberal-conservative dichotomy so often described by American political pundits but rather as a tradition-versus-change orientation. Comparing data from 1968 and 1970, researchers observed some shifting in the relative salience of these dimensions to voters. Partisanship became more prominent in 1970. At the same time, the two dimensions became more independent. These findings suggest the importance of the larger political environment in creating a context, or complex set of interrelated stimuli, for cueing appropriate subjective dispositions which voters use to perceive and organize projected traits of leaders, thus yielding diverse images.

Images of the Ideal Public Official and the Self

Given the early voting behavior research revealing the very strong relationship between party identification and candidate images, it is not surprising that researchers overlook other attitudinal constraints. One of these is the conception an individual holds of the desirable political leader, the ideal public official. To what extent does a voter's ideal of what qualities an official should possess influence his perceptions of the attributes office seekers display?

There are at least two approaches to gauging the influence of the ideal on candidate images. One is to examine various groups of people to see whether members of one or more groups desire traits in the ideal official that differ from those sought by other groups, and whether they project these same traits on the favored candidates of the group. A second approach is to divide people according to the distinct images they have of an ideal official and then to determine whether that ideal differentiates them in the ways they perceive actual candidates.

Generally following the first approach, Jackson and Miller used semantic differential rating scales to assess the congruency of ideal images of candidates for president and governor in 1972 with the images of candidates favored by groups of Illinois college students. Generally the ratings for the ideal official were closer to ratings given the candidates supported by respondents than they were to ratings of candidates not supported. This finding was more evident among the supporters of the two Democratic candidates; supporters of the Republican candidates seemed more ambivalent in their ratings of the attributes of their party's candidates. In fact, Republican voters rated the Democratic candidates closer to the ideal on several scales, even though they rated the Republican candidates closer on the average across all scales. In addition, the conception of the ideal official was generally uniform across these voting groups. For example, the only scale that strongly differentiated Nixon voters from McGovern voters was "liberal-conservative," with the latter group leaning much more heavily toward "liberal." But in the final analysis there were no major surprises. The most that can be said is that voters tended to see their preferred candidate as similar to their ideal conception. Therefore, we cannot argue on the basis of the Jackson-Miller data that voters simply formulate a conception of the ideal leader and then seek out the candidate most closely corresponding to it.[48]

Our Q-sort study (see Appendix C) of the 1971 Democratic presidential contenders casts more light on the question. That study distinguished two types of ideal presidential images. First, a general type emerged which, for the most part, represented the American political hero discussed in Chapter 3; second, a distinctly partisan, professional politician's image appeared, with party regulars laying a particular emphasis on strength.[49] Examining the image types for the four Democratic contenders in relaton to the two ideal types seems to reveal a negative response to the question of whether the ideal image precedes searching behavior.[50] In the first place, most of the respondents held very positive images of all four candidates, regardless of their images of the ideal.[51] Second, distributions of ideal image Types I and II (Table 4.2) across the candidates were similar.

Persons preferring the strong, professional politician were much less inclined to see George McGovern as a "maverick leader" and more inclined to see him as the opposite of the "great statesman." In other words, almost one-third of the people having a Type II image of the ideal president visualized McGovern as the contradiction of that ideal. We may infer that these respondents measured McGovern against their ideal and found him wanting. Note also that no respondents having an idealized image respresented by Type II perceived Edward Kennedy as a "maverick leader" (Type III).

The role that images of an ideal public official play in constraining candidate images, then, remains ambiguous. Conceptions of the ideal official probably define the kinds of traits voters are willing to accept in their candidates and, by inference, the qualities voters refuse to tolerate as well. For example, the characterization of Barry Goldwater or George McGovern as "extremist" no doubt contributed to defections from their candidacies by their own party identifiers, at least insofar as this trait is deemed undesirable by most Americans. There is a strong relationship between affective support for a candidate and the degree that his image matches that of the ideal official.

Yet there are many people who apparently have other bases for forming their positive and/or negative views of office seekers and consequently do not

**Table 4.2   The Distribution (%) of Candidate Image Types across Two Ideal
Presidential Image Types**

CANDIDATE AND IMAGE TYPES	IDEAL PRESIDENTIAL IMAGE TYPES	
	I (n = 26)	II (n = 10)
Muskie I	73	80
II	19	10
III	8	10
Humphrey I	62	70
II	23	10
III	15	20
McGovern I	46	40
II	4	10
II negative	0	30
III	50	20
Kennedy I	58[a]	70
II	15	30
III	25	0

[a] Cell n is less than 100% due to rounding.

fret over whether or not a candidate matches their image of the ideal. Research
suggests that one such basis is the image a voter has of himself, that is, an
image of what he thinks he is like, or would like to be, against which he compares contenders for public office. Unfortunately the evidence from such
research implying a direct relationship between self-images and candidate
images is both meager and contradictory. For example, in a study employing
the semantic differential to examine images in the Johnson-Goldwater contest,
respondents rated their favored candidate close to how they defined their
respective self-concepts and related political concepts.[52] However, in a study
of the Kennedy-Nixon campaign of 1960 using a similar technique, the self-
images of supporters of the candidates did not differ from one another significantly.[53] Moreover, both candidates' partisans perceived the nominees as
different from one another and from their own self-images as voters.

Although the direct evidence on the question is mixed, another research
tradition may also suggest relationships regarding the role of self-images in
political perception. Social psychologists have long been interested in the
relationship between various personality traits of followers and support for
political leaders. Such traits are basic subjective tendencies that shape the
image of self. To the extent that these traits influence political perception,
the self-image is thereby related as well. Personality traits having specific relevance to political events are more likely to have a stronger relationship to
support for political leaders than is the more general self-image.

The research conducted to date has been limited to the affective basis
of leader images. The personality trait most consistently examined is that of

authoritarianism, a trait sometimes conceptualized as a general tendency to defer to strength in leadership and related to a general authoritarian personality syndrome.[54] The typical measure of authoritarianism has been the California F scale, a measure of specific traits. Using it researchers have described a relationship between authoritarian traits and candidate support that may reflect the influence of attitudes comprising a more general authoritarian syndrome. Studies indicate that one group among whom such a relationship may exist is certain college-educated people.[55] In this subset of well-educated, authoritarian people, support for conservative candidates was apparent; indeed the perceived ideology of the candidate clearly overrides even partisanship as a factor on which to base evaluations.[56] There is reason to believe that this group constitutes a distinctive personality type, for they exhibit several related traits associated with authoritarianism: traditional family ideology,[57] religious conventionalism,[58] ethnocentrism,[59] and politicoeconomic conservatism.[60] Such a personality pattern differs from that of the egalitarian personality, said to favor liberal candidacies.[61] All in all, the research confirming the existence of authoritarian and/or egalitarian personalities and their relationship to people's evaluations of political figures has not been overly persuasive. Before credence can be placed in the proposition that distinct personality types are the source of voters' images of candidates, far more studies must be conducted.[62]

Comparative Social Judgment: The Image of the Opponent

The most interesting election campaigns usually involve a confrontation between two or more candidates. The voting decision implies an acceptance of one candidate and the rejection of others, a comparative social judgment a voter makes depending on the extent of his motivation to perceive differences between the candidates and relate those differences to his own views. Sherif and Hovland devised a theoretical posture to account for behavior in such situations.[63] Basically the theory argues that in dealing with social phenomena, a person forms an evaluative system based on a "most acceptable position"—that is, the course of action the person is most willing to follow among a range of options, some of which he deems acceptable and others unacceptable. The most acceptable position acts as an anchor, or reference point, against which to compare related positions. Social judgment is the process of making these comparisons. The positions that a person determines as acceptable to him constitute his "latitude of acceptance"; those not tolerable form his "latitude of rejection." Finally, there are also likely to be latitudes of noncommitment, or positions on which the person has no strong feelings one way or the other. Into this subjective matrix of varying latitudes flow pieces of information on questions pertaining to a person's interests. According to the theory, the individual assesses the content of each message through an assimilation-contrast process. If the informaton does not diverge sharply from the positions constituting the individual's zone of acceptance, he assimilates it as favorable to his image of things. If there is considerable divergence, however, he contrasts it with his own stock of beliefs and may reject the message as incorrect, undesirable, or irrelevant.[64]

Two elements of the Sherif-Hovland theory of social judgment are particularly relevant to understanding candidate images. First, the theory suggests that voters may have latitudes of acceptance, rejection, or noncommitment for candidates, just as they have for positions on controversial issues. When candidates win nominations for public office, it is entirely possible that voters may find none of them "most acceptable," yet all of them within a "latitude of acceptance." Social judgment theory suggests that voters then select the candidate who comes closest to most acceptable, assimilating and contrasting messages about that candidate in accordance with the process of comparative judgment. Or, voters may find none of the candidates really "acceptable." When forced either to choose one of them or just not vote, the process of comparative judgment provides a conclusion on the "lesser of two evils." In sum, the images voters form of candidates are products of successive comparisons of bits and pieces of information about all candidates assimilated into the voter's ranking of acceptable, unacceptable, and irrelevant traits.

Second, research based on social judgment theory reveals that the more involved people are in a controversy (in this case, in the electoral campaign), the larger their latitude of rejection and the smaller their latitude of acceptance and noncommitment. This suggests that whatever heightens a voter's interest in a campaign (partisanship, a candidate's appeals, the urgings of friends, the flow of news about the election in the mass media) may involve him more, move him toward commitment, tighten his zone of acceptance, and thereby prompt him to search for information about candidates that will reinforce his formulated images. It suggests also that the relatively uninvolved voter may be most susceptible to new information (if it reaches his awareness) and thus most responsive to the kinds of traits a candidate seeks to promote as attractive.

When people make social judgments that result in their voting for the "lesser of two evils," we have the phenomenon labelled by some political journalists as "negative voting." There was much talk in 1972 that the Nixon-McGovern contest for president was of this variety. The evidence, however, does not bear out such a firm conclusion. Otis Baskin's study is a case in point.[65] Using a measure of social judgment, Baskin found no indication of "negative voting." For instance, many of the Nixon supporters sampled were almost indifferent toward him, seeing him as neither better nor worse than his opponent. Although some McGovern voters in Baskin's study did not especially like the Democratic candidate, a high proportion did and voted *for* McGovern rather than *against* Nixon. Thus, McGovern's image fell within such voters' latitude of acceptance.[66]

Group Affiliations and Candidate Images

There has been relatively little study of whether differences exist among members of diverse groups in the images they have of political candidates. The studies that exist pertain primarily to image differences among people of varying age groups or to differences in political perceptions of males and females. Generally, even when people of different ages or sexes have diverse images of candidates, neither age nor sex is the influential factor in producing the variations. Rather, the relative isolation of certain groups from politics early in an election campaign (such as the inattention of some younger voters or of some housewives) results in their articulating no images at all. As the campaign evolves, however, such persons develop more interest and formulate images of office seekers not particularly at variance from those of the more consistently involved electorate.[67]

The evidence is sketchy but social class differences may well lead to variations in the way people perceive political candidates. Blumler and McQuail, for instance, found that British voters in the upper social classes—more so than members of other classes—responded to the personalities of candidates and to their policy stands but were less interested in their campaign promises.[68] At work here are probably educational and occupational differences, but to what extent and how we simply cannot discern from the available research.[69]

Of particular interest are the "new groups" emerging in American politics: the frustrated white working class[70] and the young voters just emerging from the post–World War II "baby boom." At least for the latter category, the new young voters, we can make a few observations regarding the relationship between categoric membership and their candidate images. In our 1972 Q-sort study of the Nixon-McGovern campaign (see Appendix D), a small number (18) of older voters completed the same sorts given the 100 college students. Although the sample of older voters is small and unrepresentative by almost any standard, we might still expect differences to emerge if younger voters are a distinct political group. The differences we found between younger and older voters were largely insignificant. The same image types emerged in Q-sort descriptions of the ideal president, Nixon, and McGovern from both groups. In addition, both groups used essentially the same image traits in formulating those images. The one notable difference was the older voters' slight tendency to place greater stress on partisan considerations in evaluating the candidates than did younger respondents.

Mass Communications: Stimulus Determination or Marginal Influence?

In Chapter 2 we reviewed the principal studies of voting behavior relating to the effects of the mass media on voting and the images people have of political candidates. The general rule, we said, is one of "minimal effects"; that is, mass communications influence voting behavior and voters' perceptions of candidates in certain ways (for example, by providing information about available alternatives, setting agendas for discussion, swaying the views of citizens normally uninterested in politics), but by and large the media have little independent effect on what people think and do in politics. Rather than repeating that narrative here, we think it important to mention in passing a few of the principal areas of communications research that pertain to the general problem of assessing the relative influence of mass communications as a stimulus-determined source of candidate images.

Generally, communication studies have examined the effect on politics (voting, images, perceptions) of the content, format, style, and timing of messages carrying campaign appeals. For instance, a recent study of message content—specifically, endorsements of candidates and published results of opinion polls—suggests that campaign messages have their greatest effects on undecided voters, but the direction of that influence is so unpredictable that the overall effect (once contradictory effects cancel each other) is generally marginal.[71] Again in a study of media formats—specifically, the effects of televised news—it appears that only marginal shifts occurred in voters' perceptions of where presidential candidates stand on key issues.[72] And, the word is *marginal* once more, in assessing effects of media style, specifically, the effects of duration, image size, and camera angle in the televised treatment of political speeches on the perceived credibility of George McGovern and Edward Kennedy.[73] Finally, the timing of campaign communications varies widely in effects on political perceptions, at least as indicated by studies of election night broadcasts projecting winners and losers.[74]

All of this does not imply that no studies document the potential, or even the reigning, impact of mass communications on candidate images. On the contrary, there are such studies, as illustrated by the following. With regard to message content, the degree of ambiguity in what a candidate says influences how accurately voters perceive not only his direct appeal but also his overall personal characteristics and his policy characteristics. Ambiguous content in campaign messages sometimes contributes to voters' committing themselves, for example, to a candidate who actually favors political values contrary to their own. Political advertising on television also has an effect, not so much on changing how people *feel* about candidates (their affective orientations), but on changing what they *believe* the candidates stand for (the cognitive aspect of candidate images). Political ads perceived as informative, interesting, honest, entertaining, and professionally produced can have measurable effects on voters' perceptions, even if repeated only a few times; but spot commercials of lesser quality—regardless of repetition—probably yield only marginal results.[75]

The medium of communication through which voters learn about a candidate influences what they think of him. Pool's comparison of the differential effects of radio and television in the Eisenhower-Stevenson campaign of 1952 is a landmark in this respect.[76] He found that television viewers tended to exhibit a tendency toward partisan polarization of their images of the respective candidates, but especially in the case of Eisenhower. Yet radio listeners rather uniformly exhibited more positive images of Stevenson. Moreover, television had the effect of "humanizing" the Eisenhower image by making him appear to be less of the conquering hero and more like a simple human being. However, television viewers rated Stevenson higher than did radio listeners on the attributes of "snobbish" and "domineering," seeing him as having a "superiority" complex. We should note that Stevenson's mass media presentations were typically formal speeches, whereas Eisenhower used a less formal style.[77]

Despite such examples as those cited to the contrary, most studies, especially field studies, credit campaign communications in the mass media with relatively little overall influence on candidate images. Why? There are several reasons. In the first place, there is a strong inclination for people who associate with one another regularly to prefer the same candidates in similar ways. Within their group, these persons create a climate of opinion favorable to one candidate over another. In that climate interpersonal, more than mass, communications are important.[78] Further, the way the news media report campaigns is very similar from one newspaper to another, one television network to another, and one news weekly to another, thereby affording relatively little of the raw material required for forming diverse images.[79] Finally, for voters who reach their decisions even before the start of the election campaign, selective distortion, ambiguous appeals, and confusion diminish the potential effects of campaign communications.

Attributive Processes and Candidate Images

When voters decide on a preferred candidate early in a campaign, then surely one of the sources for image formulation is the decision they have reached. Having made up their minds, many of them simply proceed to attribute desirable qualities to the preferred nominee and undesirable attributes to the opposition candidates. This is an aspect of a more general attributive process wherein the vote decision derives from one of several factors entering a person's conscious or unconscious thoughts.[80] In this process, the voter's decision occupies the status of an independent variable contributing to variations in the traits that comprise his image of candidates, the dependent variable. Other such independent variables include those we have described previously—partisanship, ideological leanings, images of ideal officeholders and office seekers, self-images, personalities, social judgments, reference group influences, and information gleaned from the mass media. Each of these factors presumably influences the type of image a person formulates of political candidates. But what is the relative influence of each in attributing images to political candidates?

To derive a preliminary (but admittedly, all too tentative) response to this question, we turn once more to the results of our Q-study of the images of the 1972 presidential candidates held by 100 college students who were voting for the first time in any election. Again, the details of that study are spelled out in an appendix. Through Q-sorts, our respondents defined their various images of the ideal president, George McGovern, and Richard Nixon. Two of those images of the candidates concern us here. One of those we label the "partisan campaigner," the image of a candidate as primarily a politician representing his party's members and programs, galvanizing his supporters to action, and taking his case to the electorate. Portions of our respondents, although not always the same people, had such an image of both McGovern and Nixon in 1972. The other image of note here was solely of McGovern, that of a "nice guy." People holding this image of McGovern emphasized his traits of sincerity, honesty, warmth, decency, and devotion to family.

The question before us is, Which of the several independent variables we have described in this chapter contributed most to the sampled college students' formation of either the "partisan campaigner" or "nice guy" image? To answer this we employ a statistical technique, *multiple regression*. Technical considerations associated with that technique need not concern us here (we discuss those more extensively in our use of multiple regression in Chapter 8). For the moment we simply note that multiple regression is a procedure for examining several independent variables (such as partisanship or ideology) to estimate their relative influence on a dependent variable, in this case, each of the two candidate images. We have used the factor loadings of persons on the Q-factors, which define each candidate image (see Appendix A) as weighted measures of the degree that our respondents hold a given image of a candidate. We employ those loadings as measures of our dependent variable, the candidate images, in a regression formula. Further, from our study we have for each respondent measures of several relevant independent variables—party identification, conservative and liberal leanings, vote intention, and the degree of holding each of four images of ideal president (described previously in Chapter 3). Table 4.3 presents the results of regression analyses for each candidate image. In the table, "Rank" indicates which variable was most important in formulating a given image. "Beta" is an indicator of that importance; and "Cumulative Variance" depicts what percentage of the variation in a given image is explained by the addition of each variable to the regression formula.

Only a few of the independent variables emerge in the regression analysis as sufficiently related to the formation of the two candidate images to warrant inclusion in Table 4.3.[81] Of those, clearly *the decision to vote or not to vote for McGovern* was the key variable. Respondents who did not intend to vote for McGovern attributed the image of "partisan campaigner" to him; those who did intend to vote for McGovern saw him instead as a "nice

**Table 4.3   Regression Analysis of Selected Nixon and McGovern Images, 1972
(n = 100)**

IMAGE	RANK	VARIABLE	BETA	CUMULATIVE VARIANCE
Nixon as Partisan Campaigner	1	McGovern vote intent	.32	.36
	2	Idealized image (Type II)	.24	.41
	3	Party identification	.29	.45
	4	Conservative ideology	−.14	.47
	5	Idealized image (Type III)	−.11	.48
McGovern as Partisan Campaigner	1	McGovern vote intent	−.48	.29
	2	Idealized image	.17	.31
	3	Liberal ideology	−.13	.33
McGovern as Nice Guy	1	McGovern vote intent	.18	.32
	2	Liberal ideology	.24	.38
	3	Idealized image (Type I)	−.16	.42
	4	Idealized image (Type II)	.22	.44
	5	Party identification	.24	.47

guy." In the case of Nixon, if people had decided to vote for McGovern they were more likely to imagine Nixon as a "partisan campaigner."

The remaining independent variables contributing to the formation of "partisan campaigner" and "nice guy" images are clearly of lesser importance. Apparently, respondents who desire their ideal president to be warm, human, and empathic (the Type II idealized image) did not find those qualities in McGovern or Nixon, for they attributed "partisan campaigner" images to both candidates. Those of a liberal persuasion were likely to find McGovern a "nice guy" and unlikely to attribute "partisan campaigner" qualities to him. Conservatives, on the other hand, did not see Nixon as partisan, although Democrats did.

The Elusive Sources of Candidate Images

The effort we have made in this chapter to explore the relationship of attributive processes by discerning the relative effects of certain variables on how people perceive candidates is little more than a suggestive exercise and far from conclusive. The interplay between stimulus- and perceiver-determined influences, which probably contributes to vote decisions and, we suspect, to subsequent crystallizations of voters' images of candidates, is a facet of candidate imagery that remains a mystery. Despite the numerous studies we have reviewed bearing directly or indirectly on the problem, we can say little more than that people formulate their images of office seekers on the basis of many considerations, some derived from communications stimuli and others from subjective predispositions. There has been insufficient research into social judgment, comparative appraisals, and attributive processes to give us much insight into how people become convinced that a candidate has this or that trait, that certain traits are desirable and others are not, or how they organize perceived traits into overall images of what a candidate stands for, how they feel about him, and how they want to respond to him. From what we have said in this and in the previous chapter, it should be apparent that we are further along in finding out how people perceive political candidates—the traits they detect and the types of images they hold—than we are in learning why they see certain qualities and formulate them into distinct images. We are, in the words of Herbert Blumer, still in the "variable analysis" stage of understanding candidate images—that is, putting together an inventory of factors that seem to influence voters' perceptions (partisanship, ideology, social class) and trying to find out precise statistical relationships between those variables and candidate images. In the process we have ignored what is essential to candidate imagery—*interpretation*, or the ways people construct a subjectively meaningful reality (or image) from the transactions of their predispositions with the campaign stimuli that bombard them.[82] Beauty, we suspect, lies in the exchange between campaign appeals and the eye of the beholder. If we are to understand candidate imagery in contemporary America, we must seek better descriptions of the interpretive sources and processes of the "beautifying" that brings them together.

Notes

1. Cf. Joseph E. McGrath and Marion F. McGrath, "Effects of Partisanship on Perceptions of Political Figures," *Public Opinion Quarterly* 26 (Summer 1962): 236–48; Gene Wyckoff, *The Image Candidates: American Politics in the Age of Television* (New York: Macmillan, 1968).

2. Erving Goffman, *The Presentation of Self in Everyday Life* (New York: Doubleday, 1959), pp. 1–16.

3. Donald Horton and Richard Wohl, "Mass Communication and Para-Social Interaction," *Psychiatry* 19 (1956): 215–29.

4. See, for example, Winfred F. Hill, *Learning: A Survey of Psychological Interpretations* (San Francisco: Chandler, 1963).

5. However, studies of the impact of computer forecasts in election night broadcasts are a step in this direction, although they are limited to assessing the impact on voting intentions; cf., for example, Kurt Lang and Gladys Engel Lang, *Voting and Nonvoting: Implications of Broadcasting Returns before Polls are Closed* (Waltham, Mass.: Blaisdell, 1968).

6. See McGrath and McGrath, "Effects of Partisanship," pp. 237–38.

7. Ibid.

8. Cf. Fred E. Fiedler, *A Theory of Leadership Effectiveness* (New York: McGraw-Hill, 1967). Indeed, McGrath and McGrath, p. 241, found only weak evidence of assumed similarity operating in self-candidate image comparisons.

9. The reader will recall from Chapter 2 that the distinction between long-term electoral forces (party identification) and short-term electoral forces (candidate image projections and issue positions) derives from the SRC/CPS studies of American voting behavior; see especially Angus Campbell et al., *Elections and the Political Order* (New York: John Wiley, 1966), pp. 78–157. On the relative importance of candidate image projections in voting decisions uncovered in a secondary analysis of SRC/CPS data, see Richard W. Boyd, "Presidential Elections: An Explanation of Voting Defection," *American Political Science Review* 63 (June 1969): 498–514.

10. McGrath and McGrath, "Effects of Partisanship," pp. 241–43.

11. Ibid., pp. 246–47.

12. Roberta S. Sigel, "Effect of Partisanship on the Perception of Political Candidates," *Public Opinion Quarterly* 28 (Fall 1964): 484–85.

13. Sigel distinguishes between partisan determination and perceiver determination. She clearly recognizes that partisan determination of images flows from party identification, an attitude or subjective construct of the perceiver, but does not clearly accept this as a particular variety of perceiver determination.

14. The precise nature of the rankings is not clearly evident in "Effect of Partisanship on the Perception of Political Candidates," but see Sigel's own description of the ideal president data in "Image of the American Presidency—Part II of an Exploration into Popular Views of Presidential Power," *Midwest Journal of Political Science* 10 (1966): 123–37.

15. Sigel, "Effect of Partisanship on the Perception of Political Candidates," p. 493.

16. Ibid., p. 492.

17. And her use of the *rho* correlation coefficient is clearly inappropriate given the nature of the data.

18. Cf. Charles E. Osgood et al., *The Measurement of Meaning* (Urbana, Ill.: University of Illinois Press, 1957), pp. 104–24.

19. Colin A. Hughes and John S. Western, *The Prime Minister's Policy Speech: A Case Study in Televised Politics* (Canberra: Australian National University Press, 1966).

20. Jay G. Blumler and Denis McQuail, *Television in Politics: Its Uses and Influence* (Chicago: University of Chicago Press, 1969), pp. 224–47.

21. Charles N. Brownstein, "Communication Strategies and the Electoral Decision-Making Process: Some Results from Experimentation," *Experimental Study of Politics* 1 (July 1971): 37–50.

22. Cf. also Robert G. Lehnen, "Public Views of State Governors," *The American Governor in Behavioral Perspective,* Thad Beyle and J. Oliver Williams, eds. (New York: Harper & Row, 1972), pp. 266–69, in which he finds that general evaluations of incumbent governors are better explained by partisan-response effects in some instances and by stimulus-response effects in others.

23. David O. Sears and Richard E. Whitney, *Political Persuasion* (Morristown, N.J.: General Learning Press, 1973).

24. Blumler and McQuail, *Television in Politics*, p. 104, found that voters preferred more "challenging forms" of televised political broadcasts, a finding hardly consistent with such distortive processes as selective exposure and selective perception.

25. Brownstein, "Communications Strategies," p. 37. For evidence of an interactive process, see James W. Dyson and Frank P. Scioli, Jr., "Communication and Candidate Selection: Relationships of Information and Personal Characteristics to Vote Choice," *Social Science Quarterly* 55 (June 1974): 77–90.

26. Herbert Blumer, "Society as Symbolic Interaction," *Human Behavior and Social Processes,* Arnold M. Rose, ed. (New York: Houghton-Mifflin, 1962), pp. 179–92.

27. See Donald E. Stokes, "Some Dynamic Elements of Contests for the Presidency," *American Political Science Review* 60 (March 1966): 19–28.

28. Cf. Denis G. Sullivan, "Psychological Balance and Reactions to the Presidential Nominations in 1960," *The Electoral Process*, M. Kent Jennings and L. Harmon Zeigler, eds. (Englewood Cliffs, N.J.: Prentice-Hall, 1966), pp. 238–64. See also Stokes, "Some Dynamic Elements of Contests"; Philip E. Converse, "Religion and Politics: The 1960 Elections," in Campbell et al., *Elections and the Political Order*, pp. 96–124; and Roberta S. Sigel, "Race and Religion as Factors in the Kennedy Victory in Detroit, 1960," *Journal of Negro Education* 31 (1962): 436–47.

29. Indeed, this popular perception of Nixon as a devoutly religious man seems to have prevailed at least through the 1972 presidential campaign, if the responses of college students in the *Q*-sort study conducted by the authors can be generalized.

30. Again, however, some attitudes, particularly those having to do with reference group affiliations, may require a stimulus to make those attitudes campaign-relevant, as with the religious factor in the 1960 presidential campaign.

31. See Muzafer Sherif, *The Psychology of Social Norms* (New York: Harpers, 1936); and Allen L. Edwards, "Political Frames of Reference as a Factor Influencing Recognition," *Journal of Abnormal Psychology* 36 (1941): 34–50.

32. See, for example, Angus Campbell et al., *The American Voter* (New York: John Wiley, 1960).

33. Richard F. Carter, "Some Effects of the Debates," *The Great Debates: Background—Perspective—Effects,* Sidney Kraus, ed. (Bloomington, Ind.: Indiana University Press, 1962), pp. 257–62; Hans Sebald, "Limitations of Communication: Mechanisms of Image Maintenance in Form of Selective Perception, Selective Memory and Selective Distortion," *Journal of Communication* 12 (June 1962): 142–49; George Stricker, "The Use of the Semantic Differential to Predict Voting Behavior," *Journal of Social Psychol-*

ogy 59 (1963): 159–67; Bertram H. Raven and Philip S. Gallo, "The Effects of Nominating Conventions, Elections, and Reference Group Identification upon the Perception of Political Figures," *Human Relations* 18 (1965): 217–30; Ralph L. Rosnow, "Bias in Evaluating the Presidential Debates: A 'Splinter' Effect," *Journal of Social Psychology* 67 (December 1965): 211–19; and Lynda Lee Kaid and Robert O. Hirsch, "Selective Exposure and Candidate Image over Time," *Central States Speech Journal* 24 (Spring 1973): 50. Nor are these findings peculiar to the United States: cf. Philip E. Converse and George Depeux, "De Gaulle and Eisenhower: The Public Image of the Victorious General," *Elections and the Political Order*, pp. 323–44; Peter Regenstreif, *The Diefenbaker Interlude: Parties and Voting in Canada–An Interpretation* (Toronto: Longmans Canada, 1965), pp. 73–76; and David Butler and Donald Stokes, *Political Change in Britain: Forces Shaping Electoral Choice* (New York: St. Martin's Press, 1969), pp. 383–87.

34. Cf. Gilbert R. Winham and Robert B. Cunningham, "Party Leader Images in the 1968 Federal Election," *Canadian Journal of Political Science* 3 (March 1970): 37–55.

35. Carter, "Some Effects of the Debates," pp. 258–63.

36. Selective perception and selective distortion otherwise seem to be most likely to occur in regard to issue positions taken by candidates rather than regarding their personal characteristics or political capabilities.

37. Herbert H. Hyman and Paul B. Sheatsley, "The Political Appeal of President Eisenhower," *Public Opinion Quarterly* 19 (Winter 1955–56): 26–39.

38. Campbell et al., *The American Voter*.

39. Sigel, "Effect of Partisanship on the Perception of Political Candidates," p. 490; and Dan Nimmo and Robert L. Savage, "Political Images and Political Perceptions," *Experimental Study of Politics* 1 (July 1971): 23–24.

40. Nimmo and Savage, "Political Images and Political Perceptions," pp. 19–22. There was a tendency among Democrats to stress stylistic aspects in their perceptions of Danforth, however.

41. Representative studies include Donn Byrne et al., "Response to Political Candidates as a Function of Attitude Similarity-Dissimilarity," *Human Relations* 22 (1969): 251–62; Larry C. Kerpelman, "Personality and Attitude Correlates of Political Candidate Preference," *Journal of Social Psychology* 76 (1968): 219–26; and Lawrence S. Wrightsman, Jr., et al., "Authoritarian Attitudes and Presidential Voting Preferences," *Psychological Reports* 8 (February 1961): 43–46.

42. William H. Flanigan, *Political Behavior of the American Electorate*, 2nd ed. (Boston: Allyn and Bacon, 1972), p. 98.

43. This argument follows from Philip E. Converse, "The Nature of Belief Systems in Mass Publics," David E. Apter, ed., *Ideology and Discontent* (New York: The Free Press, 1964), pp. 206–61. There is considerable controversy among political scientists regarding the capacities of Americans to think in ideological ways, and even about how much interest citizens have in political issues. The flavor and quality of the debate can be judged by consulting John C. Pierce and Douglas D. Rose, "Nonattitudes and American Public Opinion: The Examination of a Thesis," *American Political Science Review* 68 (June 1974): 626–49, followed by a "Comment" by Philip E. Converse and a "Rejoinder" by Pierce and Rose.

44. See, for example, Kerpelman, "Personality and Attitude Correlates," who found a strong direct relationship between a measure of politicoeconomic conservatism and a preference for Goldwater in 1964. See also John Osgood Field and Ronald E. Anderson, "Ideology in the Public's Conception of the 1964 Election," *Public Opinion Quarterly* 33 (Fall 1969): 390–98.

45. Byrne et al., "Response to Political Candidates."

46. Joseph Trenaman and Denis McQuail, *Television and the Political Image: A Study of the Impact of Television on the 1959 General Election* (London: Methuen, 1961), pp. 158–64.

47. Jerrold G. Rusk and Herbert F. Weisberg, "Perceptions of Presidential Candidates: Implication for Electoral Change," *Midwest Journal of Political Science* 16 (August 1972): 388–410. See also Herbert F. Weisberg and Jerrold G. Rusk, "Dimensions of Candidate Evaluation," *American Political Science Review* 64 (December 1970): 1167–85; Osgood et al., *The Measurement of Meaning*, pp. 104–24; and Sidney Kraus and Raymond G. Smith, "Issues and Images," *The Great Debates*, pp. 289–312.

48. John S. Jackson, III, and Roy E. Miller, "Campaign Issues, Candidate Images, and Party Identification at Multiple Electoral Levels" (paper presented at the Annual Meeting of the Midwest Political Science Association, Chicago, May 1973).

49. See Dan Nimmo and Robert L. Savage, "The Amateur Democrat Revisited," *Polity* 5 (Winter 1972): 268–76.

50. A three-factor solution for each candidate sort was arbitrarily extracted for the purpose of this particular analysis.

51. The McGovern types exhibited a pattern similar to those positive images found in the 1972 *Q*-sort study of the 1972 presidential campaign, in that one type could be described as a "Mr. Nice Guy" and the second as the "great statesman." The McGovern image Type III was the polar opposite of the "great statesman" and therefore highly negative. The fourth McGovern image was that of a "maverick leader" and was probably a positive image, as members of the New Democratic Coalition were more likely to perceive him in this fashion.

52. Leonard L. Rosenbaum and Elliott McGinniss, "A Semantic Differential Analysis of Concepts Associated with the 1964 Presidential Election," *Journal of Social Psychology* 78 (August 1969): 227–35.

53. McGrath and McGrath, "Effects of Partisanship," pp. 240–41.

54. See F. W. Adorno et al., *The Authoritarian Personality* (New York: Harper, 1950).

55. Kerpelman, "Personality and Attitude Correlates," p. 224; see also Ohmer Milton, "Presidential Choice and Performance on a Scale of Authoritarianism," *American Psychologist* 7 (1952): 597–98; and Howard Leventhal et al., "Authoritarianism, Ideology, and Political Candidate Choice," *Journal of Abnormal and Social Psychology* 69 (1964): 539–49, where the relationship was found but in every case the respondents were college students. For studies using general population samples, see Fillmore H. Sanford, "Public Orientation to Roosevelt," *Public Opinion Quarterly* 15 (Summer 1951): 189–216; Morris Janowitz and Dwaine Marvick, "Authoritarianism and Political Behavior," *Public Opinion Quarterly* 17 (Summer 1953): 185–201; and I. H. Paul, "Impressions of Personality, Authoritarianism and the *fait accompli* Effect," *Journal of Abnormal and Social Psychology* 53 (1956): 338–44; and in each instance no consistent relationship between authoritarianism and candidate ideology was found.

56. Leventhal et al., "Authoritarianism," pp. 542–46.

57. Ohmer Milton and Birt Waite, "Presidential Preference and Traditional Family Values," *American Psychologist* 19 (1964): 844–45.

58. See Kerpelman, "Personality and Attitude Correlates."

59. See Wrightsman et al., "Authoritarian Attitudes and Presidential Voting Preferences."

60. Ibid.

61. Leventhal et al., "Authoritarianism"; see also the speculation regarding the tradition of liberal-left criticism among eminent American professors and the incorporation of that tradition into self-images leading to support for George McGovern and other liberal Democratic candidates, in Everett Carll Ladd, Jr., and Seymour Martin Lipset, *Academics, Politics, and the 1972 Election* (Washington: American Enterprise Institute for Public Policy Research, 1973), pp. 21–27, 77–99.

62. Cf. Sanford, "Public Orientation to Roosevelt," pp. 207–16.

63. See Muzafer Sherif and Carl I. Hovland, *Social Judgment: Assimilation and Contrast Effects in Communication and Attitude Change* (New Haven, Conn.: Yale University Press, 1961); and Carolyn W. Sherif et al., *Atittude and Attitude Change: The Social Judgment-Involvement Approach* (Philadelphia: W. B. Saunders, 1965).

64. See Al R. Weitzel, "In Search of a 'Theory' of Campaign Communication" (paper presented at the Annual Meeting of the International Communication Association, New Orleans, April 1974).

65. Otis Baskin, "On Predicting Voting Behavior from Attitude Measures" (paper presented at the Annual Meeting of the International Communication Association, New Orleans, April 1974).

66. Note, however, that Ladd and Lipset, *Academics, Politics, and the 1972 Election,* pp. 92–96, found considerable negativism in voting decisions of college faculty in 1972, whether they voted for Nixon (35 percent) or McGovern (31 percent).

67. See Blumler and McQuail, *Television in Politics,* p. 248, on age and sex differences, for example.

68. Ibid., pp. 112–13.

69. Cf. Converse and Depeux, "Images of the Victorious General," pp. 316–17.

70. Thomas F. Pettigrew et al., "George Wallace's Constituents," *Psychology Today* 92 (February 1972): pp. 47–49.

71. Irving Roswalb and Leonard Resnicoff, "The Impact of Endorsements and Published Polls on the 1970 New York Senatorial Election," *Public Opinion Quarterly* 35 (Fall 1971): 410–14.

72. Robert D. McClure and Thomas E. Patterson, "Television News and Voter Behavior in the 1972 Presidential Election" (paper presented at the Annual Meeting of the American Political Science Association, New Orleans, 1973).

73. Thomas A. McCain and Paul Rowand, "The Effect of Camera Treatment on Political Speakers' Credibility: Network Television Coverage of the Speeches of Ted Kennedy and George McGovern to the 1972 Democratic National Convention" (paper presented at the Annual Meeting of the International Communication Association, Montreal, April 1973).

74. Lang and Lang, *Voting and Nonvoting*; and Sam Tuchman and Thomas E. Coffin, "The Influence of Election Night Television Broadcasts in a Close Election," *Public Opinion Quarterly* 35 (Fall 1971): 315–26.

75. Charles K. Atkin et al., "Quality versus Quantity in Televised Political Ads," *Public Opinion Quarterly* 37 (Summer 1973): 209–24; cf. also Dotty Robyn and Lynda Lee Kaid, "The Role of Political Advertising in a State Senate Campaign" (paper presented at the Summer Conference of the International Communication Association, Athens, Ohio, August 1973); and Kenneth G. Sheinkopf et al., "The Functions of Political Advertising for Campaign Organizations," *Journal of Marketing Research* 9 (November 1972): 401–5.

76. Ithiel de Sola Pool, "TV: A New Dimension in Politics," *American Voting Behavior,* Eugene Burdick and Arthur J. Brodbeck, eds. (New York: The

Free Press, 1959), pp. 236–61; see also Blumler and McQuail, *Television in Politics*, pp. 251–58.

77. See also Sidney Kraus, ed., *The Great Debates: Background–Perspective–Effects* (Bloomington, Ind.: Indiana University Press, 1962), for an excellent compilation of studies assessing the impact of television and a particular format on candidate images.

78. Cf. David L. Swanson, "Political Information, Influence, and Judgment in the 1972 Presidential Campaign," *Quarterly Journal of Speech* 59 (April 1973): 141; and Ladd and Lipset, *Academics, Politics, and the 1972 Election*, pp. 95–96.

79. Doris A. Graber, "Personal Qualities in Presidential Images: The Contribution of the Press," *Midwest Journal of Political Science* 16 (February 1972): 71–72; and Pool, "TV: A New Dimension in Politics," pp. 246–47.

80. See Gustav Jahoda, "Political Attitudes and Judgments of Other People," *Journal of Abnormal and Social Psychology* 49 (1954): 330–34; and Ira S. Rohter, "An Attribution Theory Approach to the Formation of Candidate Images: Part I. Theory and Research Paradigm" (paper presented at the Annual Meeting of the American Political Science Association, Chicago, September 1971).

81. In several cases the cumulative R^2 for the full model indicates that less than 25 percent of the variance is accounted for, and this is not reported here. These candidate image types are Nixon Types III and IV and McGovern Type IV. That these image types have a very idiosyncratic basis does not mean they are unimportant; they simply cannot be accounted for at any acceptable level using regression procedures.

82. Herbert Blumer, "Sociological Analysis and the 'Variable,'" *American Sociological Review* 21 (December 1956): 683–90.

Chapter Five

The "Old" and "New" Politics:
Campaign Styles and Candidate Images

During the presidential primary campaigns of 1968, various political journalists sensed that the mood and movement on the campaign trail differed from that of previous elections. Among the followers of Robert Kennedy and Eugene McCarthy the difference in mood seemed particularly striking. Four years later reporters would note the same difference among the workers of George McGovern. In these elections of the late 1960s and early 1970s, this distinctive approach to campaigning acquired the label of the "New Politics." What was the New Politics? A team of reporters from Great Britain covering the 1968 presidential campaign tried to find out. Following a lengthy interview with a speech writer of the late Robert Kennedy, they wrote: "He talked about a new style, a new compassion and commitment, a new constituency of the young, the black, the poor, and some middle-class people." Moreover, the New Politics implied novel ways of campaigning: "To many of the bright young men around Kennedy, there seemed to be no contradiction between a New Politics of commitment to radical social change and something quite different: a new sophistication in the use of modern political technology."[1]

It is to the New Politics and its alleged predecessor, the Old Politics, that we turn in this chapter. Specifically, we will examine the distinction between the two as campaign styles, their relation to candidate images, whether voters respond differently to such styles and possibly thereby to different images of political candidates associated with them. We use as our

109

example evidence from a study of the New and Old Politics in a selected congressional district.

Contemporary Campaign Styles and Candidate Images

One thing an election campaign does is to bring candidates and prospective voters into communication with one another. Several factors are involved when people communicate; one of these is style. *Style* refers to the distinctive way a communicator articulates the content of what he has to say in order to relate to his audience.[2] The overall campaign style of a candidate consists not only of how he arranges the words, sounds, or pictures of particular messages but also of the way the candidate (and his managers) defines the constituency, organizes the campaign, uses particular media to reach audiences, and emphasizes selected issues and elements of his personality.

Thus characterized a candidate's campaign style—that is, the distinctive way he goes about getting his message across to voters—is obviously crucial to his images. Our reasoning is as follows. A candidate's image (as defined in Chapter 1) consists largely of the subjective appraisals (images) voters have of him—their knowledge, feelings, and tendencies respecting him. According to Kenneth Boulding, messages influence those images to the point that, *"The meaning of a message is the change which it produces in the image."*[3] Therefore, to the extent that messages change images, style in articulating the content of messages helps determine the candidate's image with voters. We have already recognized (see Chapter 3) the importance of style in candidate imagery by giving it parallel status with distinctly political attributes of a candidate (such as his traits as a public official and a politician) in constituting the office seeker's *political* and *stylistic roles*.

Although candidates usually try to do things in distinct ways that set them apart from their rivals, general styles common to all candidates characterize certain eras. The Old Politics, much maligned by latter-day practitioners of the New Politics, was a dominant style through the early decades of this century. It was a means of contacting voters through the party organization, or "Machine." Tightly knit local party organizations under the control of political bosses (such as Tom Pendergast of Kansas City, Carmine DeSapio of New York City, Ed Crump of Memphis, Richard J. Daley of Chicago) or even loose coalitions of political hacks recruited candidates, mobilized voters, and shaped policies. Critics labelled it the politics of "influence peddling"—taking care of friends and neighbors and returning loyal, albeit often corrupt, officeholders to positions of public trust. As we discussed in Chapter 1 and as historian Richard Jensen pointed out, the style employed militarist metaphors: loyal partisan troops drilled by precinct captains comprised a phalanx of dedicated cadres, the key to winning campaigns.[4]

The political style of the last decade, known as the New Politics, has meant different things to different people. There were at least two visions of the New Politics of the 1960s: "When Democrats spoke of the New Politics, they meant the romantic insurgency of 1968 or the hope-filled ideologies of

1970-1972. But when Republicans spoke about the New Politics, they meant the specific technology which . . . was being perfected and mastered."[5] The New Politics of *technique,* according to Jensen, emerged around the presidential election of 1916. Again, as discussed in the first chapter, mercantilist metaphors replaced militarist language in describing campaigns. Thus campaigners organized elections as exercises in mass-merchandising that stressed "selling" candidates through commercial advertising. By the 1960s the technical side of the New Politics had become a major industry. Instead of looking to the "pols" (local and state party leaders experienced in loyal service to partisan causes, win or lose) for assistance in organizing campaigns, an increasing number of candidates simply hired professional campaigners. There were high-priced specialists in numerous facets of campaign technology (managing, polling, direct mail, computer technology, television, vote turnout, advertising, publicity) who used mass media as a means of placing candidates in direct communication with voters.[6] As the end of the decade approached, the names of many of these professional mercenaries (manager Joseph Napolitan, pollster Fred Currier, television producer Charles Guggenheim, and others) became as well known as the political bosses of yesteryear. And like those old pols, many of the new breed were rewarded for their contributions to victory not only with money but also with appointments to governmental office (as, for example, the appointment of Vincent Barabba, former head of a professional polling and data-gathering firm, as director of the Bureau of the Census in the Nixon administration).

The other vision of the New Politics was not limited to techniques. The new style was one of political *mood.* Here, "The essence of the New Politics was contained in the lessons learned in the peace movement and the earlier Civil Rights movements: that the old rules could be circumvented by action, mobility, drama, involvement, and confrontation; by learning all the ways in which determined People could wage guerrilla war against the Machine."[7] This conception of the New Politics owed much to the political style of the New Generation, the millions of young American who had reached voting age in the 1960s, especially those who became politically active and involved.[8] Although critical of the Old Politics of the "Machine," this new generation of political activists borrowed one important facet from it, namely, an emphasis on campaigning through personal contacts with voters. A hallmark of the candidacies of Eugene McCarthy, Robert Kennedy, George McGovern, and countless others seeking lesser offices in the New Politics style was the grass-roots organization of eager young volunteers knocking on doors from New Hampshire to California, engaging prospective voters in conversation about political issues and candidates. This vision of the New Politics stressed spontaneity, of having candidates say what they believed no matter what the benefit or cost in votes, rather than packaging their appeal in rehearsed, self-serving pronouncements.

All this, however, did not ignore the technical side of the New Politics. Instead of relying on paid television advertising, the emphasis was on such ploys as saying and doing things in highly newsworthy ways to assure maximum publicity in nightly televised news programs (thus, John Lindsay walked through the ghettos of New York City in his shirt sleeves to be "with the people"). And George McGovern, the 1972 practitioner of the New Politics, used sophisticated techniques of polling by retaining pollster Pat Caddell. In sum, there was no contradiction between the mood and technique of the New Politics.

These, then, are the so-called old and new political styles of contemporary election campaigns. The old emphasizes candidate-voter communication through party organization; the new stresses personal contact, either directly through the mass media or indirectly through highly involved volunteers serving the candidate and his ideas rather than the party organization. Do voters respond differently to such contrasting styles and, if so, do the styles contribute to differing images of political candidates? Although we cannot answer these questions conclusively (nor even as directly as we would like), we turn to a report of a research effort that provides some useful insights into voters' responses to the Old and New Politics.

Contrasting the Old and the New: An Exploration of Popular Responses

During 1971 there was an opportunity to explore how American voters react to different kinds of stimuli associated with the old and new campaign styles. It consisted of a two-stage research project. The first was an effort to fashion an instrument to measure the contrasting styles; the second was an application of that device through a sample survey of the qualified electorate in a single congressional district.[9]

Personifying the Old and New Politics

Notions such as the Old and New Politics are what social scientists refer to as *ideal types*—that is, each is a concept composed of a configuration of elements "based on observations of concrete instances of the pheonomena under study, but the resultant construct is not designed to correspond exactly to any single observation."[10] Ideal types are shorthand constructs denoting a pattern common to several discrete instances, yet not repeated in precisely the same fashion in any single case. One problem in exploring ideal types is to give them operational meaning; that is, to come up with a set of simple, observable procedures that, when people perform them, result consistently in the pattern of behavior allegedly common to the ideal. In the case of the Old and New Politics this means characterizing each style in such a way as to measure how people respond to it.

One way to obtain a characterization is to examine an election that clearly displays each style. One candidate, for example, would rely primarily on such things as the party organization, pols, traditional appeals, and perhaps his incumbency to articulate his campaign message in the "old" style; the other would utilize the mass media and grass-roots volunteers, profess commitments to clearly stated ideals, and challenge the status quo in the manner of the "new." Surveys of the perceptions and images of the contending candidates would be a clue to popular responses to the two styles. Although elections displaying both styles occur, it is frequently difficult to obtain the opportunity (in time and finances) to study them. An alternative approach that does not rely on the coincidence of actual elections and availability of research resources is to create an artificial, hypothetical election and to ask people to respond to it as though it were authentic. Dean Jaros and Gene Mason, for instance, employed this technique with considerable success in examining how residents in a Kentucky metropolitan area responded to demagogic appeals.

Jaros and Mason provided each respondent with printed information about pairs of opposing candidates in each of four hypothetical elections, eight candidates in all. In one election the fictitious candidates differed in personal characteristics (for example, one was described as "attractive" and the other as "unattractive"); a second election pitted a demagogue against a candidate simply described as disagreeing with him; the third election was between candidates with opposing issue positions; and the fourth contest matched candidates of opposed political parties. Using data generated by asking respondents how they would vote in these elections, Jaros and Mason concluded, in part, that the presence of party symbols associated with a candidate decreases the likelihood of voters' responding to demagogic appeals.[11]

A Hypothetical Election

In exploring voters' responses to the old and new campaign styles, we found the technique of the hypothetical election especially useful. The election in question pitted two fictional candidates in a contest for the United States House of Representatives. Although in the research project we gave each fictitious candidate an equally fictitious name, we shall refer to them here as the "Old" and the "New." In creating the fictitious candidates, we paired the Old and New candidates on a series of traits (pertaining to backgrounds, issue positions, campaign techniques, credibility, partisanship, and ideologies) representing the Old and New Politics. The traits also represented the two principal dimensions associated with candidate images in previous research— political and stylistic roles (see Chapter 3). References to the candidate's experience in public life and office, his reputation as a representative of a political party, and his stands on various issues reflected the political role; references to his qualities as a person and abilities as a performer, especially in using the mass media, indicated the stylistic role.

We can get some idea of the role and style contrasts of the two hypothetical candidates through the following sketches. The "Old" candidate, a four-term incumbent, had previously been a member of the state legislature. He was fifty-six years of age, educated in public schools, and had obtained a law degree. As a life-long resident of the community, he was married with one child who, by the time of the election, was also married. The candidate was active in Kiwanis and Rotary clubs, a Methodist and Shriner, former American Legion post commander, and an avid participant in all major civic affairs, charitable events, and religious activities. His seniority in Congress had given him influence on a key congressional committee; in fact, he was next in line for the chairmanship and did not aspire to any other office. Hence, his slogan was "Keep_____ on the Job for you in Congress." On issues the incumbent could be characterized as neither conservative nor liberal, although he labelled himself "a moderate liberal." His stands were pro-labor and favorable to the aged; he emphasized "equitable taxes" but not across-the-board tax cuts; he opposed legalized abortion, restrictive antipollution legislation that might hurt prosperity, and cuts in the defense budget.

Our candidate of the Old Politics had the support of the AFL-CIO, NAACP, Americans for Democratic Action, the major daily newspaper, the senior U.S. senator, the majority of members of the state legislature in his district, and all county chairmen of his political party. He would not disclose sources of his campaign funds, employed a nephew as his administrative assistant, and was a member of a law firm which his opponent charged "received benefits from government business." The congressman's campaign stressed his experience and used primarily major party organizations and large-scale rallies, although employing television and radio spots along with some direct mail. The "Old" candidate was a Democrat who believed a congressman owes loyalty "to the head of your Party, the President."

The "New" challenger had never before sought public office. He was thirty-seven years of age, a member of a wealthy family, had attended private schools, and held a graduate degree in business administration. He had moved to the community only eight years before, when he became general manager of the corporate relations division of a major company located there. He was married (his wife held an M.A. degree in child psychology and had served as a social worker) with three children—one in public school, one in private school, and one an infant still at home. He had been married briefly before, but divorced. The challenger ran on the slogan, "Person-to-Person Representation in Congress," arguing that his lack of experience was an advantage, for it would provide the opportunity for a "new approach." He was a Roman Catholic, active in the United Fund, Arts and Education Fund, Boy Scout Council, and on the board of directors of a college aid fund for underprivileged black students. If elected, he did not promise to remain in Congress, but said instead that if he could better serve the people's interests in the Senate or state legislature, he would eventually try to do so.

The younger candidate favored cuts in governmental spending and taxes, liberalized abortion laws, a disentangling of foreign commitments, even if this meant defense cuts and stricter legislation in the interests of ecology. Such stands, he said, made him a "moderate conservative." His endorsements came from weekly newspapers, a local television station, a printer's union, Americans for Constitutional Action, and a peace group. At the time of his election he was not employed, because he had resigned from his company when it refused to promote a qualified woman to an executive position (yet he had no woman in a key position in his campaign). The "New" candidate spoke often of the "challenging excitement of grass-roots politics," and had a large organization of volunteers at the lower levels of his campaign structure. His principal advisers were paid professional campaign consultants, and he made extensive use of electronic and direct-mail means of reaching the voters. He disclosed sources of his campaign funds; one such disclosure revealed a $25,000 contribution from his parents. The challenger ran as a Republican but urged, "I will vote on whether an issue is right, not simply because it's the position of the head of my party, whether he be the President, the Governor, or one of our Senators."

The contrasting traits of the two candidates were represented by twenty-eight pairs of items in a survey administered to selected respondents. For each pair, respondents examined the "Old" and "New" office seekers' characteristics and, following each pairing, stated their preferences on a five-point scale: (1) "I definitely would vote for Candidate A"; (2) "I probably would vote for Candidate A"; (3) "I am undecided and favor neither candidate"; (4) "I probably would vote for Candidate B"; or (5) "I definitely would vote for Candidate B." (The candidates were alternately identified as Candidate A or B.) The presentation of the contrasting characteristics differed substantially from procedures used in conventional surveys. The characteristics appeared in a format consistent with that which a voter normally encounters in an election campaign. For example, two of the pairs presented to the respondent consisted of photographs of the candidates (posed by models). One set of photographs was in black and white and the other in color. The photographs depicted two physically attractive males. The older appeared with white hair, in suit and vest, making a point in animated, but friendly, conversation. The challenger, clearly a younger man, had an open collar, loosened tie, rolled-up sleeves, and carried his jacket slung casually over his shoulder (a typical pose for many New Politics candidates of the 1960s). As another example, in presenting contrasting stands on issues for the two fictional candidates, interviewers gave respondents statements by the candidates in graphic visual and/or audial form. For instance, one survey item paired two newspaper clippings, another a sample of the candidates' direct-mail letters, yet another contrasted scripts and still photographs for television spot commercials, and a fourth consisted of photographs of the candidates' billboards.

It is impossible to portray the full content of each of these survey pairs, but an example will give some of the flavor of the effort. Pair 6 consisted of a statement dealing with the political ambitions and aspirations of each candidate. Respondents saw a card with the following quotation from the candidate of the Old Politics: "You have a right to expect your Congressman to build seniority and influence in Congress. This is the way you're protected and your best interest served. As long as I continue to enjoy your support, I pledge to continue to build such a record and, hopefully, eventually to be Chairman of my Committee." The other card for the pair was a statement by the challenger: "I am asking to serve as your Congressman now. But I will not pledge to stay forever in Congress. If your interests can best be served by my seeking another office at some future date, I will respond to the call of duty, whether in the Senate, the State House, or in some other aspect of our Government."

In addition to having respondents rate each of the twenty-eight pairs of traits on a five-point scale according to which candidate they would vote for in an election, the respondents were asked to perform one other procedure. After all trait pairs had been rated, interviewers handed respondents a well-shuffled deck of three-by-five-inch cards. Each card depicted one of the fifty-six total traits of the two fictional candidates, previously presented to the respondent. Following the procedures of Q-methodology (see Appendix A), each respondent arranged the fifty-six traits in a forced-sort distribution (ranging from +3 to –3 across seven piles). The instructions were as follows:

> You have now seen several pieces of information about two candidates for U.S. congressman. I am now going to hand you a deck of cards and on each card you will find information displayed identical to that you have already seen. What I would like you to do now is forget Candidate A and Candidate B. Instead, I would like you to think of the features you want most in your "ideal U.S. congressman."

> Please take these cards and sort them into seven piles. The traits you think best fit your idea of the ideal congressman should be placed in the first pile; the traits you think least fit your idea of the ideal should go in the last pile. In between these two piles you should sort the cards so that they range from what you consider to be the most desirable to the most undesirable features of the ideal congressman. Those that you have no opinion about will thus go in the middle piles.

> Feel free to change your mind at any time but, when you finish sorting, you should have as many cards in each pile as indicated in this diagram.

The diagram provided for a distribution into seven piles, of three, seven, eleven, fourteen, eleven, seven, and three statements each, with the three statements at each end being "most" and "least" desirable. Finally, in addition to ratings of the twenty-eight pairs of traits and a Q-sorting of fifty-six traits, interviewers obtained data on sociodemographic characteristics, partisanship, and so forth.

The first stage of this research aimed at developing and testing the procedure for defining the Old and New Politics. At this stage there was no desire to describe the distribution of responses in the population (the purpose of the second stage). Hence, the first sample of respondents was not a cross-section of qualified voters but a factorial, or balanced, purposive sample—that is, a sample selected to assure representation of persons in specific social and political categories deemed relevant to the research.[12] The sample design appears in Table 5.1. Each of the three sociopolitical categories (sex, age, and party) has two and three variants, respectively, and our sample represented combinations of these variants in approximately equal numbers. Thus, for

Table 5.1   Factorial Design for Selection of Pilot Respondents

SOCIOPOLITICAL CATEGORIES	VARIANTS		
Partisan Self-Image	(a) Republican	(b) Democrat	(c) Independent
Age	(d) Under 30	(e) Over 30	
Sex	(f) Male	(g) Female	

COMBINATIONS					
a-d-f	b-d-f	c-d-f	a-d-g	b-d-g	c-d-g
a-e-f	b-e-f	c-e-f	a-e-g	b-e-g	c-e-g

instance, one combination is a male Republican over thirty (the a-e-f combination); another is a male Republican under thirty (a-d-f).

Given the design in Table 5.1, there are one dozen possible combinations. For purposes of testing the instrument we obtained eight respondents for each combination (making 96 respondents) and added one combination for each involving Independents (thus producing a sample of $n = 100$). Following these procedures the sample of 100 respondents consisted of 50 males and 50 females, 50 persons under thirty-nine years old and 50 over that age, and 32 Democrats, 32 Republicans, and 36 Independents.

Interviewing took place in three metropolitan areas of a midwestern congressional district. In 1970 that district had a population of 480,000, 93 percent white. In 1970 a Democrat had been elected to Congress by defeating the incumbent, 53–47 percent. Pilot interviews took place in the late spring and early summer of 1971.

The analysis of data derived from the interview of these 100 respondents had two aims: first, to examine the results of respondents' ratings in the hypothetical contest between the two candidates and to determine what, if any, *sets of traits* define the Old and New Politics; second, to examine the results of the 100 *Q*-sorts of all traits to determine what *types of voters* orient themselves to the old and new in characterizing their ideal congressman. Since there were a number of diverse traits representing contrasting campaign styles as well as the political and stylistic roles of candidate images in this survey, we employed factor-analytic techniques (see Appendix A) to reduce a large number of variables (in this case, either the twenty-eight pairs of traits or the 100 respondents) into a smaller number of underlying major components having common characteristics.

Table 5.2 presents the results of a factor analysis of the twenty-eight pairs of traits presented to the balanced sample of respondents.[13] A brief description of each item-pair appears in the left-hand column. The five columns headed "Factors" contain the factor loadings of each item-pair on each of five factors. In simplest terms, such factor loadings represent the correlations between an item-pair and particular factor; loadings can range from

Table 5.2 Principal Components of Item-Pairs Comparing Candidates of the Old and New Politics

ITEM-PAIR	I	II	III	IV	V	COMMUNALITIES
19. Candidates' statements on what should be nation's goals.	72	13	33	20	06	68
21. Groups endorsing each candidate.	69	41	-05	18	17	71
12. Sample television spots for each candidate using still photos and scripts.	67	24	18	49	-01	78
9. Color photos of candidates' billboards.	66	24	26	35	11	70
11. Sample pieces of candidates' direct mail.	66	26	32	25	21	70
14. Color photos of each candidate.	62	22	56	25	11	82
18. Candidates' positions on taxes.	62	39	23	-01	02	59
17. Candidates' stands on ecology legislation.	61	32	36	09	14	64
15. Candidates' stands on economic issues (cost of living, employment, free enterprise).	59	33	37	09	16	63
10. Sample newspaper ad for each candidate.	56	09	42	46	-04	70
20. Candidates' sources of campaign funds.	55	43	38	34	-13	75
28. Ideological self-identification of each candidate.	35	81	-01	16	00	81
26. Political party of each candidate.	19	79	14	28	00	76
27. Candidates' statement on whether parties will influence their congressional votes.	23	64	31	10	19	61
23. Statements regarding candidates' credibility.	39	57	55	03	16	80
22. Newspaper endorsements of candidates.	35	56	42	32	04	71
24. Candidates' positions on women's rights and voting rights for 18-year-olds.	18	56	55	13	17	73
25. Newspaper stories of irregularities in each candidate's campaign.	25	55	50	11	28	70
3. Sketches of candidate's family, community, religious, and moral background.	31	39	36	34	05	50
13. Sample radio commercials played through cassettes.	38	19	68	28	06	73
5. Candidates' views on how to act as congressman.	43	11	68	18	25	75
6. Candidates' political aspirations and ambitions.	10	15	63	50	-02	68
16. Candidates' stands on foreign policy questions.	45	33	55	09	-21	67
8. Candidates' campaign slogans.	47	11	52	40	-04	66
2. Biographical and personal background sketches.	13	18	07	78	21	71
4. Candidates' political experience to date.	28	16	29	68	08	67
7. Candidates' pol/vol organizations.	21	36	40	42	03	51
1. Black-and-white photos of candidates.	16	15	09	17	88	87

Note: Decimals have been omitted from factor loadings.

+1.00 to -1.0. In interpreting Table 5.2 (decimals are omitted in the table), we look at all of the item-pairs having their highest loadings on a given factor. These item-pairs define the nature of that factor.

Consider Factor I. Eleven of the twenty-eight item-pairs have their highest loadings on this first factor. These pairs cover such items as color photographs of the candidates and of their billboards, newspaper advertisements, samples of direct mail, and issue positions. What these item-pairs have in common is that, excepting the color photographs, all employ a mass-media format (television, newspaper advertisment or story clipping, computerized letters, billboards, or brochures). And the color photographs (Pair 14) were the same as presented on direct mailers and billboards. We suspect, therefore, that this factor represents a multimedia dimension. Certainly a key component underlying the comparison of the Old and New Politics in the test instrument is a *media* dimension. It conforms closely to what we described in Chapter 3 as the dramatic performance traits associated with a candidate's stylistic role.

The second factor has eight item-pairs with highest loadings. The underlying themes seem to be those of the candidate's partisanship, party loyalty, group endorsements, ideological labelling, credibility, and personal background. (The weakest pair is Pair 3, with a loading of .39, which contributes relatively little to defining the factor.) Essentially, then, this factor contains item-pairs linked to the candidates' associations with certain groups and causes—political parties, special interests, and ideological referents. Thus we will refer to this component as the *partisan* aspect of the old and new political styles. It parallels the partisan politician traits of the candidate's political role, which we described in Chapter 3.

Factor III consists of five item-pairs. Running through this cluster of traits is a theme pitting the competing candidates' views of what a congressman is and should do. We have references to the "Old" candidate's experience, seniority, commitment to remain in Congress, and slogan about keeping "on the Job" contrasted with the challenger's desire for a "new approach," refusal to pledge remaining in Congress if elected, and assurance of person-to-person contact with constituents as a congressman. There is also an item on what Congress should do about the environment. We consider this factor as representing an *office* component in contrasting styles and view it as close to the public official element of the candidate's political role. A fourth factor has only three item-pairs, and all refer to personal background and participation in civic, social, and political groups. This is a *background* dimension comparable to the set of personal traits that add to a candidate's stylistic role. The remaining factor consists only of the black-and-white photographs of the candidates and is apparently of little consequence.

Four major components—media, partisan, office, and background factors—emerge from this form of factor analysis in which respondents compared and contrasted the Old and New Politics. Two of the traits (office and partisan) correspond to the public official and partisan politician sets of traits that comprise a candidate's political role; the remaining two (background and media) parallel the personal and performance qualities making up a contender's stylistic role. Another way to think of these factors is to consider them as separate *scales*, or sets of highly related items on a test, which assist a researcher in measuring perceptions of some phenomenon. As we shall see shortly, these four scales provide us with devices for measuring popular responses to the old and new in political campaigns. Before turning to that, however, let us consider the results of our factor analysis of the Q-sorts of the fifty-six traits submitted to respondents.

Results of the Q-Sorts

As described in Appendix A, Q-studies give us an insight into the clusters of persons who *share* similar perceptions of political phenomina, rather than the clusters of test items along which persons respond according to individual *differences*. A Q-factoring of the respondents' sorts provided five factors, or clusters, of persons.[14] By examining the types of traits people in each cluster ranked as most and least desirable in their ideal congressman, it is possible to typify these factors.

The cluster with the largest number of respondents (thirty-six) places heavy emphasis on issues. However, people in this group did not distinguish between the issue positions of the two candidates but instead ranked the positions of both candidates as things they wanted in their ideal. Apparently this set of voters simply believed a congressman should talk about economics, foreign policies, ecology, taxes, and general hopes for America regardless of what position he takes. They seemed to be saying, "We may or may not agree with you, but what we want most is that you talk about current problems and proposed solutions." These comprise the *issue oriented* voters. A second factor consists of twenty-nine persons who find most desirable the distinctive political style of the New Politics, especially as it centers on moderation in liberalism, independence of party dictates, and change. These are *New Politics* voters. A third cluster of persons closely related to the New Politics group (with seven respondents) emphasized the traits of the Republican citizen-politician—the youthful, affluent, amateur Republican seeking office. We call these the *citizens*. Thirteen of the respondents clustered on a factor reflecting the Old Politics. In their ideal they want experience, a man who studies the issues before voting, favors a strong defensive posture, economic stability, campaigns in the older fashion, has the support of respected community organizations, and is an old-line, loyal, party Democrat. They are *Old Politics* voters. A fifth cluster (with fourteen persons) consists of people who emphasized the media appeal of candidates, regardless of whether it be the old or new approach. They emphasized colorful campaign techniques—photographs, billboards, television, and attractive stimuli. This is a group responding in a visual way to all candidates, the *visual* voters.

Thus, of the five clusters, or types, of voters uncovered by this mode of analysis we find one that seems to respond to the Old Politics, a second responding to the New, and another (the citizen-politician) closely associated with the New. Of course, given the nature of the sample of respondents there is no way of knowing how such voter types might be distributed among the qualified voting population of the congressional district, but there is some indication of differing responses to differing styles. Although there were no partisan, age, or sex bases to any of the voter types, we cannot say from this balanced sample whether or not there might be such differences in the larger populations. Hence, to determine what generalized differences in response there might be to the Old and New Politics, it is necessary to turn to the second stage of the research project.

The Distribution of Popular Responses

A second survey of voters provided an opportunity to measure popular perceptions of the Old and New Politics, using a basic instrument derived from the earlier pilot project. This effort differed from the first in two significant respects. First, the sample of respondents was selected according to probability rather than balanced criteria. One-third of the voting precincts in each of the three metropolitan areas of the same congressional district used in the first stage of the project were randomly selected as sampling points; within those precincts, interviewers used systematic procedures for selecting respondents. The number of interviews in each of the metropolitan areas was in proportion to that area's share of the total adult population in all three areas. Total sample size was 400. The sample yielded a male-female ratio of 44–56, precisely that indicated by census figures for the three metropolitan areas, and a white-nonwhite ratio of 87–13, compared to the 86–14 ratio indicated by census data. The partisan breakdown of the sample was 45 percent Republican, 40 percent Democrat, and 15 percent Independent.

Second, although the basic instrument for determining how respondents rate the two fictional candidates on twenty-eight pairs of traits remained the same as in the pilot project, there were no Q-sorts in this second survey. However, a technique derived from the results of the analysis of the earlier Q-sorts provided a measure of the distribution of voter types in response to the Old and New Politics. Following a procedure developed by Stephenson, we selected each of the six traits that distinctly defined each of the issue, New Politics, Old Politics, citizen, and visual voter types. Thus, thirty of the fifty-six traits originally used in the pilot project remained. From this pool we identified the most distinctive trait for each type, providing another set of five traits. This procedure continued until the thirty traits had been divided into six sets of five traits, each set having one distinctive trait representing one of our five specific voter types. Then, during the interview respondents were asked to rank order the traits in each set according to what characteristics they found most desirable in their ideal congressman. In sum, Stephenson's procedure is a means of modifying the Q-sorting technique by having respondents rank order a small number of characteristics defining derived factors.[15] A tabulation of frequencies of item rankings makes it possible to determine the distribution of voter types in the congressional district.

What were the results? How did sampled eligible voters respond to the Old and New Politics in this hypothetical election? Let us turn first to how voters rated each candidate on the twenty-eight item pairs. To determine this we employ a technique used successfully by the American Institute of Political Communication in its study of candidate images in the 1968 presidential election (see Chapters 2 and 3).[16] Recall in that study researchers used a five-point scale for rating the candidates on various traits. The voters' *mean score* for each trait provided a way of measuring responses to selected traits. By the same token, the mean score of sampled voters using the five-point scale in this survey indicates their response to the old and new political modes. Figure 5.1 graphs the mean score on each item-pair. A mean score less than 3 indicates a

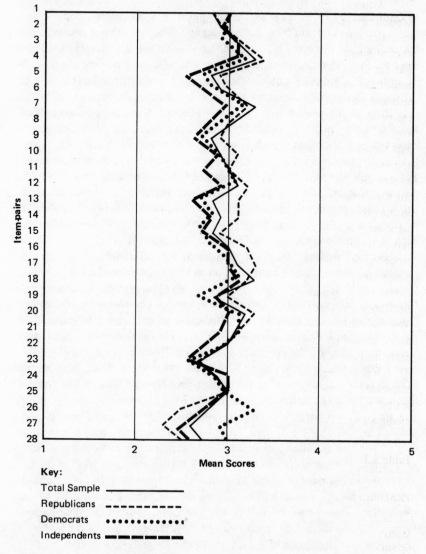

Figure 5.1   Item-Pair Mean Scores for Sample, Republicans, Democrats, and Independents

favorable response to the new political candidacy, and a mean score above 3 favors the Old Politics. The mean score across all twenty-eight item-pairs was 2.8, a response in the direction of the New Politics, but hardly clear-cut. What Figure 5.1 indicates is that the 400 respondents were largely undecided between the two styles on most items. Two of the more noteworthy exceptions are Pairs 4 and 27. In the first instance, voters indicated a preference for the background and experience of an incumbent; in the second, they approved the challenger's stand, that in voting in Congress he would be independent of dictates from the head of his political party.

Although the differences in mean scores reported in Table 5.3 among Republicans, Democrats, and Independents are not striking (and there were no major differences for various sociodemographic groups), certain patterns warrant mention. For instance, the Republican profile of mean scores parallels that for the entire sample fairly closely until we reach Pair 26, the designation of the candidates' political parties. At that point, Republican indecision ends and party members rally around their party's standard bearer, apparently irrespective of the fact that in many previous item-pairs they had leaned toward the more conventional political approach of his opponent. Republicans then reach their most decisive position (mean score of 2.3) on the next item-pair, in which their candidate proclaims he will not merely vote as party leaders dictate. Thus, partisanship was an anchor for Republican responses, but a partisanship blended with independence from party leaders. Democrats, on the other hand, lean toward the New Politics throughout most of the item-pairs, especially on such matters as statements about the congressman's role, person-to-person representation, and the candidates' views of the nation's proper goals. But on learning that the incumbent is a Democrat, party allegiance takes over momentarily. Then, independence from party dictates has an appeal and Democrats finally end the sequence of choices in the hypothetical election undecided. Independents make the strongest response favoring the New Politics, especially on items concerning the congressman's role, candidates' credibility, and the challenger's declaration that he will not follow the party lead in voting.

We get another perspective on the ratings by examining how voters responded to each of the fictional candidates on the four sets of traits, or scales, derived from the pilot project—office, partisan, background, and media scales. Table 5.3 provides assistance here also. The general indecisiveness

Table 5.3 Mean Scores by Trait Sets (n = 400 Sample)

TRAIT SET	TOTAL SAMPLE (n = 400)	REPUBLICANS (n = 181)	DEMOCRATS (n = 160)	INDEPENDENTS (n = 59)
Media	3.0	3.1	2.9	2.8
Group	2.8	2.8	2.9	2.7
Office	2.9	3.0	2.8	2.7
Background	3.1	3.2	3.0	3.0

marking all ratings extends to the four scales. However, we do get a comparison of the two political styles across the four scales. The greater response to the New Politics emerges on items pertaining to partisanship—the challenger's moderate conservatism and Republicanism, his independence from party dictates, and his credibility with groups. The older politics strikes a great responsive chord in reference to the incumbent's background, experience, and influence within his political party. Again the tendency to favor the New Politics pervades partisan identifiers across all scales (Independents are more clearly New Politics), with the notable exceptions of Republicans responding on the media and background scales. Republicans preferred the older campaign ways of the incumbent to the newer techniques of their own party's candidate. There are a number of explanations for this, but we mention two. First, given the sequence in which item-pairs were presented to respondents, the party identification of candidates did not appear until the next-to-last pair; perhaps had Republicans known which was "their" candidate, they might have reacted differently to the media and background items. Second, in the congressional election of 1970 in this district, the experienced Republican incumbent had been defeated by a Democrat using all of the media techniques associated with the New Politics; the incumbent waged a conventional Old Politics campaign. Perhaps Republicans were reacting to the ghosts of campaigns past.

For professional political consultants who believe that elections are won when technique triumphs over substance and when the *way* things are said dominates the *content* of messages, the results of the analysis of the distribution of voter types in this congressional district should be heartening. The style associated with this professional approach, the New Politics and its variants, had numerous adherents in 1971 in the sampled district. Table 5.4 presents the distribution of each voter type, both for the total sample and for Republicans, Democrats, and Independents. Note that one-fourth of the sample were clearly in the category of New Politics voters, compared with the 16 percent in the Old. Moreover, to the extent that a campaign in the style of the New Politics combines the amateurism and volunteerism of the citizen-politician, issue orientation, and visual appeal, then the persons in this sample are definite targets for the newer campaign styles.

Table 5.4 The Distribution of Voter Types

VOTER TYPE	TOTAL SAMPLE ($n = 400$)	REPUBLICANS ($n = 181$)	DEMOCRATS ($n = 160$)	INDEPENDENTS ($n = 59$)
Issue Oriented	29%	30%	29%	29%
New Politics	25	26	23	25
Citizen-Politician	10	8	9	11
Old Politics	16	15	19	17
Visual	20	21	20	18
	100	100	100	100

In several respects the findings of this probability survey of types of voters responding to campaign styles parallels the balanced survey conducted in the pilot phase of this research. For one thing, the overall distribution of voter types in this sample of 400 voters selected by probability methods is very similar to that derived in the purposively selected sample of 100 persons. Of course, we have no way of knowing at this point, but the results give us some confidence that the modification of the Q-sorting procedures for larger samples did provide a means for probing the distribution of voter types in large-sample surveys. As in the case of the pilot project, there were not notable differences in the distribution of voter types by partisan self-images of respondents or by sociodemographic characteristics. Partisanship alone does not determine how voters may respond to campaign styles. As we saw clearly in Chapter 4, the image the voter has of his ideal public official in an election and the way he compares the actual candidates with that ideal are crucial in his perceptions of the campaign.

The Old and the New Politics Revisited

We cannot avoid concluding this chapter with a marked degree of uncertainty. To our knowledge, the research reported in this chapter is the only existing effort to operationalize the notions of the Old and New Politics, test them, and measure popular responses to them. As with all such first and single efforts, it has its limitations. First, despite all efforts to avoid it, the selection of item-pairs to represent the contrasting traits of the "Old" and "New" political candidates remained largely a subjective matter. The instrument may, therefore, be insensitive to differences in the "Old" and "New" emphasized by other possible efforts. Yet retests of the instrument and replications of its use to determine the distribution of popular responses have been conducted in eleven other congressional districts, yielding largely the same results: that is, on overall and specific traits respondents rank the two political styles very closely; but as clusters of voters, regardless of party, they respond distinctly more positively to the New Politics of commitment on issues manifested through fresh faces using sophisticated techniques of communication. This is especially the case for Independents.

All of this raises again the problem introduced in Chapter 3. The images that people have of candidates within the campaign setting consist both of the comparative perceptions of the candidates' specific traits and their mental configurations of a single candidate's characteristics—emphasizing some, minimizing others, and overlooking or denying that many traits exist. When asked to differentiate between the traits of candidates, voters may find little to distinguish. Yet as groups of voters they share perceptions and evaluations of those traits they find most characteristic of a given candidate. These shared perceptions set types of voters apart and thus result in the distinctive images of political candidates and their campaign styles. At least in this sense, then, the research reported in this chapter indicates that voters are receptive to the New Politics campaigns of issue commitment wedded to sophisticated technology. But when all is said and done, does the hoopla of the election campaign really make much difference in the images people have of political candidates? This is the question we turn to in Chapter 6.

Notes

1. Lewis Chester, Godfrey Hodgson, and Bruce Page, *An American Melodrama: The Presidential Campaign of 1968* (New York: Viking Press, 1969), pp. 375–76.

2. George N. Gordon, *The Languages of Communication* (New York: Hastings House, 1969), pp. 193–94.

3. Kenneth Boulding, *The Image* (Ann Arbor, Mich.: University of Michigan Press, 1956), p. 7. (Italics are Boulding's.)

4. Richard Jenson, "American Election Campaigns: A Theoretical and Historical Typology" (paper delivered at the Annual Meeting of the Midwest Political Science Association, Chicago, May 2–4, 1968).

5. Theodore H. White, *The Making of the President 1972* (New York: Bantam Books, 1973), p. 430.

6. See Stanley Kelley, Jr., *Professional Public Relations and Political Power* (Baltimore, Md.: Johns Hopkins Press, 1966); Dan Nimmo, *The Political Persuaders: The Techniques of Modern Election Campaigns* (Englewood Cliffs, N.J.: Prentice-Hall, 1970); James M. Perry, *The New Politics: The Expanding Technology of Political Manipulation* (New York: Clarkson N. Potter, 1968); Harold Mendelsohn and Irvin Crespi, *Polls, Television, and the New Politics* (Scranton, Pa.: Chandler, 1970); and David Lee Rosenbloom, *The Election Men: Professional Campaign Managers* (New York: Quadrangle Books, 1973).

7. Chester et al., *An American Melodrama*, p. 376.

8. Frederick G. Dutton, *Changing Sources of Power* (New York: McGraw-Hill, 1971).

9. One of the authors was a principal investigator in this research project, which was conducted under the general auspices of Civic Service, Inc. Because of the confidences that must be maintained regarding certain aspects of this research, the fictitious name for the hypothetical candidates and the congressional district in which surveys were conducted are not identified beyond regional location and/or general demographic characteristics.

10. George A. Theodorson and Achilles G. Theodorson, *A Modern Dictionary of Sociology* (New York: Thomas Y. Crowell, 1969), p. 193.

11. Dean Jaros and Gene L. Mason, "Party Choice and Support for Demagogues: An Experimental Examination," *American Political Science Review* 63 (March 1969): 100–110.

12. The nature of balanced sampling criteria is discussed in William Stephenson, *The Study of Behavior* (Chicago: University of Chicago Press, 1953), pp. 66–69.

13. The principal components factor solution was rotated to simple structure by the varimax method. Criterion for extraction was an eigenvalue-greater-than-one. The five-factor solution explains 71 percent of the variance.

14. Several factor solutions were employed, but a five-factor solution, explaining 42 percent of the variance, is utilized here. The principal components solution was rotated to simple structure by the varimax method.

15. Stephenson, *The Study of Behavior*, chap. 9.

16. The American Institute for Political Communication, *Anatomy of a Crucial Election* (Washington, D.C.: American Institute for Political Communication, 1970).

Chapter Six

The Dynamics of Images of Presidential Candidates: 1952-1972

In Chapters 3 and 4 we sought to delineate the various traits that comprise the content of candidate images and to explore the complexities of image projections; in Chapter 5 we considered the relationship between candidate images and campaign styles. We now continue those general lines of inquiry by examining the changes that occur in voters' composite images of presidential candidates during election campaigns, both changes in their perceptions of image traits and the candidate- and/or perceiver-determined sources of those changes between and during election campaigns. We will look first at how changes in voters' perceptions of candidates have been measured in recent presidential campaigns. We will then examine the amount of inter- and intraelection change exhibited in the images of presidential candidates. Finally, we will consider how image changes relate to partisan self-images (a perceiver-determined factor) and to the voter's sources of communications in a campaign (a candidate-determined factor). The discussion in this chapter paves the way for our examination in Chapter 7 of one other type of dynamic in candidate images—the effects of victory and defeat on voters' perceptions of candidates.

Measuring Changes in Likes and Dislikes about Presidential Candidates

As a source of information concerning voters' images of presidential candidates in the last two decades, we turn again to a type of data we first introduced in Chapters 2 and 3, namely, the quadrennial national studies conducted by the Survey Research Center (now Center for Political Studies) of the University of Michigan. The reader will recall that in each election year, the SRC/CPS asks a representative sampling of the voting population to volunteer their likes and dislikes about the major presidential candidates. As we saw in Chapter 3, the SRC/CPS has developed a systematic way of classifying respondents—according to their answers to open-end questions probing these likes and dislikes. Among the categories of responses that refer to the *political role* traits of candidates are those pertaining to public leader attributes (leadership qualities, experience and ability, and references to the candidate's domestic and foreign policies) and to partisan qualities (references to the candidate as a party representative, to his philosophy, and to his relations with various social groups—labor unions, farmers, businessmen). Other categories contain *stylistic role* traits, specifically, those pertaining to a candidate's personal attributes, background, and character. Prior to 1972 SRC investigators classified as many as five statements each respondent made about a candidate into such categories; in 1972 they classified only three statements per respondent.

In the discussion that follows we report the total number (n) of references made about a given candidate in a particular category and the percentage of those references that were positive (%+), that is, the percentage of "likes" among all statements volunteered about the candidate. As a measure of the changes in perceptions that occur *between* elections, we compare the percentage of positive comments for the elections in question. But to determine whether or not voters' images of presidential candidates change *during* election campaigns, we use a slightly more complicated procedure. The basic SRC/CPS survey conducted prior to a presidential election is not a panel design (having a given respondent interviewed several times before the election), but instead consists of one interview per respondent conducted from early September through election day (with postelection interviews for portions of the original sample). To detect intraelection image changes we convert the SRC/CPS surveys to a quasi-panel design. For this we employ a technique John Kessel used with considerable success in his study, *The Goldwater Coalition*.[1] We simply divide the set of preelection interviews conducted in each SRC/CPS survey (1952-1972) into three representative subsamples by interview date. As did Kessel, we compared the subsamples created for each election's quasi-panel with respect to education, occupation, partisanship, and vote behavior to see if persons interviewed at one period differed from those interviewed at another. The results indicate that each election's subsamples are sufficiently representative to permit us to explore intraelection changes. However, because unequal numbers of interviews take place in each period in each election survey, we again have taken a cue from Kessel and weighted all data pertaining to the number of references to candidates (likes and dislikes) as a way of equalizing the number of interviews used for each campaign period. As we present our findings in the accompanying tables, we shall refer to four campaign periods in each election: the designation "Full" is to all preelection interviews conducted in the year's survey; the designations "First," "Second," and "Third" refer to specific campaign subperiods or subsamples.[2]

It is important to keep in mind a couple of things regarding these data. First, the standard SRC/CPS question probing likes and dislikes about each candidate, at least on the surface, taps what we described in Chapter 1 as the *affective* aspect of candidate images—that is, what people feel, or value and don't value, about a candidate. But comments about candidates cast in a like-dislike format reveal beliefs, or *cognitions*, as well as feelings. When a person says of, say, George McGovern, "I like him because he is a Democrat," he indicates that he believes McGovern is a Democrat. Hence, responses to the standard SRC question also measure information people have about candidates. People having more information about candidates are likely to have more to like and dislike and, as a result, are more likely to make several comments.

Second, and directly related to the first point, the standard SRC/CPS question places a premium on the ability of people to articulate their feelings and beliefs. Some persons discuss a candidate at length whereas others, even though they may have clear likes and dislikes, do not respond to the question. In fact, as we noted in Chapter 2, an average of 47 percent of respondents in the last two decades expressed no likes or dislikes about the candidates. Does this mean few people have affective orientations toward the candidates? We think not, for with a different type of measure of that affect we find more widespread feeling. Thus in 1972, 95 percent of respondents were able to say how warm they felt toward the two candidates when asked to rate each on a "feeling thermometer" (see Chapter 2). We suspect, therefore, that the SRC/CPS data measure the traits that a self-selected sample of political articulates see in candidates rather than measuring the views of a cross section of adults. We get an added indication that this is the case by looking at the data for 1968. Doris Graber reported that 42 percent of the SRC/CPS sample made no comments about the presidential candidates when asked their likes and/or dislikes.[3] Yet as Barbara Hinckley and her colleagues indicate, of SRC/CPS respondents who voted in 1968 (rather than of all SRC/CPS respondents), 93 percent had at least one comment to make about the major party candidates. In short, voters are much more likely than nonvoters to express feelings and beliefs about candidates. Voters rather than nonvoters are the articulators of candidate images.[4]

Interelection Changes in the Images of Presidential Candidates

Table 6.1 provides information regarding both interelection and intracampaign trends in the images of the presidential candidates of the two major political parties over the past two decades. We call the reader's attention to the data contained in the columns labelled "Full" in Table 6.1. Note that these columns contain the same data reported earlier in Tables 3.1a and 3.1b (Chapter 3), but now we look at them from the perspective of how much change occurs in elections across the six presidential elections, changes both in the affective content of the *overall images* and in perceptions of *specific image traits*.

As measured by the percentage of positive comments about a party's candidates from election to election, it is apparent that there is considerable fluctuation in the aggregate images of the party candidates. Of all comments

Table 6.1 Image Traits of Presidential Candidates by Interview Periods, 1952–1972

	DEMOCRATIC CANDIDATES								REPUBLICAN CANDIDATES							
	INTERVIEW PERIOD								INTERVIEW PERIOD							
	FULL		FIRST		SECOND		THIRD		FULL		FIRST		SECOND		THIRD	
IMAGE TRAITS	n	%+	n	%+	n	%+	n	%+	n	%+	n	%+	n	%+	n	%+
1952																
Experience/ability	722	85	274	85	297	84	352	86	1597	57	612	57	700	55	670	60
Background/character	688	64	290	63	380	63	278	67	1030	89	554	90	523	90	387	87
Personal attraction	292	55	139	55	114	54	94	59	355	94	148	97	152	94	129	88
Party representative	838	50	333	51	360	47	343	54	409	34	163	34	171	35	177	30
Issues	110	66	38	66	45	64	62	71	81	47	36	50	29	52	37	32
Domestic policies	197	64	73	61	78	69	106	57	231	67	114	78	80	56	87	55
Foreign policies	62	52	28	43	23	61	26	54	552	82	224	80	228	80	234	89
Relation to groups	169	86	75	87	61	84	78	89	112	63	50	64	51	63	49	57
Total	3078	65	1250	64	1258	64	1339	68	4357	69	1790	71	1824	68	1770	68
1956																
Experience/ability	545	59	249	59	218	58	292	58	1117	88	503	89	534	93	552	87
Background/character	776	52	379	56	333	49	325	49	1105	81	518	82	482	84	484	75
Personal attraction	370	39	186	35	165	39	136	50	830	55	411	52	330	55	395	56
Party representative	627	51	264	53	300	48	266	56	341	31	134	30	158	34	172	30
Issues	63	41	12	75	38	37	35	35	93	60	36	61	48	66	37	50
Domestic policies	210	59	68	53	102	55	120	73	305	53	145	54	139	51	117	56
Foreign policies	196	25	41	32	89	12	163	37	645	70	336	36	210	96	243	86
Relation to groups	181	91	93	90	75	93	74	99	249	56	152	65	94	51	91	38
Total	2968	52	1292	55	1329	49	1411	55	4745	70	2235	66	1995	73	2081	68

Table 6.1 (continued)

IMAGE TRAITS	DEMOCRATIC CANDIDATES								REPUBLICAN CANDIDATES							
	INTERVIEW PERIOD								INTERVIEW PERIOD							
	FULL		FIRST		SECOND		THIRD		FULL		FIRST		SECOND		THIRD	
	n	%+	n	%+	n	%+	n	%+	n	%+	n	%+	n	%+	n	%+
1960																
Experience/ability	396	64	207	64	200	63	268	67	557	91	288	91	324	92	285	89
Background/character	1033	52	527	49	597	54	570	60	529	73	267	73	301	73	313	76
Personal attraction	543	73	222	70	272	77	260	72	255	65	135	64	141	67	136	62
Party representative	314	61	159	67	158	58	236	53	357	49	178	48	199	54	232	45
Issues	63	73	22	82	46	72	52	62	79	66	38	68	50	60	44	73
Domestic policies	197	58	91	58	18	53	136	71	135	54	60	52	87	55	88	59
Foreign policies	108	47	46	50	69	41	80	55	321	79	154	81	196	78	192	77
Relation to groups	109	83	55	80	54	81	84	90	93	38	50	34	50	40	60	40
Total	2673	60	1329	59	1514	61	1686	64	2326	71	1168	71	1348	72	1350	69
1964																
Experience/ability	803	93	307	96	357	92	315	92	213	50	92	54	86	51	86	44
Background/character	1002	51	373	60	433	49	418	47	864	43	299	37	402	45	342	47
Personal attraction	299	61	123	68	136	47	105	59	224	52	98	43	99	54	79	59
Party representative	341	62	156	67	144	63	113	55	360	23	154	32	147	24	141	16
Issues	159	60	72	58	67	61	57	61	400	33	154	31	164	34	175	34
Domestic policies	598	66	237	65	262	56	232	69	793	34	281	38	341	33	345	32
Foreign policies	210	61	68	57	95	57	91	70	408	32	159	26	190	30	145	39
Relation to groups	261	94	106	90	996	94	124	94	153	14	55	20	74	12	56	13
Total	3673	69	1451	72	1590	67	1455	68	3415	36	1295	35	1503	36	1369	37

131

Table 6.1 (continued)

IMAGE TRAITS	DEMOCRATIC CANDIDATES								REPUBLICAN CANDIDATES							
	FULL		INTERVIEW PERIOD						FULL		INTERVIEW PERIOD					
			FIRST		SECOND		THIRD				FIRST		SECOND		THIRD	
	n	%+	n	%+	n	%+	n	%+	n	%+	n	%+	n	%+	n	%+
1968																
Experience/ability	578	77	201	69	245	77	267	79	595	81	230	82	280	81	237	79
Background/character	912	49	352	45	406	47	388	54	1110	50	463	52	498	51	457	49
Personal attraction	463	48	195	42	183	46	215	53	359	46	120	44	165	49	156	43
Party representative	546	42	245	42	239	40	221	43	369	44	158	46	154	46	161	40
Issues	286	36	98	25	134	32	120	44	196	52	79	54	81	51	89	51
Domestic policies	495	63	152	62	205	60	245	65	449	61	189	61	179	60	206	63
Foreign policies	301	53	127	44	134	52	119	58	428	74	139	73	212	73	171	75
Relation to groups	267	87	77	94	92	83	154	88	180	31	77	38	65	35	90	90
Total	3848	56	1457	50	1640	53	1729	60	3686	57	1455	58	1034	59	1567	56
1972																
Experience/ability	106	35	56	28	39	51	31	29	429	92	169	89	201	95	167	92
Leadership qualifications	288	10	99	13	125	10	143	07	133	56	54	56	58	57	55	56
Personal attraction	518	38	196	36	249	42	205	36	474	55	201	59	206	57	180	48
Party representative	252	23	104	25	110	22	99	22	86	32	29	31	35	43	55	22
Management of government	33	27	12	17	10	01	18	24	133	59	54	66	52	63	60	47
Philosophy of government	278	29	124	26	125	22	135	30	81	57	34	74	28	28	39	56
Domestic policies	378	37	140	31	132	27	179	42	430	47	166	45	167	51	206	47
Foreign policies	417	38	150	35	181	34	195	45	774	68	321	69	333	71	309	63
Relations to groups	235	76	84	64	91	80	124	84	183	25	71	24	69	25	90	27
Miscellaneous/other	147	33	78	37	41	32	52	23	225	62	91	68	104	53	89	57
Total	2652	35	1043	32	1103	34	1181	38	2948	61	1190	63	1253	63	1250	56

Note: n's for two of each of the three interview periods in each year (i.e., First, Second, Third) have been weighted to compensate for differences in the number of interviews which took place in each time period; consequently, the number of references on any given trait for interview periods will not total the number of references of the "Full" interview period.

Source: Survey Research Center/Center for Political Studies, University of Michigan, and the Interuniversity Consortium for Political Research.

made about Democrats in the two decades from 1952 to 1972, 57 percent were positive but the range was from a low of 35 percent in 1972 to a high of 69 percent in 1964. Over that period the Republican candidates received a 64 percent positive rating, ranging from 36 percent in 1964 to Nixon's high of 71 percent in 1960. Generally, interelection images of Republican candidates have been more stable than those of Democratic contenders. Moreover, as illustrated by Figure 6.1, Republicans have enjoyed a net image advantage over the twenty-year period (the notable exception being Barry Goldwater's image in 1964).[5]

The variation in voters' perceptions of specific image traits exceeds the amount of fluctuation in their overall images of party candidates. However, as we saw in Chapter 3, there are a few patterns that remain fairly stable across the six elections: Democrats retain a stable advantage in reference to their candidates' domestic policies and their relations to social groups; the foreign policy component of the images of Republicans is usually strong and, even in the case of Goldwater, Republican candidates enjoyed a majority of favorable comments pertaining to experience and ability. In general, the *partisan traits* associated with political roles of Democrats have been strong and stable, whereas the *governmental leadership traits* of Republicans' political roles have remained stable. But a degree of moderate stability in the political role traits of candidates from both parties has been more than offset by great variability in their stylistic role traits, especially variation in their perceived personal attributes. Kirkpatrick and his associates, employing more sophisticated procedures for examining similar data, concluded that references to the candidates as persons vary more than any of the other categories of comments.[6]

In the six elections we are considering, three candidates competed more than once: Stevenson ran in 1956 after being defeated in 1952; Eisenhower sought reelection as the incumbent in 1956; and the perennial candidate of the era, Richard Nixon, lost in 1960 but won in 1968 and was reelected in 1972 (of course, he was also on the Republican ticket as Eisenhower's running mate in both 1952 and 1956). What changes in images did these repeaters undergo? Stevenson's overall image declined considerably between 1952 and 1956. Almost two-thirds of all comments about him had been positive in 1952, but by 1956 only slightly more than a majority were favorable. On every trait in 1952, at least one-half of all comments about Stevenson were favorable; yet, in 1956 references to his issue positions generally, and foreign policies particularly, were basically negative. Sharp declines also occurred in references to his leadership (experience and ability) and stylistic role (personal attributes and background characteristics). Only along partisan traits of his political role did he hold up well.

In 1952 Eisenhower's perceived assets pertained more to his stylistic than political role (compare his personal attributes to experience and ability). But in 1956 his leadership traits were markedly positive (although partisan traits associated with his political role slipped), and there was a drop in the positive perception of his stylistic role. Still, his overall image remained

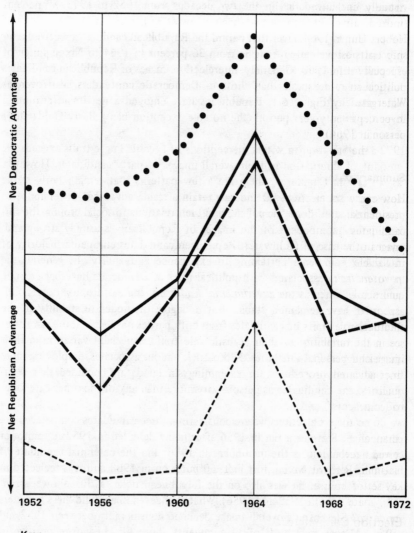

Figure 6.1 Partisan Advantage of Images of Presidential Candidates,
1952–1972, for all Respondents and by Partisan Self-Images

Source: Data adapted from Samuel A. Kirkpatrick, William Lyons, and Michael R. Fitz-
gerald, "Candidate and Party Images in the American Electorate: A Longitudinal
Analysis" (paper delivered at the Annual Meeting of the Southwestern Political Science
Association, Dallas, Texas, March 28–30, 1974).

virtually unchanged. Richard Nixon's image was up in 1960, down in 1968 (indeed, almost the same percentage of positive references as obtained by Hubert Humphrey), and slightly improved in 1972. Perceptions of the leadership traits associated with his political role were always resoundingly positive, especially references to his experience and ability; the partisan traits of that political role were just as stable, although nowhere nearly as positive. The pre-Watergate electoral rise and fall of Richard Nixon's image is told primarily in perceptions of his stylistic role—two-thirds of references in 1960 to his personal qualities were positive, by 1968 a majority were negative, and in 1972 a majority were again positive.

Summary of the Data

How can we summarize these data on interelection changes in the images of presidential candidates from 1952 to 1972? First, the popular images of the candidates have been far more dynamic than have been the partisan self-images of voters (recall the relative stability in partisan self-images as depicted in Table 2.3, Chapter 2). Second, the fluctuation in image traits exceeds that in voters' aggregate images of candidates. Third, perceptions of the personal qualities of candidates are especially changeable, more so than leadership aspects of the political role and far more so than the partisan aspects. These fluctuations in personal traits are more comparable to the type of swings we see in the direction of the vote from election to election than are any other image changes. Put simply, in elections in which one candidate has held a distinct advantage over his opponent in voters' assessments of respective personal qualities, the advantaged candidate has won; in 1968, when neither candidate had much of an advantage in personal traits, the election was close. Certainly we do not offer a comparative assessment of voters' perceptions of candidates' personal qualities as a basis for predicting election outcomes. But the perceived attributes of a candidate as a fellow human being are a key element in his image and, as we have indicated before and will again, they may be *the* key set of traits in the electoral decisions or many voters.

Election Campaigns and Changes in Candidate Images

In Chapter 2 we reviewed a large body of literature dealing with studies of voting behavior in the United States. The thrust of those studies is that in presidential elections, almost two of every three voters make up their minds by the time of the nomination of the major party candidates. The fact that relatively few voters make or change their decisions during the campaigns gives rise to the "rule of minimal change," suggesting that campaigns have relatively little effect on voting. Yet a sizable minority do reach their decisions during campaigns and, moreover, both decided and undecided voters do shift their perceptions of the candidates. The data in Table 6.1 (and later in Table 6.6, with respect to the independent candidacy of George Wallace in 1968) help us to reexamine these propositions regarding the degree of image change in presidential election campaigns.

In his study of the 1964 presidential campaign, Charles O. Jones differentiated among three phases of that contest roughly corresponding to the three months between the nominating conventions and election day.[7] Jones labelled the phases of the 1964 campaign as the *planning, testing,* and *critical* phases. The planning phase takes place in August, the candidate's plan is tested in September, and October is the month for reaching a critical decision on, and carrying out of, the most effective plan. Recent presidential campaigns suggest, however, that campaign planning probably lasts well into September, various plans are still being tested in October, and it is usually the last two or three weeks of campaigns that become critical. These are the three subperiods we have used to divide the SRC national preelection samples in each of our six presidential contests, and although they do not match Jones's phases as closely as we might like, we can at least consider them rough approximations of his campaign stages.

From the columns for each subperiod of the various campaigns depicted in Table 6.1, one pattern is fairly clear. Generally, the overall image of a candidate emerges early in a campaign and changes only in marginal ways thereafter. The 1952 campaign exemplifies this pattern. In the planning phase 64 percent of comments about Stevenson and 71 percent about Eisenhower were positive, with the image of the former improving and of the latter declining only slightly as the campaign progressed. The 1956 campaign differed only in that the low point of Stevenson's image came during the second, testing, period when Eisenhower's image was higher than at any other time. In 1960 there was a steady rise in the percentage of positive comments regarding John Kennedy, but the overall increase was only 5 percent, and his image was never as positive as Nixon's at the Republican candidate's lowest point. Again, in 1964 Johnson's overall image changed relatively little (there was a slight decline), and a negative image dogged Goldwater throughout the campaign.

The strongest case that overall images can change markedly in presidential campaigns is that of Hubert Humphrey in 1968. Much like his rise in the Gallup and Harris polls, Humphrey improved fairly steadily in his overall image and experienced a 10 percent increase in positive comments over the course of the campaign. The image of third-party candidate George Wallace in 1968 (see Table 6.6) tailed off in the testing and critical phases of the campaign. And, in 1968 there was no consistent pattern in the rise and fall of respondents' appraisals of Nixon. As with Humphrey's image in 1968, that of George McGovern improved in 1972 but scarcely enough to overcome its clearly negative tone at the campaign's start. Nixon's overall image declined in 1972, dropping to the low in positive comments he received at the close of the 1968 campaign.

Although the overall images of presidential candidates during campaigns remain fairly stable, campaigns do modify perceptions of specific image traits. Each of the six campaigns reflects this tendency to some degree. For example, in 1952 there was a general decline in specific facets of both Eisenhower's perceived political and stylistic roles. Note the drop in appraisals of his domestic policy stands and the slippage in his personal appeal (see Table 6.1).

Yet perceptions of his experience and ability, background and character, and foreign policies remained firm and, indeed, improved in some instances. In 1956 Stevenson entered the campaign with only a minority of positive references to his personal attractiveness, yet as the campaign moved into its latter phases there was a substantial improvement in this image trait. In 1960 almost 40 percent of all comments made about John Kennedy were to his background and character. Since on this single attribute he gained steadily during the campaign, it is reasonably clear that a large portion of his improvement in overall image can be attributed to improved stylistic quality. In fact, 55 percent of all references to Kennedy were to stylistic role (background, character, and personal attributes). In the first period of the campaign 56 percent of these references were favorable; by the close of the election 64 percent were positive, a vivid demonstration of the impact of the perceived Kennedy style.

In 1964 professional pollster Thomas Benham and the Opinion Research Corporation conducted surveys on behalf of the Republican nominee. Benham found that at the start of the campaign, people were more inclined to ascribe favorable qualities to Johnson than to Goldwater. They found Johnson, for instance, "warm and friendly," "progressive and forward-looking," and Goldwater a man who "acts without thinking." The effects of the campaign, observed Benham, were to heighten negative qualities in Johnson's image and lower many favorable ones, while increasing perceptions of Goldwater's both favorable and negative traits.[8] Table 6.1 suggests similar results. There we see evidence of a steady decline in the percentage of positive references regarding Johnson's background and character and, although the decline was not as steady, to his personal attributes as well. There was steady improvement in appraisals of Goldwater's background, character, and personal attributes. In fact, during the last interviewing period the percentage of positive references to Goldwater's personal attributes matched the 59 percent afforded Johnson. Johnson also suffered some slippage in his partisan role as indicated by the 12 percentage-point decline in positive comments about him as a party representative. Interestingly enough, although Goldwater received severe journalistic criticism for his foreign policy stands, both candidates improved in voters' perceptions of this trait.

As with John Kennedy's in 1960, a good portion of the improvement of Hubert Humphrey's image in 1968 grew out of increasingly positive assessments of his stylistic role (background, character, and personal attributes). In addition, shifting perceptions of Humphrey's leadership traits (his experience and ability and his stands on foreign policies) also worked to his benefit, far more than they had for Kennedy in 1960. Perceptions of Nixon's image traits changed very little in 1968. We can observe a steadily improving Humphrey image pitted against a moderately positive, but stable, Nixon image. When the campaign opened, a bare majority of references to Humphrey were positive, contrasted with the 58 percent favorable mentions of Nixon. By the critical phase of the campaign, 60 percent of Humphrey's references, compared to 56 percent of Nixon's, were positive.

Unlike three of our major party candidates destined to lose (Stevenson,

Goldwater, and Humphrey), and unlike John Kennedy who won, George McGovern did not improve in the voters' eyes on those traits associated with stylistic role (traits limited primarily to the category of personal attributes in the 1972 SRC survey). The highest percentage of references made about McGovern were to his personal attributes, but these were generally negative and improved scarcely at all during the campaign. Assessments of his leadership qualities declined throughout the campaign. Marginal improvement in his image came from increasingly favorable perceptions of his relations to groups and his domestic policies. Nixon received high marks for his experience and ability and for his foreign policies throughout the campaign but suffered declines in references to his personal attributes and partisan role.

Do campaigns, then, make a difference in the images of presidential candidates? We think so, at least in some respects. The general pattern is clearly one of overall image stability. People form a composite impression of likes and dislikes which, when averaged out across the respondents in a nationwide survey, does not change very much during a campaign. Yet Humphrey's improvement in 1968 is a notable exception to this stability. Moreover, in relation to the specific traits we defined in Chapter 3 (leadership, partisanship, personal qualities, and dramatic performance), we suspect there are sizable shifts in perceptions. At least as measured by such perceived qualities as personal attributes, background and character, and experience and ability, we have noted greater shifts in image traits than in the overall images of candidates. Generally, perceptions of candidates' stylistic roles are most changeable, and when perceived personal attributes such as background and and character comprise a major aspect of the candidate's image (as with Kennedy and Humphrey), substantial improvements in popular responses to the candidate's style take on added importance.

Partisan Self-Images and Changes in the Images of Presidential Candidates

From our discussion in Chapter 2 of the early voting studies and the description of voting behavior that has grown out of the various studies of the SRC/CPS, we know that partisan self-images influence how people perceive the qualities candidates publicize in campaigns. We would expect, then, that interelection and intracampaign changes in candidate images would be, at least in part, a function of the voters' party loyalties. We can explore this possibility by referring to Table 6.2.

Let us first consider interelection changes. We noted earlier that from 1952 to 1956 there was a general decline in Stevenson's overall image coupled with an improvement in Eisenhower's. As indicated by the "Full" column in Table 6.2, the decline in positive references to Stevenson in 1952 touched all partisan groups—Democrats, Independents, and Republicans. The greatest decline was among Independents, whose percentage of favorable comments about Stevenson fell from 63 to 40 between elections. Eisenhower, however, did not experience a comparable improvement among all three groups of partisans; his overall image among Democrats was less positive in 1956 than it had been earlier. This doubtless reflects partisan conflicts in Eisenhower's first term, which highlighted for Democrats the fact that he was a Republican president.

One of the most notable failures of party loyalists to hold positive images of their party's standard bearer comes with the candidacy of George McGovern. Prior to 1972 no Democratic candidate had failed to garner at least 73 percent favorable comments from his fellow Democrats. But in 1972 less than a majority of the comments made about McGovern by Democrats were favorable. Even among weak Democrats, prior to 1972 no Democratic candidate had failed to receive 60 percent favorable commentary. Yet McGovern could do no better than 40 percent positive references (in fact, as with John Kennedy in 1960, McGovern's image was stronger among independent Democrats). Now, contrast McGovern's image among Democrats with Goldwater's among Republicans in 1964. Republicans were clearly less critical of their party's most unpopular candidate in two decades than Democrats were to be of George McGovern in 1972. Almost two-thirds of Republicans' comments about Goldwater were favorable; only among weak Republicans did he not achieve a favorable rating (the only time, in fact, in the two decades that a set of independent, weak, or strong Republican identifiers did not have a majority of favorable comments about the candidate of the GOP).

Basically, then, the interelection stability in candidate images, which we found for all respondents earlier, holds in a predictable way for people having various partisan self-images. Year in and year out, Democrats (except in 1972) and Republicans think well of their own party's candidate and ill of the opposition's. As expected, Independents are far more fickle, principally when it comes to evaluating Democratic candidates. Independents made 63 percent positive comments about Stevenson in 1952, but only 40 percent in 1956; their comments about Kennedy and Johnson were clearly positive, but only 44 percent of what they said about Humphrey and 29 percent of their references to McGovern were favorable. On the whole Independents have been more charitable and consistent toward Republican candidates. Only in the case of Goldwater did they turn away. Although Independents were more responsive to George Wallace in 1968 than were other partisan groupings, less than a majority of references made by Independents to Wallace were favorable (see Table 6.6).

What of the relationship of partisan self-images to campaign changes in the images of presidential candidates? Because of the limited number of references made to any specific image trait of a candidate in a given campaign period, we are limited in this discussion to changes in overall images of the candidates. In this respect, shifts in overall candidate images are *inversely related* to strength of partisan self-images. For example, in 1952 the improvement in Stevenson's image and slim decline in Eisenhower's can be attributed in part to Stevenson's improved standing and Eisenhower's slippage among Independents. Four years later, however, comments about Stevenson among

Table 6.2 Images of Presidential Candidates by Interview Periods and Partisanship of Respondents, 1952–1972

PARTISAN SELF-IMAGE OF RESPONDENT	DEMOCRATIC CANDIDATES								REPUBLICAN CANDIDATES							
	FULL		INTERVIEW PERIOD						FULL		INTERVIEW PERIOD					
			FIRST		SECOND		THIRD				FIRST		SECOND		THIRD	
	n	%+	n	%+	n	%+	n	%+	n	%+	n	%+	n	%+	n	%+
1952																
Democrat	1558	80	611	81	644	79	710	81	1820	54	687	54	794	55	796	55
Strong	857	86	325	85	384	86	347	86	824	42	296	43	389	44	327	35
Weak	701	74	286	77	260	68	363	77	996	65	390	62	405	65	469	69
Independent	710	63	302	62	283	63	294	69	1099	71	387	74	431	69	428	66
Republican	765	36	319	36	314	34	311	39	1360	87	580	88	568	85	502	88
Weak	359	43	149	45	143	41	158	45	643	83	273	85	255	80	269	84
Strong	406	29	170	28	171	29	152	46	717	90	306	92	313	88	233	94
1956																
Democrat	1466	73	656	76	628	70	721	76	1764	49	573	52	785	52	860	47
Strong	785	80	375	82	312	77	299	81	868	40	400	41	341	38	472	39
Weak	681	66	281	68	316	63	322	69	896	61	353	63	444	62	388	58
Independent	648	40	239	43	314	38	322	43	1177	76	490	75	553	77	531	73
Democrat	239	60	93	52	105	62	135	66	338	60	159	61	139	63	165	53
Independent	192	34	55	50	111	32	82	21	361	77	136	72	193	79	135	79
Republican	217	19	91	31	98	19	105	30	478	57	195	58	221	88	231	84
Republican	852	26	407	30	304	23	352	23	1559	92	803	90	640	93	643	90
Weak	376	33	192	34	159	30	148	38	728	87	422	85	265	90	295	85
Strong	476	19	215	27	215	17	204	12	831	96	381	96	375	96	348	95

Table 6.2 (continued)

PARTISAN SELF-IMAGE OF RESPONDENT	DEMOCRATIC CANDIDATES								REPUBLICAN CANDIDATES							
	INTERVIEW PERIOD								INTERVIEW PERIOD							
	FULL		FIRST		SECOND		THIRD		FULL		FIRST		SECOND		THIRD	
	n	%+	n	%+	n	%+	n	%+	n	%+	n	%+	n	%+	n	%+
1960																
Democrat	1369	76	742	74	677	80	858	77	942	49	524	50	441	44	597	52
Strong	700	85	366	86	400	85	361	84	419	31	229	32	223	30	216	26
Weak	699	67	376	63	277	73	497	71	523	63	295	64	218	59	381	67
Independent	505	60	207	56	343	59	356	70	483	75	200	77	353	75	268	70
Democrat	168	75	61	64	113	75	152	92	105	43	34	32	83	53	80	35
Independent	182	64	72	65	131	60	120	70	170	78	70	81	136	76	68	76
Republican	155	38	74	41	99	38	84	29	208	89	96	90	134	88	120	90
Republican	778	33	373	31	487	35	432	31	879	92	433	92	539	93	461	92
Weak	360	41	147	39	267	44	200	36	392	89	175	87	279	91	188	89
Strong	418	25	226	26	220	25	232	26	485	95	258	95	260	96	273	94
1964																
Democrat	1995	83	779	85	846	80	820	83	1631	20	584	19	716	19	685	20
Strong	1090	88	465	90	422	85	466	91	868	13	346	15	366	15	354	10
Weak	905	76	314	78	424	75	354	73	763	27	238	25	350	24	331	32
Independent	821	63	321	67	381	62	293	61	765	37	305	33	348	38	281	38
Democrat	358	73	131	78	154	69	151	75	342	24	117	23	161	29	134	19
Independent	226	67	106	81	107	62	63	57	200	34	101	27	81	30	68	49
Republican	237	44	84	32	120	53	79	38	223	58	84	54	106	59	79	61
Republican	823	40	347	47	351	40	314	34	987	63	402	59	432	64	375	64
Weak	453	54	180	68	193	48	182	49	493	48	191	41	219	55	191	45
Strong	370	24	167	25	158	32	132	12	494	77	211	76	213	73	184	84

Table 6.2 (continued)

PARTISAN SELF-IMAGE OF RESPONDENT	DEMOCRATIC CANDIDATES INTERVIEW PERIOD								REPUBLICAN CANDIDATES INTERVIEW PERIOD							
	FULL		FIRST		SECOND		THIRD		FULL		FIRST		SECOND		THIRD	
	n	%+	n	%+	n	%+	n	%+	n	%+	n	%+	n	%+	n	%+
1968																
Democrat	1997	73	755	65	802	69	951	78	1452	34	614	36	613	35	630	32
Strong	1058	84	307	82	398	81	575	87	677	19	214	17	273	21	340	18
Weak	939	60	448	54	404	58	376	65	775	47	400	46	340	46	290	50
Independent	978	44	388	43	448	40	398	48	1117	60	360	53	554	63	446	57
Democrat	351	56	136	37	145	50	161	69	340	36	120	32	157	43	143	30
Independent	272	42	117	53	118	39	114	40	357	62	122	51	173	62	144	67
Republican	354	32	135	39	185	33	123	28	420	76	118	75	224	77	159	74
Republican	799	25	305	21	328	23	371	28	1131	82	478	90	475	82	494	80
Weak	483	29	161	25	199	29	234	31	630	76	214	83	264	75	299	74
Strong	316	18	144	17	129	15	137	23	501	91	264	95	211	91	195	88
1972																
Democrat	1043	47	438	49	421	49	461	51	1121	46	463	49	450	52	479	48
Strong	347	60	144	59	134	63	152	70	387	34	151	31	162	42	173	35
Weak	696	40	294	44	287	42	309	42	734	52	312	54	288	58	306	56
Independent	945	29	332	29	329	36	388	37	967	60	434	68	391	63	388	51
Democrat	279	53	102	53	98	56	153	57	322	40	135	44	106	45	159	28
Independent	249	24	84	20	104	29	122	26	292	67	100	72	147	69	115	59
Republican	417	15	146	16	127	27	113	21	353	74	199	81	138	70	114	74
Republican	703	15	273	16	323	17	270	16	813	81	311	86	349	82	327	77
Weak	393	18	141	16	188	21	165	18	450	75	161	81	195	76	188	73
Strong	310	11	132	08	135	12	105	13	363	88	150	91	154	91	139	83

Note: *n*'s refer to total number of references about a candidate. *n*'s for two of each of the three interview periods in each (i.e., First, Second, Third) have been weighted to compensate for differences in the number of interviews which took place in each time period; consequently, the number of references for any given partisan category for interview periods will not total the number of references for the "Full" interview period.

Source: Survey Research Center/Center for Political Studies, University of Michigan, and the Interuniversity Consortium for Political Research.

strict Independents grew more unfavorable as the campaign progressed. In 1960 strict Independents, independent Democrats, and weak Democrats were sources of the improvement of Kennedy's image. Again, in 1964 strict Independents help account for the decline in Johnson's image and the slight improvement of Goldwater's. In 1968 Nixon's image improved among strict Independents, but overall so did Humphrey's. The proportion of positive references to Humphrey also increased among weak Democrats and independent Democrats. For George McGovern the prime sources of image improvement did not lie with Independents (although he became slightly better received by that group) but instead among strong Democrats. Thus 1972 is an exception to the general pattern of greater change among weaker partisans (compare also Goldwater's improvement among strong Republicans in 1964). Perhaps strong partisans, initially dissatisfied with the nominee of their party, return to the party fold as the election nears and, in the process, rationalize their decision by thinking happier thoughts about their party's standard bearer. Nixon's image in 1972 declined noticeably among Independents and to a lesser degree among both weak and strong Republicans, but the decline produced no serious defections to McGovern.

In stating that campaigns make a difference, then, we must add that they do so for some types of people. Most apparent in our six presidential elections were changes among Independents, either strict Independents or covert identifiers. These are the types of voters that the studies we reviewed in Chapter 2 predict would be most swayed by a campaign. But in campaigns in which a party has nominated a candidate whom many loyalists find unattractive, the candidate may still improve his image among weak and/or strong party identifiers, a pattern we have noted for Goldwater, Humphrey, and McGovern. In short, *the uninvolved and the disenchanted are more inclined to shift their images*, just as they are more inclined to shift their voting intentions and behavior.[9]

Communication Sources and Candidate Images in Presidential Elections

Two decades have passed since Kenneth Boulding, in his little book entitled *The Image*, called on political scientists to develop a cross-disciplinary study of "eiconics," the science of message-image relationships.[10] He argued that the relationship between messages about the world communicated to people and their images of the world is reciprocal, one affecting the other. As one facet of our investigation of the images of presidential candidates, we want to explore the relationship between the communication sources people use to learn about presidential campaigns and their images of the candidates. Although from our data we cannot say how candidates' messages affect voters' images and vice versa, we can say that voters' images of candidates are associated with voters' sources of information about elections. Moreover, we can see what campaign changes in the overall images of candidates relate to exposure to different communications media.

Sources of Campaign Information

What kinds of communications sources are available to voters and which do they use? We get an idea by examining Table 6.3, which portrays the percentages of respondents in SRC/CPS surveys, 1952–1972, who followed the presidential campaigns in four types of mass media (television, newspapers, radio, and magazines), the number of such media they used, and whether they were contacted by either of the two major political parties (and thus used party contact as a communication source). In Table 6.3 the column designated with an "F" refers to the total sample, and those designated "1," "2," or "3" pertain to each of the subsamples for interview periods, as employed earlier.

Several points stand out in the data reported in Table 6.3. First, we see clearly that television has become the primary source of campaign information for most people. From a bare majority who viewed television in its early days of the 1952 campaign, its usage has increased to the point that nine of every ten respondents followed the 1972 election on television. In contrast, as sources of campaign communications newspapers, radio, and magazines declined in use over the two-decade period. In 1972 only 57 percent reported reading about the campaign in newspapers, only one-third used magazines, and only four in ten followed the campaign on radio. Other studies have documented the pervasive use of television by Americans to learn about politics, at least for national elections. For instance, asked in a 1972 Roper survey from what source they became "best acquainted with the candidates for national offices," two-thirds said television, 29 percent replied newspapers, and only 8 and 5 percent, respectively, selected radio or magazines. The same study, however, reports that in elections at the city and county level, newspaper reading exceeded television viewing as a source of learning about candidates by 43 to 28 percent, a pattern dating back to at least 1964. In statewide elections, newspapers and television serve to an equal degree as sources of campaign information.[11]

Second, as our data indicate, there is very little variation in the rates of usage of each medium throughout the presidential campaign. Approximately the same percentages of respondents report following the campaign in the various media from one period to another.

Third, an average of about one in ten Americans have not followed presidential campaigns through any of the mass media in the last two decades. Moreover, 15 percent have normally employed only one of the four media. In contrast to the one-fourth of Americans who follow campaigns through the media either not at all or who use only a single medium, about one-third have used two or three media and 14 percent have employed all four. As with the types of media used, there is very little variation in the number of media used relative to campaign periods.

Finally, the bulk of Americans simply are not contacted by political parties during presidential campaigns. Prior to 1964 an average of 80 percent of Americans reported no party contact. From 1964 onward there has been a noticeable, but slight, increase in the percentage of persons reporting contact

Table 6.3 Sources of Campaign Communication in Presidential Elections by Interview Periods, 1952–1972 (results in percentages)

										ELECTION AND INTERVIEW PERIOD														
COMMUNICATION SOURCE	1952				1956				1960				1964				1968				1972			
	F	1	2	3	F	1	2	3	F	1	2	3	F	1	2	3	F	1	2	3	F	1	2	3
Types of Media:																								
Television	51	51	54	46	74	75	74	70	87	87	88	86	90	91	91	88	89	90	88	90	88	87	86	92
Newspapers	79	80	80	77	68	69	68	69	80	79	83	80	80	79	80	81	74	70	76	75	57	58	54	59
Radio	69	69	69	70	45	47	46	42	43	43	44	44	52	51	50	56	42	40	39	45	43	41	46	43
Magazines	40	42	40	38	31	32	30	30	42	40	45	42	44	44	44	44	34	29	35	35	33	32	31	35
Number of Media:																								
None	15	16	15	19	8	8	9	8	10	9	10	13	11	10	10	12	17	19	15	18	6	5	4	8
One	13	12	12	14	19	18	18	22	12	13	11	10	11	14	11	10	13	14	14	12	21	23	22	17
Two	27	26	29	25	32	31	32	33	27	27	26	28	28	26	30	26	28	30	28	27	33	31	35	35
Three	32	33	32	29	28	30	27	26	33	34	33	31	33	32	32	36	29	27	31	28	27	28	27	25
Four	13	13	13	13	13	13	13	12	18	17	20	18	17	17	16	17	12	10	12	14	13	13	12	15
Party Contact:																								
None	83	88	86	90	82	81	85	82	78	80	75	77	63	67	64	64	71	74	76	65	71	68	71	75
Democratic	4	4	4	2	6	7	5	6	8	8	9	6	15	15	16	18	12	10	7	18	12	13	11	11
Republican	5	5	5	3	6	6	5	6	9	8	10	11	18	18	20	18	17	16	17	17	8	8	7	11
Both Parties	4	3	5	5	6	6	5	6	5	4	6	6	–	–	–	–	–	–	–	–	7	8	8	5

Note: Percentages refer to proportion of respondents interviewed in postelection surveys stating they knew of the campaign from each source; however, interview periods (1, 2, 3) refer to preelection interviews and, thus, for purposes of determining intracampaign changes, postelection interviewees are grouped according to the data of the preelection interviews. "F" refers to the overall campaign in each election year.

Source: Survey Research Center/Center for Political Studies, University of Michigan, and the Interuniversity Consortium for Political Research.

by one or the other major political party. Yet, despite the increase less than one-half of Americans are typically contacted by a political party during presidential election years. Upwards of one-third of respondents have been contacted in recent presidential elections, and Republicans have generally done slightly more than Democrats. In 1964 Republicans contacted 18 percent of respondents, the highest contact rate in the two decades being studied. In spite of the highly trumpeted grass-roots campaign planned by McGovern forces in 1972, only about one in ten respondents indicated they had been reached by the Democrats. As with other sources of information about the campaign, there is little intraelection variation in contact rates.

What this suggests is that for most Americans, the likelihood is that the messages contributing to their images of presidential candidates are conveyed primarily through television, to a lesser degree by newspapers, and relatively little by radio, magazines, and direct contact with political parties. (Of course, we report no data here on how much association with friends, neighbors, co-workers—the personal influence discussed in Chapter 2—might contribute to popular images of candidates.) Do people derive different overall images of presidential candidates and do they perceive different image traits as a result of the source of their campaign messages? To answer these questions we can turn to Tables 6.4 and 6.5.

Information Sources and Perceived Image Traits

It is convenient to begin with the "Full" column in Table 6.4, which contains for each of six presidential elections the total number of references and percentage of positive comments about the candidates by people using and not using a specific variety of mass media (television, newspapers, radio, and magazines), the number of media they use, and whether or not political parties contact them. We noted earlier that about two-thirds of the comments made about Adlai Stevenson in 1952 were positive, a rating that differs very little for users and nonusers of the electronic media, television and radio. He did not do that well among readers of newspapers or magazines, but fared considerably better among people who did not read about the campaign at all. Persons following the campaign in only one medium made a higher proportion of favorable comments about Stevenson than the average; people using four media made a noticeably lower percentage of favorable comments. And, as we might expect, people contacted by the Democratic party were more positive, either because of the contact or because they were already favorably disposed when contacted (especially if the parties followed the common campaign practice of canvassing in areas where their supporters were likely to be found). In contrast with Stevenson's image, there was little difference in the Eisenhower image from one communication source to another.

Continuing through the remaining five presidential campaigns, the reader will find relatively little variation in candidate images by communication sources. Of course, there are a few differences, but scarcely so sizable as to warrant the proposition that varied media exposure had a major effect on the

Table 6.4 Images of Presidential Candidates by Interview Dates and Campaign Communication Sources, 1952–1972

COMMUNICATION SOURCE	DEMOCRATIC CANDIDATES								REPUBLICAN CANDIDATES							
	FULL		FIRST		SECOND		THIRD		FULL		FIRST		SECOND		THIRD	
	n	%+	n	%+	n	%+	n	%+	n	%+	n	%+	n	%+	n	%+
1952																
Television	1739	64	924	66	723	62	654	67	2331	69	1215	69	1018	68	809	71
No television	1036	65	523	62	401	66	501	70	1579	69	837	72	619	68	678	63
Newspapers	2503	63	1326	62	1013	61	1029	68	3476	69	1845	72	1443	68	1314	67
No newspapers	333	80	147	82	153	78	143	78	512	66	256	65	222	67	198	68
Radio	2053	64	1095	63	811	62	857	68	2827	69	1477	70	1160	68	1138	69
No radio	788	68	376	67	351	67	319	71	1169	69	622	73	515	68	375	63
Magazines	1388	60	765	59	555	59	523	65	1941	73	1035	76	816	71	706	70
No magazines	1443	70	699	70	602	68	384	72	2042	65	1062	67	849	64	807	64
One medium	223	78	79	76	92	77	118	84	378	66	155	67	150	68	168	62
Two media	899	58	261	70	358	65	366	68	1155	67	433	70	505	64	468	69
Three media	1208	63	512	58	472	65	493	70	1635	68	683	74	668	67	602	64
Four media	573	58	257	64	237	54	220	61	787	72	325	69	325	74	318	76
No media	314	66	118	69	122	70	123	64	470	68	202	67	183	69	199	67
Democratic contact	123	72	76	70	47	64	37	100	165	68	90	61	81	78	30	50
Republican contact	178	62	78	67	79	69	55	64	211	72	134	73	83	69	53	79
No party contact	2402	65	1248	65	972	63	1014	68	3437	69	1798	71	1421	67	1356	67
1956																
Television	2454	53	989	51	1114	49	1094	55	2587	71	1202	73	1163	73	1628	66
No television	506	50	217	53	206	49	288	49	926	74	408	72	395	75	468	74
Newspapers	2428	51	1067	55	1066	47	1009	47	3565	72	1576	72	1563	73	1700	67
No newspapers	543	56	244	55	254	58	218	50	007	72	491	73	432	72	399	72
Radio	2616	54	635	55	621	50	621	51	2295	73	1707	71	924	73	896	67

Table 6.4 (continued)

	INTERVIEW PERIOD															
COMMUNICATION SOURCE	DEMOCRATIC CANDIDATES								REPUBLICAN CANDIDATES							
	FULL		FIRST		SECOND		THIRD		FULL		FIRST		SECOND		THIRD	
	n	%+	n	%+	n	%+	n	%+	n	%+	n	%+	n	%+	n	%+
1956 (continued)																
No radio	1415	58	681	55	693	48	761	56	2436	72	1783	74	1068	74	1199	68
Magazines	1265	51	578	54	553	48	573	52	1787	72	898	68	782	72	800	68
No magazines	1688	54	564	57	757	51	800	55	2764	71	1248	71	1198	74	1281	68
One medium	384	57	160	59	167	59	200	50	646	70	301	68	257	74	322	69
Two media	899	52	370	54	407	49	420	55	1508	73	636	75	659	74	730	68
Three media	1059	49	489	55	451	43	456	56	1517	71	688	71	646	73	640	69
Four media	590	52	251	57	255	52	263	51	756	70	530	74	339	71	335	64
No media	101	57	31	77	41	49	44	57	205	74	86	71	94	74	75	75
Democratic contact	198	54	102	56	87	57	70	37	319	72	162	72	135	70	126	69
Republican contact	190	43	89	46	69	42	113	44	285	73	147	73	107	72	139	73
No party contact	2345	53	994	57	1067	49	1099	55	3618	72	1672	67	1624	74	1487	77
1960																
Television	2341	61	1169	60	1297	60	1567	63	2048	71	1033	71	1176	72	1195	68
No television	216	53	113	48	121	51	116	76	176	70	87	66	97	76	120	73
Newspapers	2219	59	1098	58	1282	61	1399	61	1948	71	962	71	1148	72	1147	69
No newspapers	323	64	184	66	135	51	228	77	271	69	158	64	120	81	160	70
Radio	1204	60	606	59	694	59	702	67	1020	69	499	68	620	70	573	67
No radio	1346	60	670	59	721	60	946	61	1199	73	617	73	651	75	742	70
Magazines	1281	56	601	55	801	55	770	60	1191	73	544	73	782	74	682	69
No magazines	1255	64	672	63	610	65	846	66	1023	69	567	68	499	69	609	71
One medium	231	67	137	68	105	60	124	77	211	71	107	64	72	69	80	70
Two media	650	60	324	58	347	62	472	63	577	73	306	72	297	77	370	64

148

Table 6.4 (continued)

| | DEMOCRATIC CANDIDATES | | | | | | | | REPUBLICAN CANDIDATES | | | | | | | |
| | FULL | | FIRST | | SECOND | | THIRD | | FULL | | FIRST | | SECOND | | THIRD | |
COMMUNICATION SOURCE	n	%+	n	%+	n	%+	n	%+	n	%+	n	%+	n	%+	n	%+
1960 (continued)																
Three media	1033	60	519	59	576	61	674	62	905	70	445	70	544	70	521	70
Four media	603	57	283	56	392	56	331	61	539	71	246	71	361	71	308	70
No media	152	67	66	59	99	73	96	75	140	74	64	73	82	68	80	85
Democratic contact	233	63	118	63	141	60	116	69	193	62	93	65	133	67	76	32
Republican contact	284	52	122	52	179	50	212	57	266	74	115	70	180	73	164	88
No party contact	1861	61	980	60	953	61	1203	66	1604	71	855	72	839	73	951	66
1964																
Television	3135	69	1233	72	1367	67	1230	68	2942	36	1109	35	1296	35	1179	37
No television	308	69	106	75	127	69	144	66	271	33	89	29	120	33	120	36
Newspapers	2855	67	1094	69	1230	66	1164	66	2717	38	1016	37	1170	37	1132	38
No newspapers	593	78	245	85	267	75	212	77	501	26	182	21	248	28	172	26
Radio	1692	67	623	68	709	65	738	70	1610	37	587	39	672	38	712	35
No radio	1740	70	704	76	784	69	633	66	1593	34	601	31	744	34	584	39
Magazines	1579	61	604	63	701	62	619	59	1586	40	627	41	698	38	611	42
No magazines	1836	75	721	79	786	73	741	75	1599	31	566	29	708	31	668	31
One medium	345	79	167	81	160	79	96	74	277	28	116	28	114	32	83	32
Two media	970	72	385	79	426	68	372	72	892	32	325	26	417	34	336	33
Three media	1311	67	461	70	577	68	577	65	1238	38	418	38	540	36	545	41
Four media	705	59	303	58	316	58	297	65	767	41	325	44	312	42	312	37
No media	281	60	124	77	111	64	110	72	238	34	100	33	100	41	90	27
Democratic contact	605	67	229	62	283	67	221	72	577	33	230	37	264	35	208	25
Republican contact	778	61	304	62	341	60	305	59	758	45	290	41	348	41	282	45
No party contact	2644	71	1021	75	1147	60	1062	71	2435	34	896	33	1064	35	1006	34

INTERVIEW PERIOD

Table 6.4 (continued)

	DEMOCRATIC CANDIDATES								REPUBLICAN CANDIDATES							
COMMUNICATION SOURCE	FULL		FIRST		SECOND		THIRD		FULL		FIRST		SECOND		THIRD	
	n	%+	n	%+	n	%+	n	%+	n	%+	n	%+	n	%+	n	%+
1968																
Television	3089	56	1126	48	1345	54	1377	61	3012	58	1157	58	1359	59	1264	56
No television	276	51	101	69	125	42	118	54	265	53	87	47	119	53	119	54
Newspapers	2723	54	962	50	1222	51	1188	59	2667	58	948	55	1236	60	1119	57
No newspapers	659	62	267	49	252	62	320	66	624	53	201	61	248	54	275	49
Radio	1528	58	538	54	603	54	757	63	1472	56	511	57	636	58	673	53
No radio	1813	54	681	46	867	52	715	59	1785	59	727	56	842	60	691	58
Magazines	1409	51	475	50	615	46	644	56	1476	59	488	61	678	59	643	56
No magazines	1966	59	733	50	855	58	861	65	1866	58	780	55	820	59	779	57
One medium	388	57	133	53	167	50	173	65	380	58	164	67	164	58	157	64
Two media	1047	59	404	40	493	62	413	62	991	57	416	52	427	57	389	58
Three media	1265	54	441	55	540	46	573	61	1217	59	417	58	585	63	502	55
Four media	628	53	219	50	252	50	305	56	668	56	224	60	284	56	312	55
No media	539	57	384	41	187	55	251	59	498	51	342	69	189	49	215	43
Democratic contact	590	57	308	62	251	52	308	62	573	57	70	24	252	59	289	51
Republican contact	651	48	238	48	275	41	299	55	736	60	297	61	334	62	301	57
No party contact	3162	57	1195	51	1355	56	1413	62	1935	56	1135	57	1302	57	1272	54
1972																
Television	2082	35	798	32	857	36	986	39	2308	62	936	64	950	64	1017	58
No television	180	33	70	31	90	40	70	30	218	56	89	63	106	54	76	36
Newspapers	1463	34	598	31	549	34	701	38	1608	61	773	62	660	63	705	58
No newspapers	800	38	285	32	396	39	368	39	893	64	342	69	414	68	384	55
Radio	1034	38	393	34	452	40	477	42	1138	60	460	62	488	39	498	55

Table 6.4 (continued)

COMMUNICATION SOURCE	DEMOCRATIC CANDIDATES								REPUBLICAN CANDIDATES							
	FULL		FIRST		SECOND		THIRD		FULL		FIRST		SECOND		THIRD	
	n	%+	n	%+	n	%+	n	%+	n	%+	n	%+	n	%+	n	%+
1972 (continued)																
No radio	1213	32	487	29	493	33	587	36	1346	64	555	65	557	68	562	62
Magazines	996	41	372	28	347	36	481	43	998	60	383	66	410	61	461	56
No magazines	1356	36	516	33	609	37	608	37	1603	65	629	63	630	65	641	57
One medium	406	40	147	39	198	38	162	40	467	60	201	60	199	62	198	58
Two media	745	33	279	27	305	37	374	35	776	65	310	70	339	67	352	58
Three media	686	35	298	31	272	34	313	39	763	60	334	61	295	63	336	58
Four media	391	39	149	33	148	38	213	44	433	59	158	62	148	58	217	55
No media	468	50	175	36	159	30	279	35	417	73	195	59	170	68	310	52
Democratic contact	304	36	137	31	91	33	156	47	340	51	160	56	114	50	150	45
Republican contact	193	26	75	23	83	41	86	17	201	71	87	69	84	69	80	78
No party contact	1493	35	545	34	642	36	739	37	1678	62	634	65	728	64	775	58

INTERVIEW PERIOD

Note: *n*'s for two of each of the three interview periods in each year (i.e., First, Second, Third) have been weighted to compensate for differences in the number of interviews which took place in each time period; consequently, the number of references on any given communication source for interview periods will not total the number of references for the "Full" interview period.

Source: Survey Research Center/Center for Political Studies, University of Michigan, and the Interuniversity Consortium for Political Research.

overall evaluations of the candidates. Generally, Democratic candidates have had more positive images among people using either one or no medium than among heavy media users. Perhaps this merely reflects the traditionally greater Democratic appeal to voters with lower levels of income and formal education, factors associated with fewer opportunities to follow campaigns in media other than the all-pervasive television. Where party contact makes any difference, the relationships are predictable. Respondents contacted by the Democratic party in 1960 were less favorable to Richard Nixon than were people generally, and those contacted by Republicans in 1964 favored Goldwater in their comments more than did persons in the sample at large.

To explore whether these patterns in overall candidate images extend to specific image traits, we look at Table 6.5. Of the four image traits comprising the candidate's political stylistic role, which we derived in Chapter 3 (that is, governmental leadership, partisanship, personality, and dramatic performance), we are able to provide data from the SRC/CPS surveys for the first three. We have reclassified the SRC categories as follows: *party/group* references include all those partisan attributes mentioned by respondents pertaining to liking or disliking a candidate because of people in a political party, party characteristics, the candidate as a party representative, and groups associated with the candidate; *government* traits include references to the candidate's management and philosophy of government and to his domestic and foreign policies; *personal* traits in this instance relate to background, character, experience, ability, and personal attributes. (This reclassification also assures a sufficient number of references to calculate percentage of positive references appearing in each cell of Table 6.5.)

Tendencies rather than definite conclusions appear. For example, on the party/group set of traits for Democratic candidates, magazine readers have been less favorable than persons using other media. The 1972 campaign is an exception; references to McGovern on the party/group dimension were about the same for all media users other than television viewers, who were markedly more positive about the candidate. Again, on party/group traits for Democrats, people following the campaigns in a single medium and those contacted by the Democratic party have been more favorable in their commentary. Aside from party contact there is little variation for Republican candidates on party/group image traits. Evaluations of candidates of both parties vary slightly on government traits as expected, but there are no other differences in tendencies on government traits.

The considerable variation we have seen before in the personal traits attributed to candidates, both between and during election campaigns, is not as clearly evident for the various types of communications sources. For the most part, the percentages of favorable comments about any of a candidate's personal traits are about the same regardless of what medium, or how many, voters use. (There are some differences in personal assessments related to party contact, but, excepting the ratings of Humphrey in 1968, these are expected.) The principal variations relate to the number of media people used

Table 6.5 Images of Presidential Candidates by Communication Sources, 1952–1972 (results in percentages)

IMAGE TRAITS	DEMOCRATIC CANDIDATES										REPUBLICAN CANDIDATES									
	MEDIA TYPES[a]				MEDIA NUMBER				PARTY CONTACT		MEDIA TYPES				MEDIA NUMBER				PARTY CONTACT	
	TV	N	R	M	1	2	3	4	D	R	TV	N	R	M	1	2	3	4	D	R
1952																				
Party/Group	52	51	55	45	81	43	51	41	65	38	38	40	39	44	40	36	39	44	31	33
Government	63	57	56	52	88	54	55	50	63	59	74	75	75	78	71	77	67	77	72	77
Personal	72	71	71	69	72	72	71	69	78	76	73	72	72	76	71	69	73	75	73	76
1956																				
Party/Group	58	57	58	56	74	59	53	58	62	41	36	37	39	38	33	40	45	30	32	56
Government	45	40	29	39	44	44	42	32	42	35	74	76	83	79	73	75	76	78	79	82
Personal	52	52	52	51	49	50	50	56	52	47	76	76	76	76	78	79	75	75	75	73
1960																				
Party/Group	66	64	67	56	81	66	68	58	63	48	47	47	45	52	55	45	51	44	38	51
Government	58	55	56	45	78	57	51	57	61	47	72	71	69	74	76	69	72	73	62	74
Personal	60	59	59	58	62	66	60	59	63	54	78	79	77	78	79	84	76	78	71	79
1964																				
Party/Group	76	73	72	61	86	83	74	58	65	61	20	20	20	21	25	18	18	25	16	27
Government	65	61	62	67	71	67	62	60	62	59	32	34	33	37	23	29	36	36	28	38
Personal	69	68	69	59	80	71	68	59	70	60	46	48	49	49	36	42	48	52	45	51
1968																				
Party/Group	56	52	56	44	64	63	51	50	60	35	40	41	38	42	40	41	41	39	40	54
Government	54	51	56	46	55	57	51	49	56	44	65	67	62	68	63	62	67	66	60	65
Personal	57	56	58	55	55	58	57	56	50	53	58	58	58	57	63	57	59	57	55	59
1972																				
Party/Group	70	43	47	44	57	48	43	41	51	50	30	26	30	27	27	23	27	27	12	35
Government	25	38	40	37	36	44	31	32	32	31	62	58	59	61	62	64	61	55	54	73
Personal	29	36	31	30	24	27	31	57	21	19	72	72	71	71	71	74	71	73	64	77

[a] Media types are abbreviated as follows: "TV" refers to television; "N" refers to newspapers; "R" refers to radio; "M" refers to magazines.

Source: Survey Research Center/Center for Political Studies, University of Michigan, and the Interuniversity Consortium for Political Research.

to follow the campaigns of 1964 and 1972. In 1964, for instance, the commentary about Johnson was clearly more favorable from people using a single medium than those using four; in contrast, people using four media were more positive than those employing one in their comments about Goldwater. This 1964 pattern is consistent with the general trend observed earlier—that is, Democratic images are stronger among users of fewer media, those voters likely to be of lower income and education. But respondents' perceptions of McGovern's personal qualities in 1972 were a clear exception. Only among users of four media did positive comments about McGovern's personal qualities reach a majority, let alone exceed it—an exception for which there is no ready explanation.

Information Sources and Image Change

Finally, what intracampaign changes in candidate images are associated with communication sources? In the columns of Table 6.4 pertaining to campaign periods, there are no trends common to all six presidential elections or to the two major contenders in any given election. There are distinct trends for some candidates. Thus, Stevenson's image in 1952 improved steadily across the three campaign periods among nonusers of television and with people using one or three media; Eisenhower's image declined among nonusers of television and radio, newspaper and magazine readers, and users of one or three media, while improving among people who followed the campaign in all four media. In contrast, in 1956 Stevenson's image generally declined from the first to second campaign period regardless of the voters' communications sources, then improved in the third period; only among users of four media was there a steady decline. Eisenhower's image declined slightly in the final period of the 1956 campaign, perhaps in reaction to his handling of two international crises at the time—the invasion of Suez by allied forces of Israel, Britain, and France and the movement of Soviet troops into Hungary. Unfortunately we lack sufficient evidence to discover whether the communication of news of such critical events significantly alters candidate images.[12]

Political folklore says that in 1960, John F. Kennedy's image improved sharply relative to Richard Nixon's as a result of a series of dramatic televised debates between the two candidates. Published research on the effects of these debates is much less conclusive, generally finding that Kennedy's stock did rise among many who were undecided before the debates (especially as a result of the first debate), but that among those who had already made up their minds, especially Democratic and Republican party identifiers, the debates made little difference. Moreover, there was some evidence that people reacted differently to the debates according to whether they viewed them on television (where Kennedy is said to have gained) or listened on the radio (which assisted Nixon).[13] The data in Table 6.4 provide little evidence of a dramatic reversal in candidate images. The first of the debates came at the close of the first interview period, and the others took place in the second period. Among television users, Kennedy's image rose slightly and Nixon's declined between these periods. Nixon's improvement among radio listeners was scarcely noticeable, no more so than his improvement among nonlisteners. As it turns out, the big improvement in the Kennedy image came among people who did not follow the campaign on television at all, and those who did not hear about the campaign in any of the media!

Each of the remaining elections provides its own unique pattern. In 1964 Johnson's image deteriorated and Goldwater's improved among people who did not use the electronic media, among single media users and those who used three media, and among those contacted by Republicans. People who used all four media thought less of Goldwater as the campaign drew on. In 1968 Humphrey did as well in some instances among people who did not follow the campaign in certain media as he did among people who did (for example, newspaper readers and nonreaders). For the most part, the image of Richard Nixon reached its highest level in late September and early October, with declines thereafter. He suffered steady declines among nonreaders of newspapers, magazine readers, and people who used few or all of the media. George Wallace's image (see Table 6.6) dropped among persons exposed to all types of media, but it improved among people who did not follow the campaign at all. Like perceptions of Humphrey in 1968, those of George McGovern in 1972 became more favorable among television viewers, newspaper readers and nonreaders, and persons contacted by Democrats; his stock also rose among magazine readers and people using a wide variety of media. Nixon's change in image followed a trend only slightly different from that of four years earlier. In sum, looked at from the perspective of what sources people use to learn about presidential campaigns, changes in candidate images are only marginally detectable, and no patterns transcend individual campaigns and candidates.

How Much Do Candidate Images Change?

In this chapter we have seen that the images of candidates in the presidential elections of the last two decades are dynamic, but that changes both between and during campaigns take place within a larger context of stability. We have observed patterns that transcend various presidential elections and aspirants, but patterns having exceptions that are both election-specific and candidate-specific. Among such patterns we can say that samples of articulate Americans have been fairly generous in their overall assessments of presidential candidates—excepting Barry Goldwater, George Wallace, and George McGovern. The percentage of voters' voluntary positive comments about contenders generally exceeds that of negative remarks. And even in the cases of Goldwater and Wallace, a majority of the comments about the candidates' personal attractiveness were favorable.

Over the course of the six presidential elections, the images of Democratic candidates have been most consistently positive in respect to one principal trait—partisanship. References by voters to the Democratic candidate's political party and to the social groups he and the party represent generally promote positive images. In a less consistent fashion, people have assessed

Table 6.6 Images of George Wallace as a Presidential Candidate, by Interview Date, 1968

| | TOTAL SAMPLE | | SEPT. 6-NOV. 4 | | | INTERVIEW PERIOD | | | | | |
| | | | DEMOCRATS | INDEPENDENTS | REPUBLICANS | SEPT. 6-25 | | SEPT. 26-OCT. 25 | | OCT. 16-NOV. 4 | |
	n	%+	%+	%+	%+	n	%+	n	%+	n	%+
Image Traits:											
Experience/ability	352	16	18	18	07	137	30	137	14	174	14
Background/character	1126	22	19	24	21	382	25	481	22	525	21
Personal	297	62	56	71	62	101	53	118	65	148	62
Party representative	221	26	27	30	18	96	23	108	29	78	24
Issues	256	56	49	68	52	103	60	99	60	123	51
Domestic policies	1029	46	41	51	57	451	52	436	45	437	44
Foreign policies	147	51	52	56	57	36	66	72	50	65	48
Relation to groups	288	22	18	25	26	127	28	122	16	122	26
Total	3716	35				1433	40	1570	34	1672	34
Respondent's Partisanship:											
Democrat	1636	31				598	42	697	30	745	27
Strong	762	24				223	35	343	24	352	20
Weak	874	37				375	46	354	36	393	34
Independent	1168	41				460	42	533	38	481	42
Democrat	390	38				159	33	158	34	180	43
Independent	359	45				159	44	151	41	154	51
Republican	419	39				142	51	224	40	147	33
Republican	881	34				362	32	322	35	441	34
Weak	530	34				199	31	186	33	282	36
Strong	351	33				163	33	136	38	159	30

Table 6.6 (continued)

| | SEPT. 6–NOV. 4 | | | | INTERVIEW PERIOD | | | | | |
| | TOTAL SAMPLE | | DEMOCRATS | INDEPENDENTS | REPUBLICANS | SEPT. 6–25 | | SEPT. 26–OCT. 25 | | OCT. 16–NOV. 4 | |
	n	%+	%+	%+	%+	n	%+	n	%+	n	%+
Communication:											
TV	2960	34				1121	39	1277	34	1313	32
No TV	273	30				98	49	120	93	162	46
Newspapers	2685	34				957	38	1192	34	1182	33
No newspapers	573	35				259	48	212	35	273	29
Radio	1416	35				516	41	564	36	688	31
No radio	1804	33				699	39	833	32	735	33
Magazines	1401	31				434	35	615	31	653	29
No magazines	1801	35				785	43	749	33	784	34
One source	326	32				146	45	142	32	132	26
Two sources	1010	35				392	42	460	35	417	32
Three sources	1204	32				467	37	522	32	534	31
Four sources	624	34				184	36	255	36	314	32
No sources	550	42				262	41	191	38	264	46
Democratic contact	554	26				110	41	239	27	280	23
Contact D and R	283	30				89	34	127	32	117	27
Republican contact	459	34				94	37	291	36	274	30
No contact	3034	35				1195	40	1272	34	1365	35

Note: *n*'s for interview subsamples have been weighted to compensate for differing numbers of interviews during each time period.

Source: Survey Research Center/Center for Political Studies, University of Michigan, and the Interuniversity Consortium for Political Research.

Democratic candidates positively for their domestic policy stands, a governmental leadership trait associated with the candidate's political role. By contrast, Republican candidates have received generally positive assessments of selected personal traits—personal attributes, background, and character—and have added to that positive perceptions of such leadership qualities as experience and ability. But we have noted that perceptions of candidates' personal traits vary considerably from election to election. And, although there tends to be a partisan coloring to how voters perceive any candidate, a presidential aspirant (an Eisenhower or a Kennedy) may have a distinct appeal to weak partisans and Independents because of his perceived qualities as a fellow human being.

As for alterations in the images of presidential candidates during campaigns, the rule of minimal change applies, but people still shift selected perceptions of the candidates and of specific traits. We have seen clear evidence of changes in voters' overall evaluations of candidates in a campaign (such as Kennedy's in 1960 and Humphrey's in 1968) and in their perceptions of specific traits, even if these alterations have not affected their overall assessment greatly (as with perceptions of Johnson in 1964). Of various traits voters perceive, their perceptions of candidates' personal qualities are most changeable during campaigns, just between elections. When personal traits comprise a major portion of a candidate's image, changes in how people respond to that candidate as a person can lead to a substantial shift in their overall assessment of him (again, the Kennedy candidacy is a case in point).

We have noted—as the classic voting studies insisted—that Independents of various shades are most likely to change their images of candidates during campaigns. If, as we keep being reminded by various political observers, the bonds of party are loosening and independence is on the ascendency, we might expect more fluidity in candidate images in future presidential elections.[14] But, contrary to expectations deduced from voting studies, we have also detected image changes even among persons holding strong partisan self-images. We suspect these are people who start out unhappy with the party's nominee, but who sort themselves out from other partisans as the campaign progresses and ultimately return to the party fold, gradually improving their appraisals of their nominee in the process.

Finally, we have found that relatively few changes in the images of political candidates, both between and during elections, stem from voters' following the campaigns through different kinds of communication sources, at least as we have been able to measure them. The changes that we have been able to document have been specific to traits of individual candidates, but there is no general pattern of image change relative to differing types or numbers of media. We are obviously reluctant to assert that mass communications are unimportant in the imagery of presidential candidates, for to do so would fly in the face of findings from other studies as well as our own intuition. We caution the reader that this is one of many areas where students of political communication have the most tentative things to say.[15]

Within an overall pattern of stability, then, the images of candidates between and during presidential elections do change. They change for specific candidates, in specific traits, among specific types of voters. There is at least one other variety of change that can occur in candidate images—the changes that take place in people's perceptions of candidates as a result of their winning or losing elections. We move to that question in the following chapter.

Notes

1. John H. Kessel, *The Goldwater Coalition* (Indianapolis: Bobbs-Merrill, 1968), pp. 255–97.

2. The dates of interviews for each presidential election survey, 1952–1972, are as follows:

YEAR	FULL	FIRST	SECOND	THIRD
1952	Sept. 15–Nov. 2	Sept. 15–30	Oct. 1–16	Oct. 17–Nov. 2
1956	Sept. 17–Nov. 5	Sept. 17–Oct. 2	Oct. 3–19	Oct. 20–Nov. 5
1960	Sept. 12–Nov. 7	Sept. 12–30	Oct. 1–19	Oct. 20–Nov. 7
1964	Sept. 7–Nov. 2	Sept. 7–Oct. 3	Oct. 4–17	Oct. 18–Nov. 2
1968	Sept. 6–Nov. 4	Sept. 6–25	Sept. 26–Oct. 15	Oct. 16–Nov. 4
1972	Sept. 8–Nov. 6	Aug. 8–Oct. 3	Oct. 4–17	Oct. 18–Nov. 6

3. Doris A. Graber, "Personal Qualities in Presidential Images: The Contribution of the Press," *Midwest Journal of Political Science* 16 (February 1972): 46–76.

4. Barbara Hinckley et al., "Information and the Vote: A Comparative Election Study," *American Politics Quarterly* 2 (April 1974): 131–59. Hinckley et al. report that voters are more likely to perceive and to comment on the traits of presidential candidates than on candidates for lesser offices. Thus, 93 percent of voters commented about presidential candidates in 1968 compared to 86 percent for gubernatorial and 85 percent for U.S. senatorial candidates.

5. Figure 6.1 is adapted from data presented by Samuel A. Kirkpatrick et al., "Candidate and Party Images in the American Electorate: A Longitudinal Analysis" (paper delivered at the Annual Meeting of the Southwestern Political Science Association, Dallas, Texas, March 28–30, 1974).

6. Ibid.

7. Charles O. Jones, "The 1964 Presidential Election: Further Adventures in Wonderland," *American Government Annual 1965–1966* (New York: Holt, Rinehart and Winston, 1965), pp. 1–30.

8. Thomas W. Benham, "Polling for a Presidential Candidate: Some Observations on the 1964 Campaign," *Public Opinion Quarterly* 29 (Summer 1965): 185–200.

9. Philip E. Converse, "Information Flow and the Stability of Partisan Attitudes," *Public Opinion Quarterly* 26 (Winter 1962): 578–99.

10. Kenneth Boulding, *The Image* (Ann Arbor, Mich.: University of Michigan Press, 1956).

11. Television Information Office, *What People Think of Television and Other Mass Media, 1959–1972* (New York: The Roper Organization, 1973).

12. An attempt to relate critical events to how people perceive various presidents has been made by John E. Mueller, *War Presidents and Public Opinion* (New York: John Wiley, 1973), chap. 9.

13. For a summary of studies of the effects of the televised debates, see Dan Nimmo, *The Political Persuaders* (Englewood Cliffs, N.J.: Prentice-Hall, 1970), pp. 161–62. One of the major studies of such effects is Carter's effort based on responses of a panel consisting of sixty Democrats and sixty Republicans. He found that Kennedy improved in such image attributes as industriousness and experience as a result of the debates; both candidates were perceived as "tough" in light of the debates. There were no substantial changes, however, in such perceived personal attributes as hard-working, honesty, intelligence, and imagination. See Richard F. Carter, "Some Effects of the Debates," in Sidney Kraus, ed., *The Great Debates* (Bloomington, Ind.: Indiana University Press, 1962), chap. 14.

14. See Walter Dean Burnham, *Critical Elections and the Mainsprings of American Politics* (New York: W. W. Norton, 1973); Louis Harris, *The Anguish of Change* (New York: W. W. Norton, 1973); and Everett Carll Ladd, Jr., *American Political Parties: Social Change and Political Response* (New York: W. W. Norton, 1970), chap. 6.

15. A comprehensive review of the studies indicating the effects of mass communications on elections is Maxwell E. McCombs, "Mass Communication in Political Campaigns: Information, Gratification, and Persuasion," in F. Gerald Kline and Phillip J. Tichenor, eds., *Current Perspectives in Mass Communication Research* (Beverly Hills, Calif.: Sage Publications, 1972), pp. 169–94.

Chapter Seven

Winners and Losers: The Effects of Victory or Defeat on the Images of Political Candidates

Not everyone running for public office wins. Many candidates seem to be called, few chosen. Do people think differently of a candidate after the election than they did during the campaign? And, are there differences in the postelection images of the victor and the vanquished? These are questions directly related to the more general phenomenon of changes in candidate images, which we discussed in the preceding chapter. In this chapter we shift our attention from the electoral campaign to its outcome and ask what changes in the images of political candidates flow from electoral victory or defeat. We begin our exploration by reviewing what other researchers have found. Then we consider the presidential election of 1972, employing research designed specifically to examine postelection shifts in voters' perceptions of candidates.

Congratulations and Condolences Effects

Studying the effects of electoral outcome on popular images of the contenders is important not only because by pursuing it we learn more about the general phenomenon of perceptual change, but it is also of practical political relevance, especially to students concerned with the workings and persistence of democratic regimes. In political terms the question is, Can a democratically elected official—emerging from an unusually heated, often bitter, conflict-laden political campaign—rally support and respect among his opponents' followers simply because there are image benefits attached to being victor? V. O. Key, Jr., a perceptive student of electoral politics, suggested that an important feature of elections, "whatever else they decide, is their production of acceptable decisions on the succession to power in the state. At the leadership levels that quality of elections manifests itself when the losers surrender the seals of authority to the winners of the popular majority. The development of norms, expectations, and restraints that enable those with authority to surrender it in response to popular decisions is a rare phenomenon among the rulers of men." Key also found that the mass of those who have supported a losing candidate reconcile themselves to the defeat. From surveys he cited the following evidence: (1) after an election people reach the conclusion that the outcome probably will make little difference to the country, anyway; (2) many persons report having voted for the winner when they actually did not, perhaps out of a reluctance to admit being on the losing side; and (3) some people simply want to be good sports about the whole thing and accept defeat gracefully. For whatever reason, says Key, "People accomplish a psychological adjustment to the loss of an election."[1] That adjustment contributes to a stable political climate wherein leaders are able to rally popular support in crises, and in which the routine, day-to-day workings of government can proceed.

Published studies of the effects of electoral victory or defeat on popular perceptions of political candidates revolve around two questions implied in Key's observations—do postelection changes occur and, if so, what is the source of the psychological adjustment? Two of the earliest studies of the problem appeared well before Key's observations. In separate investigations Korchin[2] and Paul[3] found that after an election there was a tendency for people to be more favorable toward the winner, although not so toward the defeated candidate. Both researchers labelled such a postelection restructuring of the winner's image the "fait accompli effect." In other words, faced with an accomplished fact—such as the knowledge that the winner of a presidential election would be president for four years—both supporters and opponents accept it and react to the winner as an officeholder and a power wielder, not merely as an office seeker.

Where the winner, but not the loser, improves his image after an election, the fait accompli effect seems to offer a reasonable explanation. But what if the defeated candidate also improves his image? Does this happen? Research indicates that it does. George Stricker examined the pre- and postelection images of Richard Nixon and John Kennedy in 1960. Using semantic differential ratings, Stricker found that his subjects prior to the election evaluated the two candidates in distinctly different ways, respondents being more favorable toward the candidate of their voting choice than toward his opponent (as we might expect). After the election, however, both the winner and the loser received more favorable evaluations than they had received before. Moreover, respondents after the election tended to see the two candidates as possessing more similar images than they had earlier.[4] In a study similar to Stricker's, Raven and Gallo had their subjects (students in upper-division sociology and psychology courses) rate Richard Nixon and John Kennedy on a twenty-item semantic differential test, one day prior to the 1960 presidential election and one month later. They found that the students rated both candidates more favorably after the election than prior to it, and that the image improvement took place whether partisans were rating their own candidate or that of the opposing political party.[5]

Faced with the fact that perceptions of *both* winning and losing candidates were more favorable after the election than before it, how did researchers explain this phenomenon? In the case of the winner, the fait accompli effect was a partial answer. Raven and Gallo added to fait accompli what they considered another effect of election outcome. Based on other studies they had conducted, they argued that presidential nominating conventions have a polarizing effect on voters' perceptions of candidates. That is, when contenders are nominated and thus identified to the mass of citizens as representatives of opposing parties, voters' partisan self-images greatly influence their images of contending candidates. After the election a *depolarizing* effect sets in. Winning partisans continue to favor their candidate, and losing partisans grow more favorable toward the winner because his party affiliation simply is less important to them once the heat of the campaign is over. Thus, the fait accompli and depolarizing effects enhance the winner's image. And, since winning partisans also become more favorable to the losing partisan (because his negative party affiliation no longer seems as salient), depolarization actually benefits both winners and losers.[6]

Of the studies reviewed thus far, all examined primarily a single aspect of a candidate's image—affect. All focused essentially on favorable versus unfavorable evaluations of candidates. For example, the studies of Stricker and of Raven and Gallo in employing the semantic differential concerned themselves mainly with the evaluative element in voters' semantic differentiation (see Chapter 3) rather than with ratings of potency or activity. Two recently published studies are more broadly based. Anderson and Bass had two groups of subjects (one composed of thirty-one students and the other of twenty-one industrial workers) rate various candidates on three sets of semantic differential scales before and after election day. The candidates rated were presidential contenders (Lyndon Johnson and Barry Goldwater), Michigan gubernatorial candidates (George Romney and Neil Staebler), and two Michigan senatorial candidates (Phillip Hart and Elly M. Peterson). Of the three sets of semantic differential scales used, one represented evaluation, activity, and potency scales (they were the same scales previously used by Stricker); a second set tapped the respondents' perceptions of a candidate's "political image" (a series of traits comparable to what we have referred to several times as political role); and a third set sought ratings of the candidate's "personal image" (a dimension comparable to stylistic role in our discussions).

By factor analyzing respondents' pre- and postelection ratings of the candidates, Anderson and Bass discovered two major components of candidate images. The first was an *evaluative* component consisting of ratings of the candidates on such items as "good-bad," "wise-foolish," and "considerate-inconsiderate." This component reflected primarily affective orientations toward the candidates and turned out to be roughly parallel to the "personal image," or stylistic role of candidates. The second component Anderson and Bass dubbed a *political assertiveness* factor. It included such ratings as "influential-noninfluential," "decisive-indecisive," and "powerful-powerless." It was primarily cognitive in emphasis (reflecting beliefs about candidates' assertiveness) and comparable to political image, or the political role of the candidate. What, then, were the pre- to postelection changes in these components? According to Anderson and Bass, "the ratings of the winning candidates increased significantly on both the evaluative and political factors, while the ratings of the losing candidates showed significant decreases only on the political factor. Such results are quite in contrast to the . . . studies that found significant increases in the evaluative ratings of both the defeated and winning candidates."[7]

Anderson and Bass offer relatively few new insights into why winning candidates receive more favorable postelection evaluations whereas losers do not. The fait accompli and depolarization effects apparently operated; however, Anderson and Bass simply attributed changes in the political assertiveness component to the fact that all three winning candidates were incumbents seeking reelection, whereas the losers were all seeking offices of president, governor, and senator for the first time. Incumbents reelected retained and enhanced their political assertiveness. But the defeated candidates not only lost the offices they sought, they also surrendered their previously held positions. Thus, left without offices of governmental and party influence, their political assertiveness waned, or so it apparently seemed to the respondents surveyed.

The second recent study that attempted to ascertain something more than simply changes in voters' evaluations of candidates was conducted by Blumler and McQuail. They examined the changing perceptions of a representative sampling of voters in the 1964 general elections in Great Britain. More specifically, they had respondents rate the leaders of the political parties—Sir Alec Douglas-Home of the Conservatives, Harold Wilson of the Labourites, and Jo Grimond of the Liberals—on several semantic differential items. Their respondents rated the party leaders before, during, and after the campaign. A factor analysis revealed three major components of these ratings—evaluative, strength, and activity factors congruent with the semantic differential scales. The major shifts in perceptions came through the reassessments of the strength of the rival party leaders. With respect to the two major party leaders, "The net effect of the 1962 election was to depress Sir Alec's—and to enhance Mr. Wilson's—public reputation for strength."[8] Blumler and McQuail offered the view that a campaign and its outcome present a challenge primarily for a leader to demonstrate his strength, confidence, and persuasiveness. Rarely are there sufficient moral tests of a leader to contribute to changes in moral evaluations. Hence, changes in perceptions of the leader's potency are those most likely to result from elections.

How can we summarize the major conclusions of these studies of the effects of victory or defeat on the perceptions of political candidates? First, it is apparent that *voters become more favorable to winners,* due to the effects of fait accompli and depolarization. Perhaps voters are congratulating the victors and themselves for having voted for him and, if they did not vote for him, at least rationalizing the outcome. Second, the *results are mixed respecting perceptions of the losers.* Some studies indicate a more favorable postelection image of the loser, who benefits from depolarization, condolences, and sympathy. Third, as a result of winning or losing, *perceptions of different image traits seem to change.* Political role traits are more likely to change after the election than are stylistic role traits. This finding is in sharp contrast to the inferences we drew in Chapter 6 about intracampaign image changes, which appear to consist largely of shifting perceptions of personal attributes. Hence, let us tentatively assert that campaigns contribute to changes in stylistic appraisals of candidates, and electoral outcomes work to change political appraisals. In any event, as suggested, voters make psychological adjustments following elections. Two studies of the 1972 presidential election may help us find out what kinds of voters change and in what ways.

Victor and Vanquished: Pre- and Post-Election Images of Nixon and McGovern

We saw much earlier (Table 2.3) that the proportion of persons in 1972 who thought of themselves as Democrats outweighed that of Republican self-images by a 53 to 34 margin. Since Richard Nixon was reelected to a second term in the White House with 61 percent of the popular vote, we can assume that not only did Nixon attract Independents but also that a substantial portion of Democrats defected from their party in 1972. In previous chapters we suggested a few reasons why—for example, the role of issues in the election (Chapter 2) and the comparative images of Nixon and McGovern (Chapters 3 and 6). Chapter 8 explores the relative influence of a number of factors, including candidate images, on the outcome of the 1972 presidential election. Here, however, we want to look at the effect of the outcome, which was of landslide proportions, on the images of the major party candidates.

On the basis of the studies just reviewed, what kinds of effects might we expect in relation to candidate images after the 1972 election? First, we might expect that Nixon's image would be more favorable following the election, reflecting the effects of fait accompli, depolarization, and the new lease on his incumbency. The effect of McGovern's lopsided defeat on his image is harder to predict. If depolarization and/or condolences are at work, we might expect McGovern's postelection image to be more positive than were voters' preelection perceptions of him. Yet, although people might evaluate him more favorably out of sympathy, we might expect their estimates of his political strength to decrease.

Beyond what we might expect for Nixon's image and McGovern's when we treat each candidate alone, we can also ask what we might anticipate regarding the relative change in the images of the two candidates when compared with one another. Few of the studies we have looked at help us make these estimates, but there is a body of theory derived from the study of attitudes and attitude change that may be of assistance. The theory has many variations (theories of cognitive dissonance,[9] balance,[10] and congruence[11]), among which there are both sharp and subtle differences. At the risk of oversimplification, for our purposes we shall lump them together under the label of *consistency* theory.[12] Essentially, consistency theory says that when perceivers have inconsistent or contradictory views of an object-of-perception in a situation (recall our discussion in Chapter 3), that inconsistency produces discomfort or tension. The inconsistencies may arise from incompatible beliefs (cognitions), feelings (affects), or tendencies (conations) toward an object-of-perception, or from combinations of these three. The resulting tension motivates the perceiver to reduce the conflict by changing one or more of his perceptions.

In relation to the images of political candidates, consistency theory suggests a number of possible changes in the postelection shifts in comparative perceptions of Nixon and McGovern in 1972. To illustrate, we will note a few of these shifting perceptions. Consistency theory predicts that if a person acts or commits himself regarding an object, that action influences how much he will change his view of the object. For instance, if he did not like a candidate, yet for one reason or another voted for him, after the election the voter might come to think more highly of the candidate (especially if his voting choice won). Or, suppose a voter had favorable images of both Nixon and McGovern prior to the 1972 election but voted for the loser. Consistency theory suggests that the voter's image of McGovern would remain favorable while his image of Nixon would decline, as he reduces the conflict between having voted for an appealing loser but having failed to vote for an equally appealing winner. Similarly, take the voter who liked neither candidate before the election. If he voted for Nixon, having supported a winner sets up a conflict between not liking Nixon but liking being a winner. The probability is that he would think more highly of Nixon after the election. But if he voted for McGovern and does not like being a loser, he might see Nixon, the winner, more positively (a variation on the fait accompli effect).

Of course, since the candidates represent their respective political parties, partisan self-images enter the calculus. The Democrat who prefers Nixon to McGovern but is hardly wild about either is in a state of conflict. He might, after voting for Nixon, become more positive toward the winning Republican candidate. But it might also be that, after the election, his partisanship makes him uncomfortable with "four more years" of Republican Nixon and as a result his postelection image of Nixon becomes even more unfavorable and perhaps his image of McGovern becomes more positive. In sum, the relative strength of the respective beliefs, feelings, and leanings influence the amount of postelection reconstruction of images that a person undergoes.

Let us now shift our attention to data and studies of Nixon and McGovern images in 1972. Then we will take a backward look at explanations derived from previous studies and consistency theory to help us understand shifts in candidate images.

Pre- and Postelection Images of Presidential Candidates: A Survey Analysis

In Chapter 2 we described a technique employed by the Survey Research Center/Center for Political Studies to obtain information on how people feel about political candidates. People rate respective candidates on a "feeling thermometer," a scale that ranges from 0 to 100 degrees. If they feel warm toward a candidate, they may give him a rating at the upper end of the scale, say 70 degrees or above; if they are cool, they may rate him below 40 degrees; and if they are noncommital, a rating near 50 degrees is not unusual. In the 1972 SRC/CPS presidential election study, a panel of respondents rated both Richard Nixon and George McGovern in pre- and postelection interviews. We have coded those ratings into seven levels of degrees. Table 7.1 displays the distribution of pre- and postelection ratings for the two presidential candidates.

Table 7.1   Pre- and Postelection Ratings of Presidential Candidates, 1972

THERMOMETER RATINGS	NIXON		McGOVERN	
	PRE-ELECTION	POST-ELECTION	PRE-ELECTION	POST-ELECTION
0–15°	08.6%	08.6%	21.5%	24.3%
16–28	00.5	00.3	01.0	01.2
29–43	09.3	11.3	17.8	20.1
44–55	10.2	09.7	16.9	15.7
56–69	12.4	11.6	11.5	11.2
70–84	19.3	19.6	13.5	13.7
85–97	39.7	38.9	17.8	13.8
Total	100.0%	100.0%	100.0%	100.0%
	$n = 2143$		$n = 2105$	

Source: Survey Research Center/Center for Political Studies, University of Michigan, and the Interuniversity Consortium for Political Research.

It is apparent that feelings toward Nixon both before and after the election were considerably warmer than they were toward McGovern. Almost 40 percent of respondents rated Nixon above the 85-degree level prior to and after the election; approximately one-fifth of the sample rated McGovern below 15 degrees in pre- and postelection interviews. Table 7.1 indicates that there were also changes on the candidates' ratings following the elections, more so in the loser's than in the winner's case. Both before and after the election Nixon's average, or mean, rating lay in the 56–69-degree range. At each degree range, there were only slight pre- to postelection variations in thermometer ratings. McGovern, in comparison, received less favorable ratings after the election. His mean rating fell from the 44–55 to the 29–43-degree range. Cooler ratings increased in proportion, warmer ones decreased. Overall, the pre- to postelection ratings of Nixon correlated with a coefficient of .72; the McGovern correlation was .69.

To get a better idea of pre- to postelection changes in affective orientations toward the 1972 presidential candidates, as measured by the "feeling thermometer," we turn to a comprehensive study undertaken by Cigler and Getter.[13] Using the raw thermometer ratings respondents gave to each candidate (rather than relying on a seven-level classification, as we have in Table 7.1), Cigler and Getter compared the mean ratings of Nixon and McGovern before and after the election. The mean ratings of both candidates fell following the election, but Nixon's rating hardly declined—from 65.5 to 65.3—whereas McGovern's dropped from 48.9 to 45.4. Thus, the changes in mean ratings constituted striking evidence of neither a congratulation nor a condolence effect for the respective candidates.

However, mean ratings of the samples can obscure the image changes that took place. Therefore, Cigler and Getter shifted their attention to the percentages of respondents that did and did not change their ratings, and by how much. They found that approximately two-thirds of those sampled changed their evaluations of each of the two candidates after the election, with only 37 and 31 percent, respectively, not changing their evaluations of Nixon and McGovern. Of change toward Nixon, 34 percent felt warmer toward him following the election; 29 percent felt cooler. Only 28 percent warmed to McGovern after the election, but 42 percent felt cooler.

Recall that two of the previously published studies (those by Korchin and by Paul) reported that their subjects evaluated winners, but not losers, more favorably after the election. On the other hand, the studies of Stricker and of Raven and Gallo revealed more favorable postelection evaluations for both winners and losers. The findings of Cigler and Getter do not reconcile these conflicting results. To be sure, the Stricker and Raven-Gallo view receives no support from the fact that only 9 percent of the SRC/CPS respondents surveyed increased in warmth toward both candidates (16 percent even rated both candidates *lower*). Yet only 20 percent rated Nixon higher and McGovern either the same or lower, compared to 19 percent who rated McGovern higher and Nixon either the same or lower—scant confirmation of the Korchin-Paul thesis. We might conclude that a great deal of postelection shifting of perceptions seems to occur, but not always in a consistent way. If we treat those changes only in aggregate ways, the favorable-unfavorable shifts for each candidate cancel each other and, as a result, we lose sight of the fact that various subgroups of the population change their evaluations of the candidates. The Cigler-Getter data indicate clearly that some people congratulate the winner and console the loser; others congratulate the winner but don't console the loser; still others console the loser but don't congratulate the winner; and a surprising proportion become more negative toward both candidates.

Based on the SRC/CPS surveys, what kinds of voters changed their evaluations of Nixon and McGovern in 1972? The Cigler-Getter analysis suggests that changes are more likely a function of whom respondents voted for than of voters' demographic and socioeconomic characteristics, partisan self-images, ideological leanings, or positions on key election issues. People who voted for Nixon, as we might expect on the basis of consistency theory, evaluated him more favorably after the election and judged McGovern more negatively. McGovern voters—in the potential conflict of having preferred a loser—resolved any inconsistency in the direction of their committed act; that is, they became more hostile toward Nixon following the election and slightly more favorable toward McGovern. Clearly the fait accompli effect discovered by Korchin and by Paul, in which supporters of the loser evaluate the winner more favorably, did not materialize in 1972.

Cigler and Getter make one other point, concerning people's comparative evaluations of candidates, which is particularly relevant to our discussion. That is, Is the candidate they rate more favorably before the election the same candidate they are more positive toward afterward? More than three-fourths of respondents were consistent in their comparative ratings in 1972. Only 3 percent who rated Nixon higher than his opponent in preelection interviews rated McGovern higher than Nixon after the election; another 3 percent were more favorable to McGovern than Nixon before the election, but changed to Nixon over McGovern in postelection interviews. Older voters were more consistent in rating Nixon higher; younger voters rated McGovern more consistently. Those having more formal education were more consistent than those having less. And, Republicans and Independents were more consistent in their comparative evaluations than were Democrats. But voting behavior was most clearly related to consistency of comparative ratings: McGovern voters made their postelection ratings consonant with their votes, that is, rating McGovern higher than Nixon. Nixon voters rated Nixon higher than McGovern in postelection interviews, thus reinforcing their voting decision.

In sum, the Cigler-Getter findings from their study of pre- and postelection affective responses to Nixon and McGovern in 1972 provide interesting contrasts with those of earlier investigations. They uncovered no evidence of congratulation and condolence effects, or what they refer to as a "bandwagon" effect for the winner and "sympathy" for the loser. Rather than a fait accompli effect they suggest that supporters of each candidate simply reinforced their preelection evaluations. Backers of the winner became slightly more pro-Nixon and anti-McGovern, and McGovern supporters grew more hostile to the winner and more favorable to the loser. Instead of a depolarization effect there was, in short, polarization.

Studies employing the "feeling thermometer" to assess the effects of victory or defeat on the images of political candidates necessarily limit themselves to measuring the affective aspects of images. Moreover, they are little help in determining what changes occur in voters' perceptions of the political and stylistic role dimensions of candidate images. Therefore, to round out our picture of the effects of the outcome of the 1972 election on the images of the victor and the vanquished, we offer the results of a study conducted by the authors.

Pre- and Postelection Images of Presidential Candidates: A Q-Analysis

We have reported elsewhere (Chapters 3, 4, and 5) the results of various *Q*-studies, the methods and techniques of which we spell out in appendixes at the end of this book. The reader will recall that in such studies, people define their images of an object-of-perception (in this case, a candidate) by sorting or ranking opinion statements according to their degree of agreement or disagreement with those statements. The persons' sorts are correlated and factor analyzed to produce clusters of people who have ranked the statements in similar ways. Each cluster, according to William Stephenson, who devised *Q*-technique, can be thought of as people who share a common image.[14] *Q*-studies, then, permit us to explore image types as well as image traits, a distinction we developed in Chapter 3.

During a two-week period before the presidential election in November 1972, we asked people to sort fifty-two statements describing various attributes of presidential candidates. The same people sorted the statements again after the election. We selected the fifty-two statements from several hundred obtained earlier by asking students in political science courses on two college campuses to write short paragraphs describing their "ideal president," Richard Nixon, and George McGovern. We asked the students to write about the four image traits that characterize a candidate's political and stylistic roles; that is, we asked them to describe the ideal president or one of the candidates as a public official, party leader, person, and dramatic performer. From all of these statements we selected thirteen to represent each of the four image traits and yield a *Q*-sample of fifty-two statements. For example, one of the statements in the public official category was, "He is a good administrator." "He articulates what the party stands for and always tries to show how each action or proposal is moving toward that goal" exemplifies the politician trait set. One of the personal attributes was, "He has the highest degree of honesty, integrity, and intelligence." For the performer category one statement was, "He is cool, calm, and collected."

Thirty-six subjects sorted the statements three times before and three times following the election—one sort to define their preferred characteristics of the ideal president, one for Richard Nixon, and one for George McGovern.[15] The subjects consisted of eighteen college students who, until 1972, had never before voted in a presidential election and eighteen college graduates, each of whom had voted in several presidential elections. The eighteen subjects in each set came from three university communities—six "first-time" and six "experienced" voters from each campus—located in a midwestern state, a border state, and a southern state. Obviously, we make no claim that such a sample represents all voters or even those with exposure to college, anymore than the findings of Stricker or of Raven and Gallo were intended to be generalized to a universe of all voters. (This study was part of a larger research effort, and we describe the rationale for our sample selection in Appendix D.)

Pre- to Postelection Perceptual Stability

Shifts in our subjects' perceptions of political candidates following the election took place against a background of remarkable stability. An indicator of this stability is the fact that respondents sorted the fifty-two statements in approximately the same ways in two interview periods separated by as much as six weeks. Consider the mean correlations between pre- and postelection sorts.[16] For instance, on sorts pertaining to the ideal president the correlation was .77; coefficients for the Nixon and McGovern sorts were lower, but still a robust .65 and .64, respectively. (Compare these coefficients with the .72 and .69 correlations of SRC/CPS respondents for Nixon and McGovern on the "feeling thermometer," a measuring device more likely to produce pre- to postelection stability.) Coefficients for all respondents on the ideal president sort were statistically significant; those for all but two respondents on the Nixon sort and one on the McGovern sort were also significant.[17]

Drawing inferences from the voting studies reviewed in Chapter 2, we could have expected our respondents to be fairly stable in their perceptions from pre- to postelection periods simply because those voting studies have generally found that the belief and value systems of persons having several years of formal education (in this case, college experience) are fairly consistent, organized, and integrated over time. But we would also have expected less stability in perceptions of first-time voters than in those of more experienced voting citizens. According to Angus Campbell's classic study, *The American Voter,* "The stronger the individual's sense of attachment to one of the parties, the greater his psychological involvement in political affairs."[18] Positively related to the strength of partisan self-images and degree of involvement is greater stability in political beliefs, feelings, and tendencies. And, "Since most individuals hew to a single party throughout their lives, strength of party identification increases with age." Thus, our youthful first-time voters are "less securely bound to the existing party system"[19] and, therefore, we might expect weaker partisan self-images among first-time voters than among experienced voters, contributing to less stability in political images.

Our data, however, indicate no significant differences in the pre- to postelection stability of the images of first-time and experienced voters. For first-timers, the mean correlation coefficient for pre- and postelection sorts on the ideal president was .75, and for the experienced it was .80. On Nixon sorts the correlations were .61 and .67, respectively, for first-time and experienced voters; and the coefficients were .59 and .66, respectively, for McGovern sorts.[20] In fact, extensive analysis of first-time versus experienced voters revealed few image differences.

We get another indication of the overall stability in the images of the winner and loser and some hint of changes that did occur by examining how well each party's candidate fit the conception of ideal president held by respondents prior to, and following, the election. Before the election the mean correlation of subjects' Nixon sorts to their ideal was .23; the correlation of

McGovern sorts to the ideal was also .23. Following the election the mean correlation of subjects' postelection sorts on the ideal to Nixon sorts was almost the same, .22, but on McGovern it had dropped to .08. In short, the losing candidate was farther from respondents' image of the "ideal president" than he had been earlier. Oddly enough, among Nixon voters their winning candidate was also less ideal following the election than before. The mean preelection correlation among Nixon supporters between their ideal and Nixon was .75, and between their ideal and McGovern it was −.12; after the election the ideal to Nixon correlation was .28, and to McGovern it was −.14. Similarly, among his supporters McGovern proved less ideal after the election than before: mean preelection correlation between sorts on "ideal president" and George McGovern was .55, and between the ideal and Nixon −.36; postelection correlations of ideal to McGovern dropped to .24, while for Nixon the correlation improved to .10. On the surface, then, among their respective supporters both Nixon and McGovern were less ideal after the election.

It is worth noting that prior to the election, the correlation between the ideal and Nixon among his supporters was positive; between the ideal and McGovern it was negative. Among McGovern supporters the correlation between the ideal and Nixon was negative, and positive between the ideal and McGovern. In short, we see scant evidence of the alleged phenomenon of "negative voting," whereby both candidates should be negatively correlated with the ideal. Across the entire sample, preelection sorts of both candidates were positively correlated with the ideal, hardly an indication that voters throught they were choosing between the lesser of two evils.[21]

Pre- to Postelection Images

To ascertain the pre- and postelection images of our subjects, we Q-factor analyzed each of the three sets of thirty-six preelection sorts for "ideal president," Richard Nixon, and George McGovern, and each of the three sets of postelection sorts.[22] Each factor is a cluster of persons sharing an image, and we can describe that image by examining the factor scores of Q-statements to determine which items generally were ranked "most" and "least" characteristic by all persons in a cluster. For pre- and postelection sorts on the ideal president, we extracted a single factor for each set of sorts; for convenience we refer to the preelection "ideal" image as PREID and to the postelection image as PSTID. For each of the pre- and postelection sets of sorts on Richard Nixon and George McGovern, we extracted two factors. We label the two images derived from preelection Nixon sorts as PRENIX 1 and PRENIX 2, and those derived from postelection Nixon sorts as PSTNIX 1 and PSTNIX 2; the two images from preelection McGovern sorts we label PREMC 1 and PREMC 2, and the postelection images are PSTMC 1 and PSTMC 2. Table 7.2 presents the factor loadings for each respondent on various factors.

But prior to and following the election there was marked consensus on the image of the ideal president. By correlating factor scores of statements

for PREID and PSTID we can measure the similarity of the pre- and postelection images of the "ideal president." That correlation is a striking .94. Four statements most clearly defined PREID and PSTID, the consensus image of the "ideal president" transcending pre- and postelection periods. Two of the *most* characteristic qualities of the ideal emphasized the president's political role as a public official: "He is a good administrator" and "He arrives at decisions through careful consideration and analysis of all available information." The other two most characteristic statements reflected personal qualities: "When he is wrong he admits it" and "He is calm, analytical, and cautious, yet bold and decisive in carrying out his plans." Three statements appear on both pre- and postelection factors as *least* characteristic of the ideal president. One is from the party leader category: "He should be elected as a result of his party allegiance because talk is cheap and all candidates promise great things." The other two indicate a dislike for qualities of a television performer: "His voice, speech patterns, expressions, and cool appearance are more important than the mere words of his speech" and "He is proof that Madison Avenue advertising techniques make television appearances more effective." Looking at other key statements, it is apparent that the stable image of an ideal president emphasized experience as a public official and warm personal qualities as most characteristic, politician and dramatic qualities as least desirable.

Of the two images revealed by preelection sorts on Richard Nixon, twenty-nine respondents shared PRENIX 1, and seven held PRENIX 2. The images derived from postelection Nixon sorts revealed seventeen persons with PSTNIX 1, and nineteen with PSTNIX 2. Although the number of people holding each image changed after the election, the two pre- and postelection images correspond well. The PRENIX 1 to PSTNIX 1 correlation is .79; the PRENIX 2 to PSTNIX 2 correlation is .68.

Prior to the election PRENIX 1 subjects emphasized favorable qualities as a public official and as a person most characteristic of Richard Nixon. Distinguishing statements included three that we have already seen characterizing the ideal president: "He is a good administrator," "He arrives at decisions through careful consideration and analysis of all available information," and "He is calm, analytical, and cautious, yet bold and decisive in carrying out his plans." Another most characteristic attribute refers to a personal quality: "He is ambitious." Ranked least characteristic of Nixon by persons with a PRENIX 1 image were "He is imaginative, experimental, and hip," "He is natural and sincere and does not appear to be trying to impress people," "He exhibits warmth and personal appeal on television," and "His personal magnetism and physical attractiveness are positive assets." PSTNIX 1 is similar; however, it emphasizes Nixon's qualities as a public official even more strongly. Thus, statements about his being a good administrator and a blend of prudence and decisiveness are ranked most characteristic, but joining them are "His words, actions, and manner always reflect the dignity and honor of the office" and "He is a statesman and a leader who explains to the people as much as possible the reasons behind his actions or proposals." Found least characteristic of Nixon by PSTNIX 1 subjects are politician qualities

Table 7.2 Factor Loadings for Pre- and Postelection Q-Factors

SUBJECT	PREID	PSTID	PRENIX		PSTNIX		PREMC		PSTMC	
			1	2	1	2	1	2	1	2
01	63	53	44	06	73	15	18	38	16	05
02	58	57	34	15	40	33	51	08	50	13
03	69	65	32	05	01	56	41	54	30	47
04	74	66	46	58	10	63	65	22	73	26
05	70	44	44	−55	74	04	08	70	23	45
06	72	61	17	67	−29	56	54	22	59	08
07	73	71	53	−50	77	−06	09	64	47	43
08	66	69	73	07	53	33	61	27	62	08
09	75	67	52	−15	59	07	19	66	−10	64
10	70	71	57	−09	22	56	40	29	49	22
11	75	74	51	−56	69	12	16	66	17	54
12	75	76	62	35	23	66	75	−43	71	−25
13	47	81	−35	30	76	08	56	−55	13	58
14	78	65	11	68	−23	60	60	−17	61	−18
15	61	52	34	51	−05	63	55	−11	45	−05
16	69	63	27	32	00	37	57	06	63	33
17	77	74	62	−32	61	21	−25	58	12	67
18	83	89	45	−41	67	−03	−33	68	12	76

Table 7.2 (continued)

SUBJECT	PREID	PSTID	FACTOR LOADINGS							
			PRENIX 1	PRENIX 2	PSTNIX 1	PSTNIX 2	PREMC 1	PREMC 2	PSTMC 1	PSTMC 2
19	77	77	41	-67	74	-12	-20	70	19	77
20	76	84	58	-15	26	56	76	-16	80	-12
21	56	71	83	-13	48	61	61	36	51	36
22	78	75	55	28	24	66	55	09	76	00
23	63	45	71	12	31	71	61	-03	66	20
24	72	79	81	-05	84	24	53	29	57	32
25	47	49	66	-10	55	31	27	57	35	60
26	72	64	74	19	50	58	59	25	31	40
27	59	74	38	67	-02	12	75	-04	78	17
28	81	81	65	07	59	47	28	54	52	38
29	60	53	60	33	43	63	79	00	62	-12
30	57	64	55	08	22	61	34	53	65	27
31	73	60	59	-22	44	29	-43	50	-20	67
32	67	86	70	-23	66	11	25	33	34	24
33	78	79	58	28	32	70	50	56	75	29
34	59	45	48	20	15	68	47	36	31	57
35	79	73	56	-42	70	-01	67	37	60	57
36	68	65	38	21	-56	52	28	62	-53	46

Note: Decimals have been omitted from factor loadings.

("He should be elected as a result of his party allegiance because talk is cheap and all candidates promise great things") and dramatic qualities ("His voice, speech patterns, and cool appearance are more important than the words of his speech"). In sum, PRENIX 1 and PSTNIX 1 persons had a positive image of Nixon and were particularly responsive to his qualities as a public official.

Both PRENIX 2 and PSTNIX 2 are markedly more negative. The PRE-NIX 2 image regarded Nixon as ambitious, basically a party politician, one who "earnestly wants to be liked and respected," and "proof that Madison Avenue advertising techniques make television appearances more effective." Moreover, persons sharing this view rated the statement that "When he is wrong he admits it" least characteristic of Nixon. To the degree that there are differences in the image represented by PRENIX 2 and PSTNIX 2, they lie in the tendency of the PSTNIX 2 subjects to find least characteristic of him such positive personal qualities as "imaginative," "experimental," "sincere," and "caring."

The PREMC 1 to PSTMC 1 correlation is .92; the PREMC 2 to PSTMC 2 correlation is .89. Thus, the two distinct images of the defeated candidate were even more stable than were those of the victor. Twenty subjects held the PREMC 1 image and twenty-four, the PSTMC 1; sixteen persons shared PREMC 2 and twelve, PSTMC 2. Three attributes are most characteristic of McGovern, according to people holding images of PREMC 1 and PSTMC 1: "The central quality which gives depth and substance to all the others is his quality of caring," "He has the highest degree of honesty, integrity, and intelligence," and "He is a statesman and a leader who explains to the people as much as possible the reasons and actions behind his proposals." References ranked least characteristic are that he should be elected because "talk is cheap . . . ," that he is a product of "Madison Avenue," and that "He is a middle-of-the-roader." Overall, those PREMC 1 and PSTMC 1 subjects are positive toward McGovern, particularly emphasizing his positive personal qualities as most characteristic.

Persons sharing PREMC 2 and PSTMC 2 emphasize four personal qualities of McGovern as most characteristic: "He earnestly wants to be liked and respected," "He is a good family man," "He is imaginative, experimental, and hip," and "He is ambitious." They sorted three statements as least characteristic of McGovern: "He arrives at decisions through careful consideration and analysis of all available information," "He sticks to his decisions once they are made," and "He makes only those promises he has the ability to keep." People holding these images responded to McGovern as a good person but were skeptical of his prudence and decisiveness.

In summary, we uncovered a consensus image of the ideal president and two distinct images for each candidate. Despite the fact that he was the victor, the two images of Nixon after the election were basically the same as before; and for the defeated candidate, the two images are even more per-sistent. Although the images remained the same for both victor and vanquished, there were shifts in our subjects from one image to another; that is, the collec-tive images persisted across the election, but individuals changed their images.

Pre- to Postelection Shifts

Table 7.3 displays the numbers of subjects by voting groups, with each of the pre- and postelection Nixon and McGovern images. We have divided the voters in five ways: (1) first-time voters, (2) experienced voters, (3) partisan self-images (both direction and intensity), (4) ideological self-images, and (5) whom they voted for in the election (in all instances it turned out to be the same candidate preferred in preelection interviews). Overall, twelve subjects who had a PRENIX 1 image shifted to PSTNIX 2, and three moved from PRENIX 2 to PSTNIX 1. On McGovern images, seven respondents moved from PREMC 2 to PSTMC 1, and three from PREMC 1 to PSTMC 2. Shifts on Nixon images were statistically significant; no shifts on McGovern images were significant. Among subgroups, shifts on Nixon images by Democrats, moderate and strong identifiers of both parties, liberals, and McGovern voters were statistically significant.[23]

Given the makeup of our sample, we would not argue that these same subgroups in the electorate at large changed their images of Nixon. Yet pre- to postelection correlations among partisan subgroups on the SRC/CPS "feeling thermometer" suggest that the stronger the partisan self-images of voters, the more likely they changed their evaluations of Nixon after the election. The pre-post coefficient for strong Democrats was .65, compared to .67 among weak Democrats, .74 among independent Democrats, and .69 among strict Independents. For strong Republicans the correlation was .49, and for weak Republicans it was .51. Pre-post correlations of McGovern ratings offer a different pattern, but even there strong Democrats and strong Republicans displayed lower coefficients than did other partisan groups.

Finally, aside from the shifts from PRENIX 1 to PSTNIX 2, there is further evidence that principal changes in our respondents' images of the presidential candidates were confined to images of the victor. Subjects scored significantly *lower* on the postelection positive image of Nixon than they had on the positive image before the election; they scored significantly *higher* on the Nixon negative image following the election.[24] Thus, the images of the two candidates remained the same across the election period. However, the general evaluation of the winner across the sample grew more negative, whereas that of the loser was relatively unchanged. Moreover, specific subjects—who tended to be moderate-to-strong Democratic, liberal, McGovern voters—shifted from a positive to negative image of the victorious incumbent.

Changes in Perceptions of Image Traits

As noted earlier our *Q*-statements reflected four image traits—qualities as a public official, party leader, person, and dramatic performer. In sorting the statements our subjects were unaware of these categories. Yet we are interested in knowing whether the four categories were distinctive traits of candidate images to our respondents. As a means of finding out, we factor analyzed our data to obtain clusters of statements rather than of people.[25]

Table 7.3 Pre- to Postelectoral Shifts of Principal Voter Groups, Using Two-Factor Solutions

VOTER GROUPS	PRENIX		PSTNIX		PREMC		PSTMC	
	1	2	1	2	1	2	1	2
Total sample (*n* = 36)	27	9	18	18	20	16	16	12
First-time voters (*n* = 18)	11	7	8	10	10	8	11	7
Experienced voters (*n* = 18)	16	2	10	8	10	8	13	5
Republicans (*n* = 14)	11	3	12	2	4	10	5	9
Democrats (*n* = 18)	13	5	4	14	13	5	15	3
Independents (*n* = 4)	3	1	1	3	3	1	4	0
Independent-weak partisans (*n* = 22)	15	7	12	10	12	10	13	9
Moderate-strong partisans (*n* = 14)	12	2	6	8	8	6	11	3
Liberals (*n* = 19)	13	6	4	15	15	4	16	3
Middle-of-the-road (*n* = 6)	5	1	5	1	1	5	4	2
Conservatives (*n* = 11)	9	2	9	2	4	7	4	7
Nixon voters (*n* = 18)	15	3	15	3	5	13	7	11
McGovern voters (*n* = 16)	10	6	2	14	14	2	15	1
Nonvoters (*n* = 2)	2	0	1	1	1	1	2	0

FACTORS

Although we found few statistically significant postelection differences for any image trait (there are statistically significant differences regarding one of Nixon's traits), changes did occur and their direction is worth noting.[26] Prior to the election Nixon's qualities as a public official were ranked significantly higher than his personal qualities. Clearly this and our other analyses indicate that the most distinctive aspect of the Nixon image lay in his experience as an incumbent. After the election, however, ratings on Nixon's public official trait did not differ significantly from those on the party leader, personal, or television traits. For McGovern both prior to and after the election, personal style was dominant over his perceived political role attributes. Before the election his personal qualities stood out significantly more than his qualities as either a public official or a party politician. At this point his attributes as a television performer also ranked significantly higher than his party leadership potential. Following the election, style remained the most visible dimension of his image. His qualities as a person and performer still ranked significantly higher than his characteristics as a public official or party leader; and, between personal and performer attributes, the former were significantly dominant among our subjects.

After Which Election?—The Effects of Victory or Defeat as Election-Specific

In contrast to many of the previous studies of the effects of victory or defeat on the perception of political candidates, neither the Cigler-Getter nor our study of the images of presidential candidates in 1972 reveals a large number of changes in overall perceptions and distinctive candidate images. In certain respects this stability is not too surprising. Certainly, we would not expect marked changes in the images people have of an ideal president between pre- and postelection periods. Our subjects were agreed and persistent in their image of the ideal, at least sufficiently so as to sort fifty-two Q-statements in highly correlated ways, both before and after the election, to generate a single, common factor. Moreover, we should probably not expect the images of the 1972 presidential candidates to change much as a result of victory or defeat. For one thing, images of Richard Nixon, as a result of his long political career and the fact that he was the incumbent, were probably formed in the minds of most of our respondents well before the election and therefore unlikely to change. Moreover, the Nixon victory may well have been a foregone conclusion for most people; for them the simple announcement of that victory on election day probably had little effect. Although images of Senator McGovern were probably not as well articulated among the general population before the election, among our sample of voters with relatively high levels of formal education—and thereby more likely to be informed and aware of alternative candidates—those images were more likely to exist. Again, since preelection polls indicated his defeat a foregone conclusion, we might have expected little postelection change in the images of McGovern.

The qualities actually emphasized in the images of the ideal president, Richard Nixon, and George McGovern might also have been accurately anticipated. For their ideal, people subscribed to the notion still found in many high school and college civics and government textbooks (see Chapter 3), that a president must be far more than a "politician"; indeed, he must operate above partisan strife as an efficient, deliberate "statesman." As for images of Nixon, it is reasonable to expect that an incumbent president will be primarily known as a public official—especially in Nixon's case, with his campaign slogan to "Re-elect the President," we might have expected this to be the most distinctive positive dimension of his image. And, given McGovern's more recent rise to national prominence, we could guess there to be a greater emphasis on personal qualities and dramatic performance in determining what was "most characteristic" of him.

As a result, we could have predicted stability in the evaluations and images of ideal and actual candidates in 1972. But despite the stability in overall images, we discovered changes in the images shared by specific types of respondents that, on the basis of previous studies, we simply could not have expected. Earlier studies of the effects of victory or defeat would not have predicted that a winning candidate might be evaluated more negatively. Moreover, drawing on the political scientist's contemporary understanding of voting behavior, we would anticipate that the voters least likely to have changed their images would have been the moderate-to-strong partisans (whether Democrats or Republicans) having clear-cut ideological leanings. Yet these were precisely the persons who did change, at least to the extent that they shifted from a reasonably positive image of Richard Nixon, which emphasized his qualities as a public official, to a markedly more negative one after his victory, stressing an ambitious, partisan, insincere, and "packaged" side. Equally interesting is the finding that among voters whom previous studies have found most fickle in their orientations toward political objects (that is, the inexperienced, independent or weak partisans, and middle-of-the-roaders), we found no significant shifts in images.

At best we can only speculate as to why in our Q-study more than one-third of our subjects changed their images of Richard Nixon, and why these were primarily experienced, moderate and strong Democrats, liberals, and McGovern voters. Perhaps Democrats of fairly intense partisanship who were committed to vote for McGovern even in the anticipation of likely defeat simply could not visualize his qualities as a public official or party leader, but could live comfortably with the notion that he was a good person. In viewing Nixon, the conflict of these partisans between their performances and expectations was even more severe. The opposition party's candidate was likely to win. In something of a fait accompli effect, therefore, they searched for positive qualities in Nixon even before the election, found them in his role of a public official, indeed as the incumbent president, and labelled them "most characteristic," rather than dwelling on his shortcomings as a person or politician.

In short, in the anticipated loser they discerned personal qualities that could justify their supporting him; in the inevitable winner were political, or "statesman," qualities close to the ideal president, which allowed them to accept his "inevitable" victory. Once the election was over and they were no longer saddled with a prospective loser, they did not need to resolve a conflict between preferences and expectations. As consistent Democratic partisans they could still regard their party's defeated candidate as a good person, but they had no pressure to salve the pain of the inevitable. Instead, they had Nixon "to kick around" once more. This they did by adopting a more negative image of the winner, one stressing not his qualities as a public official but a host of negative personal and politicianlike attributes. In other words, there was a postelection fait accompli effect in reverse. Supporters of a candidate seemingly doomed to defeat adjusted to the situation by finding positive things in the man whose victory was assured; then after the election and without the necessity of resolving a conflict between preferences and expectations (and without the necessity of having to justify a vote for the winner, since they had voted for the loser), they could return to criticizing the candidate they opposed, but thought would win, in the first place.

In Chapter 6 we found that many of the changes in voters' perceptions of presidential candidates between elections and during campaigns have qualities that are unique to specific elections, campaigns, candidates, and image traits. Of course, these unique changes appear within an overall context of considerable stability and of a few consistent patterns of variations (such as the fact that perceptions of personal attributes are more mercurial than are other qualities of candidates). We suspect that much the same holds for image changes resulting from electoral outcomes. The evaluations most voters make of a presidential candidate probably do not change much after the election, and when they do, previous research indicates that Key's admonition—that a psychological adjustment permits a reconciliation of voters to the winner—is usually correct. But this general pattern of change does not always occur, as studies of the presidential election of 1972 indicate. The '72 election outcome did not reconcile divergent images of the candidates, nor was there evidence (either in the Cigler-Getter or our analysis) of a striking psychological adjustment accruing to the advantage of Richard Nixon. If anything, people distinguished more sharply between the two candidates after the election than they had before, and either did not change their evaluation of the winner or viewed him in even more negative terms. Thus, as far as changes in the images of the presidential candidates in 1972 are concerned, the electoral outcome did not moderate the sociopolitical conflict of the times, but instead may have exacerbated it.

Notes

1. V. O. Key, Jr., *Public Opinion and American Democracy* (New York: Alfred A. Knopf, 1961), pp. 478–79.

2. J. Korchin, "Reconstruction of Attitude Following a National Election: The *fait accompli* Effect," *American Psychologist* 3 (1948): 272 (abstract).

3. I. H. Paul, "Impressions of Personality: Authoritarianism and the *fait accompli* Effect," *Journal of Abnormal and Social Psychology* 53 (1956): 338-44.

4. George Stricker, "The Operation of Cognitive Dissonance on Pre- and Post-election Attitudes," *Journal of Social Psychology* 63 (1964): 111-19. See also his earlier study, "The Use of the Semantic Differential to Predict Voting Behavior," *Journal of Social Psychology* 59 (1963): 159-67.

5. Bertram H. Raven and Philip S. Gallo, "The Effects of Nominating Conventions, Elections, and Reference Group Identification upon the Perception of Political Figures," *Human Relations* 18 (August 1965): 217-29.

6. Ibid. See also Arnold Thomsen, "What Voters Think of Candidates Before and After Election," *Public Opinion Quarterly* 2 (1938): 269-74.

7. Lynn R. Anderson and Alan R. Bass, "Some Effects of Victory or Defeat upon Perception of Political Candidates," *Journal of Social Psychology* 73 (1967): 234-35.

8. Jay G. Blumler and Denis McQuail, *Television in Politics* (Chicago: University of Chicago Press, 1969), chap. 12.

9. Leon Festinger, *A Theory of Cognitive Dissonance* (Evanston, Ill.: Row-Peterson, 1957); Jack W. Brehm and Arthur R. Cohen, *Explorations in Cognitive Dissonance* (New York: John Wiley, 1962).

10. Fritz Heider, "Attitudes and Cognitive Organization," *Journal of Psychology* 21 (1946): 102-12; Fritz Heider, *The Psychology of Interpersonal Relations* (New York: John Wiley, 1954).

11. Carolyn Sherif et al., *Attitude and Attitude Change* (Philadelphia: W. B. Saunders, 1965).

12. Various consistency theories are discussed in Robert B. Zajonc, "The Concepts of Balance, Congruity and Dissonance," *Public Opinion Quarterly* 24 (1960): 280-96; and Daniel Katz, "Attitude Formation and Public Opinion," *Annals* 367 (September 1966): 150-62.

13. Allan J. Cigler and Russell Getter, "After the Election: Individual Responses to a Collective Decision" (paper presented at the 1974 Annual Meeting of the Southwestern Political Science Association, Dallas, Texas, March 28-30, 1974).

14. William Stephenson, *The Play Theory of Mass Communication* (Chicago: University of Chicago Press, 1967), p. 41.

15. The scale for scoring and distribution of the fifty-two statements was:

	MOST CHARACTERISTIC					LEAST CHARACTERISTIC			
Score	+4	+3	+2	+1	0	−1	−2	−3	−4
Number of statements	4	5	6	7	8	7	6	5	4

16. Correlation coefficients were first transformed into a Fisher's z, mean coefficients calculated on the basis of Fisher's z and then retransformed.

17. Employing the expression $1\sqrt{n}$ as the standard error of a zero order correlation, with $n = 52$ statements, coefficients exceeding .35 are significant at the .01 level.

18. Angus Campbell et al., *The American Voter* (New York: John Wiley, 1960), p. 143.

19. Ibid., p. 497.

20. The difference in coefficients between first-time and experienced voters is significant at the 0.5 level. However, heeding Kerlinger's admonition that, due to the lack of independence in ipsative measures such as Q-sorts, mean differences at the 0.5 level should be viewed warily, researchers are therefore

advised to opt for the .01 requirement, which we do. See Fred N. Kerlinger, Q-Methodology in Behavioral Research," in Steven R. Brown and Donald J. Brenner, eds., *Science, Psychology and Communication* (New York: Teachers College Press, 1972), pp. 3–38.

21. As another means of testing for pre- to postelection changes, especially in image attributes, we conducted a *t* test and *F* test on the mean raw scores assigned to each statement by subjects in their sorts. Only on Nixon sorts did even a single item appear significant: "His perserverence, firmness, coolness, and aggressiveness clearly project a 'take charge' image." The item was scored significantly higher before than after the election. We can only wonder what score it might have received after the impeachment proceedings following the Watergate revelations.

22. For each set of Q-sorts we generated several factor solutions using several criteria. The single-factor solution for each of the ideal president sorts and two-factor solutions for all others proved most informative. In multifactor solutions, rotation to simple structure is by the varimax method. Percentages of total variance explained for each sort are as follows: preelection ideal, 48; postelection ideal, 47; preelection Nixon, 42; postelection Nixon, 45; pre- and postelection McGovern, 43.

23. But changes on Nixon images were significant only at the .05 level as measured by the McNemar test. Shifts among Democrats, liberals, and McGovern voters were significant at the .01 level, and among moderate and strong partisans at the .025 level.

24. In calculating these changes we took the factor loading of each of our respondents on each candidate's pre- and postelection factors and treated it as the respondent's test score on that factor. Thus, for instance, Respondent #1 had a PRENIX 1 score of 44 and a PSTNIX 1 score of 72. Using these scores we calculated *F* ratios to compare the pre-post performance of respondents on the candidate images. The only two significant *F* ratios were for Nixon images: the PRENIX 1 to PSTNIX 1 ratio was 2.87, significant at the .01 level; and the PRENIX 2 to PSTNIX 2 ratio of 2.08 was significant at the .05 level.

25. That is, instead of a Q-type factor analysis we conducted what convention refers to as an R-type analysis, which Stephenson refers to as a System 4 analysis. See William Stephenson, *The Study of Behavior: Q-Technique and Its Methodology* (Chicago: University of Chicago Press, 1953), pp. 51–52.

26. In this portion of the analysis we employed single-factor solutions for each of the pre- and postelection Nixon and McGovern sorts. We then conducted an analysis of variance using the standardized factor scores for the thirteen statements in each of the four categories of image traits.

Chapter Eight

Political Images and Voting Behavior

This volume has brought together an interrelated series of studies pertaining to the images of political candidates. We have reviewed the research of those earlier voting studies. In addition, using data derived from our own research we have looked at various dimensions and sources of candidate images, contrasting campaign styles of office seekers, changes that occur in images during election campaigns, the relation of mass communications to the conceptions people have of candidates, and the effects of winning or losing on the images of political figures. Before summarizing our main findings in these areas and suggesting lines of future research, there remains one additional question to consider—that is, How much of an independent effect do the voters' perceptions of competing candidates have on the choice they subsequently make at the polls? In seeking an answer to this question we shall do two things: first, we reexamine in greater detail than we did in Chapter 2 what relevant studies have said about how candidate images influence voting behavior; second, we offer data from the 1972 presidential election that measure the relative influence of candidates, parties, issues, and ideologies on the votes of a nationwide sample of Americans. We conclude this text with a summary of where we are and where we hope to go in the study of candidates and their images.

Candidate Images and Voting Behavior: Hints from Contemporary Research

At various points in our discussion we have described how voters' images of political candidates in partisan elections relate closely to those voters' partisan self-images. It is reasonable, therefore, to expect a party's identifiers to perceive their party's candidate positively and the opposition's standard bearer negatively. We saw in Chapters 3, 4, and 6 that this is not always so, but certainly in presidential elections it has been a typical pattern. The information in Table 8.1 illustrates how this pattern is related to voting and suggests what effect partisanship has on the party identifier who does not like his party's candidate and on the Independent who does not have a partisan self-image.

Table 8.1 contains the same type of information we are familiar with from our discussion in Chapter 6. Based on the quadrennial presidential election surveys of the Survey Research Center/Center for Political Studies, it displays the percentages of positive references made to each political party's nominees in response to questions probing voters' likes and dislikes about the candidates. The table portrays these percentages separately by voting behavior for Democrats, Independents, and Republicans. Taking 1952 as an example, among Democrats voting for Adlai Stevenson, 90 percent of their comments about their candidate were positive, whereas only 41 percent of their remarks about Dwight Eisenhower were positive; among Independents, 84 percent of the comments regarding Stevenson by his voters were positive, whereas 83 percent of comments made among Eisenhower voters were favorable to Ike.

A few of the main points in Table 8.1 are noteworthy. First, as expected, among party identifiers loyal to their party's candidate, the proportion of positive comments about their choice is high: among Democratic identifiers the percentage of positive comments about Democrats has ranged from 70 in the case of George McGovern in 1972 to 90 in 1952; among Republicans the range of percentages of positive references has been from 77 in 1964 (references to Barry Goldwater) to 95 in the case of Richard Nixon in 1960. Conversely, party loyalists have predominantly negative comments about the opposition's candidate. For instance, only 12 percent of the remarks about Goldwater among Johnson Democrats were positive, and among Nixon Republicans in 1972 only 15 percent of remarks about McGovern were positive. Admittedly these are extreme cases, but among Stevenson Democrats even in Eisenhower's peak year of general popularity, 1956, only four in ten of comments about Ike were positive.

Second, party identifiers who defect generally have more positive images of the opposition's candidate than of their own party's. More than 60 percent of references made by defecting Democrats about the opposition candidate have been positive in the last six presidential elections, while in no election has the Democratic candidate received a majority of positive comments from the defecting group. There is a similar pattern among Republican defectors, but with one notable exception—Republicans voting for McGovern in 1972 made very few references to him, and less than half of those were positive.

Third, the voting of Independents is congruent with their overall images of competing candidates. Thus, Independents voting Democratic have expressed positive images of the Democratic candidates, negative of the Repub-

Table 8.1 Candidate Images by Voting Behavior for Democrats, Independents, and Republicans: 1952–1972 Presidential Elections

ELECTION YEAR AND PARTY PRESIDENTIAL CANDIDATES

PARTY AND VOTING BEHAVIOR	1952				1956				1960				1964				1968						1972			
	D		R		D		R		D		R		D		R		D		R		I		D		R	
	n	%+	n	%+	n	%+	n	%+	n	%+	n	%+	n	%+	n	%+	n	%+	n	%+	n	%+	n	%+	n	%+
Democrats:																										
Voted Dem.	869	90	868	41	941	80	1038	40	884	87	543	34	1336	88	1107	12	1100	84	711	20	742	12	465	70	435	27
Voted Rep.	271	48	434	79	236	44	386	75	232	38	228	82	169	46	169	62	190	33	250	72	239	43	284	18	332	71
Voted AIP	—	—	—	—	—	—	—	—	—	—	—	—	—	—	—	—	106	36	80	50	166	84	—	—	—	—
Independents:																										
Voted Dem.	216	84	188	26	176	70	228	52	217	80	150	51	371	80	313	17	107	53	161	26	158	21	217	64	207	31
Voted Rep.	340	43	578	83	373	28	708	84	188	36	253	88	198	32	222	66	380	30	491	79	417	32	342	18	403	76
Voted AIP	—	—	—	—	—	—	—	—	—	—	—	—	—	—	—	—	80	19	98	46	174	86	—	—	—	—
Republicans:																										
Voted Dem.	18	88	28	50	25	84	32	53	39	82	34	47	199	73	190	16	51	72	57	46	64	36	22	41	34	50
Voted Rep.	639	33	1091	88	703	25	1271	92	654	30	744	95	498	26	647	77	551	22	804	85	572	29	532	15	582	83
Voted AIP	—	—	—	—	—	—	—	—	—	—	—	—	—	—	—	—	28	18	27	63	49	69	—	—	—	—

Note: Percentages refer to proportion of all positive comments made about Democratic or Republican presidential candidates (or, as in 1968, the candidate of the American Independent Party) by partisans or independents voting for a particular candidate; *n*'s refer to all comments, positive and negative, made about a candidate by a particular category of voters.

Source: Survey Research Center/Center for Political Studies, University of Michigan, and the Interuniversity Consortium for Political Research.

lican; those voting Republican have been positive about the Republican contender, negative about the Democrat. A minor variation on this theme occurred in 1960, when a majority of positive references to Nixon persisted among Independents voting for John Kennedy.

Finally, in our single case of a third-party candidate, George Wallace, there seems to have been a predictable association between voters' images of and support for him, regardless of party stripe. In 1968 Wallace's image among his Democratic, Republican, and Independent voters was clearly positive; it was just as notably negative among groups voting for either of the two major party candidates.

In sum, the data in Table 8.1 suggest a close association between the *affective* orientations of voters toward the candidate and the direction of their choices. Published and unpublished studies of voting behavior, usually confined to presidential elections, reinforce this conjecture. One of the earliest is the first major volume in the voting studies of the Survey Research Center/Center for Political Studies, *The Voter Decides.*[1] That analysis, based on a nationwide sample of adults, was an extensive interpretation and analysis of the presidential election of 1952. One of its principal aims was to identify the major forces that comprise the voter's "motivation" to choose one candidate over another. The inquiry focused on three components of that motivation—party identification (or partisan self-image, as we have labelled this phenomenon in our discussions) issue partisanship, and candidate partisanship. *The Voter Decides* demonstrated that these three forces directly relate to candidate preference; for example, the more strongly a person considered himself a Republican, and/or liked the Republican candidate, and/or preferred Republican issue postures, the more likely he voted for Eisenhower. The less distinct his stand on any of these three factors, the more likely he chose the opposition party's candidate (or, in many instances, the less likely he voted at all).

Here, then, was an early statement of three of the key variables—party, candidate, and issue orientation—that enter into the description of voting behavior, which we reviewed in detail in Chapter 2. In that description the measurement of the distribution of partisan self-images, or party identifications, permits an estimate of the normal vote, that is, the split in the two-party vote if people simply vote their partisanship. The voter's stands on issues and his likes and dislikes of the candidates influence both nonpartisans and moderate partisans, thus giving rise to defections from party loyalties and deviations from the normal vote.

But the normal-vote description of electoral behavior, although a direct descendent of *The Voter Decides,* is no direct copy of the older notion. For one thing, in the normal-vote model party identification (partisan self-image) assumes a different and more transcendent role than either candidate or issue orientation. In *The American Voter* (the SRC's analysis of the 1952 and 1956 presidential elections) the authors treat party identification as a predisposing attitude, usually formed early in life, that acts as a perceptual screen for each voter, thereby influencing how he sees and evaluates candidates, issues, and political parties.[2] In short, partisan self-images influence other political images. Party identification in the normal-vote description is thus further removed from the voting act (and thereby a long-term factor, as we described in Chapter 2) than other short-term forces specific to each election. *The American Voter* identified six such short-term components: (1) the voter's image of the Republican candidate, (2) his image of the Democratic candidate, his views on (3) domestic and (4) foreign policy issues, (5) his image of the two political parties as managers of government, and (6) the voter's group-related views such as "Democrats are good for unions." More recent studies have added an ideological component—how the voter places himself on a liberal-conservative continuum.[3]

The question that interests us regarding long- and short-term components is the relative influence of voters' images of candidates on electoral decisions and other facets of political behavior. Generally, as Peter Natchez points out in his thoughtful review and critique, published studies of the SRC/CPS indicate that, "The most powerful components of the electoral decision process are those which capture attitudes toward the competing candidates."[4] In the early presidential elections following World War II, neither domestic nor foreign policy issues determined election outcomes, although influencing some voters. Also, attitudes toward the political parties and other political groups did not produce striking shifts in voters' preferences from election to election. Although voters are now more conscious of issues, their images of candidates still influence their votes, probably more so than do their attitudes toward issues. For example, using SRC/CPS data from the 1960 and 1964 presidential elections, Natchez and Bupp found that issues clearly had come to be of considerable influence on voting, but candidate image remained a more important force.[5]

Along with the efforts of Natchez and Bupp, studies of more recent presidential elections indicate that issues are growing more crucial in shaping voting behavior, yet these studies also note that candidate image remains a principal short-term force. In 1968, for instance, the Comparative State Election Project (CSEP) focused on voting behavior in southern states (combining several statewide surveys with a nationwide sampling). By comparing respondents' positions on selected issues with their perceptions of the candidates' stands on the same issues, the analysis explained a considerable portion of the variation in voters' preferences for Nixon, Humphrey, and Wallace. Although issues were clearly important in 1968, a key component of the voting decision was thus the *voters' images of the candidates' issue stands* (correct or incorrect), along with their perceptions of other candidate qualities.[6] And, an analysis of the 1972 presidential election also revealed that how warm or cold voters felt toward the candidates (as measured by the "feeling thermometer" described in Chapter 2) and voters' perceptions of the candidates' issue stands compared with their own positions were major factors in helping people make up their minds.[7]

Natchez and Bupp have observed that the long-term component of voting—partisan identification—is "causally" associated with candidate image, so that "people do not perceive candidates through neutral eyes" but "seem to take their focus from previously established identifications."[8] Yet we have seen from Table 8.1 (as well as from data reported in preceding chapters) that partisans often have more positive images of the opposition party's candidate than they do of their own. Richard Boyd examined the presidential elections of 1956 to 1964 to determine what happens in such cases where voters' images of the candidates are inconsistent with their partisan self-images. Using SRC data he derived measures of consistency between a person's voting decision, party identification, and images of the candidates, political parties, and selected policy issues. Comparing the rates by which voters defected from their standing partisanship with the consistency between their partisan self-images and their images of candidates, parties, and issues, he found that voters' images of the candidates provided the principal statistical explanation for voting defection.[9] In sum, their likes and/or dislikes of the nominees were more likely to make them defect than how they felt about the issues or the major political parties.

Contemporary research indicates that candidate image is the most important short-term force contributing to partisan preference and defection; it also suggests that how people perceive candidates is the principal determinant of whether or not they will split their tickets, and how. In 1972 Miller and Jackson undertook a probability sampling of registered voters in three counties of southern Illinois to discover what factors related to voters' splitting their tickets between the two major political parties in contests for the presidency (Richard Nixon versus George McGovern) and the Illinois governorship (Richard Ogilvie versus Daniel Walker). Respondents evaluated each of the four candidates using various seven-point semantic differential scales (liberal-conservative, honest-dishonest, qualified-unqualified, sincere-insincere, successful-unsuccessful, friendly-unfriendly, and calm-excitable). Respondents also rated traits of their "ideal president" and "ideal governor" on the same scales. Miller and Jackson derived measures of the proximity between respondents' evaluations of each presidential candidate and their image of the "ideal president"; they also calculated proximity measures for the gubernatorial candidates. Candidate images (that is, how closely each voter's image of any candidate approximated the voter's image of the "ideal" officeholder) enabled Miller and Jackson to predict straight and split-ticket voting with considerable success. They correctly forecast the votes of more than three-fourths of their respondents and found that candidate images, rather than strict partisanship, were especially helpful in predicting the behavior of strong and moderate party identifiers.[10]

Finally, people's images of candidates affect at once how they vote and how they look at political objects, thus performing a function similar to partisan self-images. For example, Donald Stokes found that popular images of the two major political parties are affected primarily by what people like and dislike about the candidates those parties nominate for president. He found that in 1956 voters' perceptions favored the Republican over the Democratic party by 8 percent, largely as a result of Eisenhower's being considerably more popular than Stevenson; by 1964, however, the contrast in voters' perceptions of Johnson and Goldwater resulted in a 5 percent advantage for the Democratic party when people spoke of their likes and dislikes about the two major parties, a shift of 13 percent. Said Stokes, "A more eloquent statistical comment on the personal contribution which candidates for President can make to electoral change could hardly be given."[11] He noted also that "attractive" qualities of candidates consist of more than the attributes office seekers try to emphasize in their campaigns. People must be disposed to regard such attributes as "attractive" if there is to be a successful appeal. Thus, not only did Eisenhower campaign as a former military conqueror, but the electorate of the 1950s was also resonant to such an image. Moreover, Stokes pointed out that whatever qualities a candidate projects reach voters after being filtered through the mass media and personal conversations. Hence, the effects of candidate images on such things as voting (straight and split-ticket voting and partisan defections) and perceptions of other political objects result from an intricate interplay of publicized and communicated candidate attributes and voters' predispositions, the complex message-image transaction we introduced earlier in Chapter 1 and elaborated in Chapters 3 and 4.

Thus, running through the bulk of the major research relating various components of voting decisions to electoral behavior is the view that candidate images constitute the most powerful of the various short-term forces. We do not know, however, the relative effect of partisan self-images in relation to candidate images. Nor have we precise measurements of the relative effects of party, candidate, and issue orientations on recent voting behavior. We also have no indication from published voting studies what specific dimensions of candidate images most influence voting. Are the things we believe about candidates (cognitions), our feelings (affects), or our dispositions (conations) most influential? Within this context of past studies and the unanswered questions they raise let us now examine the interplay of image forces in the behavior of voters in the 1972 presidential election.

Voting Behavior in the 1972 Presidential Election

Contemporary research specifies several components that make up voting decisions. Data from the SRC/CPS study of the 1972 American national election help us measure five of these psychological components and their various dimensions: (1) partisan self-images, (2) ideological self-images, and (3) issue orientations of the electorate, as well as their (4) candidate images and (5) images of political parties. By correlating and factor analyzing these measures and employing a regression analysis we will describe the relative influence of each component on voting behavior in 1972.

Measuring the Images

Of the five major components of the voting decision (listed above), three pertain to the voter's self-image; that is, what kind of partisan the voter thinks he is, what kind of liberal or conservative he says he is, and what stands he takes on specific policy questions. Our interest in these components is in the voter's perceptions of self, not in his view of the candidates or the political parties.

In Chapter 2 we described how the SRC has been measuring the distribution of *partisan self-images* in America for more than two decades by asking cross sections of American citizens a series of questions pertaining to whether they "usually think" of themselves as Republicans, Democrats, Independents, or something else. This practice provided the seven categories of identification reported earlier in Table 2.3. We have used these categories for the 1972 election to create a seven-point scale (ranging from a score of "1" for strong Democrat through "4" for strict Independent to "7" for strong Republican) as our measure of partisan self-images. Also in the 1972 presidential election survey respondents placed themselves on a seven-point scale designed to measure their *ideological self-images*. The categories we employ, rated from "1" to "7," are extremely liberal, liberal, slightly liberal, moderate/middle-of-the-road, slightly conservative, conservative, and extremely conservative.

Finally, we employ a measure for the voter's *issue orientations*. Respondents in the 1972 survey, using a seven-point scale, revealed their positions on several policy issues. We have selected nine issues for our analysis covering a diverse gamut of social, economic, foreign policy, civil rights, and civil liberties matters: (1) whether the federal government should guarantee every person a job and a good standard of living; (2) whether income taxes should be progressive or everyone pay the same portion of their income regardless of how much they make; (3) the question of legalizing the use of marijuana; (4) the use of bussing to achieve racial integration in public schools; (5) governmental versus private coverage of medical and hospital expenses; (6) immediate withdrawal from Vietnam versus efforts to achieve complete military victory; (7) total versus no governmental action to battle inflation; (8) protection of the legal rights of those accused of committing crimes; and (9) whether the federal government should make every effort to improve the social and economic position of blacks and other minority groups.[12] On each seven-point scale used to measure the respondent's self-placement on an issue, the score of "1" represented a position favorable to governmental action, progressive taxation, legalization of marijuana, bussing, withdrawal from Vietnam, or protection of the rights of the accused. A score of "7" indicated a position on each issue at the opposite extreme. By combining respondents' self-placements on each of the nine issues we derived a composite, seven-point index of each voter's issue orientations.

In employing the 1972 presidential election study to examine the relative impact of candidate images, we relied on several different measures to tap their cognitive, affective, and conative aspects (see Chapter 1). For the *cognitive* aspect (that is, belief about each candidate regardless of how one feels about him or tends to react toward him) we have employed two sets of measures. First, in addition to being asked to identify what kind of liberal or conservative each respondent considered himself to be, the respondent—again, a seven-point scale—designated what he believed to be the respective ideological leanings of Richard Nixon and of George McGovern. Second, on each of the nine issues listed above, each respondent also designated where he believed each candidate's stand was, according to the same seven-point scale he used to declare his own position. As in the case of respondents' self-placement on issues, we used a composite index to summarize their perceptions of each candidate's stands across all nine selected issues.

For the *affective* aspect of candidate images (how people feel about contenders and what they like and dislike about them) we have also used two measures. First, we use the respondents' ratings of Nixon and McGovern on the "feeling thermometer" that taps the coolness or warmness of feelings on a scale of 0–100 degrees.[13] Second, using the responses of people to the standard SRC question asking for likes and dislikes about each candidate, we derived a "Candidate Index," a thirteen-point scale on which the higher the score, the more pro-McGovern the respondent's affect. We realize (as we argued in Chapter 6) that responses to the SRC's questions concerning likes and dislikes about the candidates tap cognitive as well as affective aspects of voters' perceptions. We think, however, that combining all the positive and negative comments emphasizes the affective quality of the responses.[14]

The *conative* aspect of images is more complex and warrants more detailed consideration. In an article published almost three decades ago, M. Brewster Smith provided a systematic differentiation among three aspects of images—cognitions, affects, and conations. Although he spoke of "attitudes" rather than "images," Smith stressed that a person's beliefs and feelings about any political object have much to do with what he *proposes* to *have done* about it. He treated a person's preferred action, or "policy orientation," as the conative element of his attitude.[15] Following Smith and others, Harding, Proshansky, Kutner, and Chein later examined the conative aspect of prejudice, but in doing so broadened the notion to include people's "action orientations" as well as their ideas about "what should be done." They reported studies measuring conation as the social distance perceived by people between themselves and members of various ethnic groups. This idea of distance or proximity we incorporate into our measurements as the conative aspect of images.[16]

Thus, with respect to such a political object as a candidate, the conative aspect is more than the voter's policy orientation in the sense of his intention to vote for or against the candidate. It refers also to *how close or distant* a voter perceives himself to be in relation to the political object; that is, a proximity along specific dimensions (or with reference to specific matters) of sufficient salience to trigger for him an idea of what he proposes to do about the object. The 1972 survey provides two kinds of information from which to construct such measures of proximity.[17] First, since respondents placed themselves and each candidate on a seven-point ideological scale, it is possible to compare the absolute difference between the respondent's self-location and where he places a candidate. We have derived a seven-point scale along which the higher a person's score, the more distant he considers his

ideological position from that of Nixon or McGovern. Similarly, since respondents rated themselves and each candidate on nine salient policy issues,
we have been able to calculate an issue proximity index for each candidate
indicating, as the score increases on a seven-point scale, an increasing distance
between the respondent's issue stand and his perception of the candidate's
stand.

In measuring images of the Republican and Democratic political parties,
our procedures paralleled those for measuring candidate images. Our measures
of the cognitive aspect of each party's image consist of (1) the respondent's
perception of each party's position on the seven-point liberal/conservative
scale and (2) his perception of each party's stand on each of the nine policy
issues (again combined into composite scales for each party). We measure the
affective aspects of party images using a "Party Index" (derived in the same
way as the Candidate Index) by relying on respondent's stated likes and dislikes about each party. And, we employ ideological and issue proximity measures similar to those used for candidates to measure the conative aspect by
relying on respondents' placements of each party on the liberal/conservative
scale and on the nine policy-issue scales.

Thus, in the tradition of voting studies that focus on the relationship
of psychological variables to the vote, we have derived twenty-three measures
of the relative influence of five major components of the voting decision.
Three of these refer to self-images (partisan, ideological, and issue self-percep-
tions); eleven to the cognitive, affective, and conative aspects of candidate
images; and the remainder to the same three aspects of party images. A
summary of our various measures and the abbreviated labels we use for them
in reporting our findings appears in Table 8.2.

We fully recognize certain problems associated with such measures.
For one, we arbitrarily posit that our Candidate Index measures affect, when
clearly, volunteered likes and dislikes about a candidate (from which the index
is derived) tap cognitions as well. Similar difficulties inhere in our efforts to
measure other aspects of candidate and party images. In addition, we realize
that in treating voters' perceptions of the issue stands of candidates or parties
as aspects of candidate and party images, we can be charged with defining
images to encompass everything (we shall report tests of our arbitrariness and
inclusiveness that help provide a balanced interpretation of our final results).
Finally, we acknowledge that there are methodological difficulties in dealing
with data derived from vastly different question formats in surveys. We have
tried to minimize these by striving at least for some comparability of scales,
but by no means do we deny that difficulties remain.[18] We think, however,
there are gains as well as losses from employing multiple measurements and
that our research question calls for their use.[19]

Candidates, Parties, Issues, and Ideologies

In presenting our findings from this secondary analysis of data generated by
the 1972 presidential study, we first want to see if any of our twenty-three
image variables exhibits any relationship to the vote decision; for this purpose

Table 8.2   Summary Measures of Self-, Candidate, and Party Images: 1972

Respondents' Self-Images

1. Partisan self-images (PSI)
2. Ideological self-images (ISI)
3. Issue orientations, or self-placement on issues (SPI)

Candidate Images

Cognition:

1. Respondent's perception of Nixon's ideological position (NLC)
2. Respondent's perception of McGovern's ideological position (MLC)
3. Respondent's perception of Nixon's issue stands (NI)
4. Respondent's perception of McGovern's issue stands (MI)

Affect:

1. Respondent's rating of Nixon on feeling thermometer (NT)
2. Respondent's rating of McGovern on feeling thermometer (MT)
3. Index of respondent's likes and dislikes of the candidates, i.e., Candidate Index (CI)

Conation:

1. Respondent's proximity to Nixon's perceived ideological position (NLCP)
2. Respondent's proximity to McGovern's perceived ideological position (MLCP)
3. Respondent's proximity to Nixon's perceived issue stands (NIP)
4. Respondent's proximity to McGovern's perceived issue stands (MIP)

Party Images

Cognition:

1. Respondent's perception of Republican party's ideological position (RLC)
2. Respondent's perception of Democratic party's ideological position (DLC)
3. Respondent's perception of Republican party's issue positions (RI)
4. Respondent's perception of Democratic party's issue positions (DI)

Affect:

1. Index of respondent's likes and dislikes of the political parties, i.e., Party Index (PI)

Conation:

1. Respondent's proximity to Republican party's perceived ideological position (RLCP)
2. Respondent's proximity to Democratic party's perceived ideological position (DLCP)
3. Respondent's proximity to Republican party's perceived issue positions (RIP)
4. Respondent's proximity to Democratic party's perceived issue positions (DIP)

we shall examine the simple correlations between each variable and the vote. Second, it is desirable to test the degree of arbitrariness in our classification of the twenty-three variables into political self-, candidate, and party image components; this we can accomplish through factor analysis. Finally, we are interested in determining which of the twenty-three image variables are most important as explanations of vote decisions, a task for multivariate regression.

Table 8.3 contains the simple correlation coefficients (r) of each of the twenty-three image variables and the vote.[20] (The vote variable we scored as "0" for Nixon and "1" for McGovern.) In addition, Table 8.3 contains the intercorrelations of the twenty-three image variables. Looking first at correlations between each of the image measures and the vote, it is apparent that

Table 8.3 Intercorrelations of Voting Behavior and Measures of Self-, Candidate, and Party Images: 1972

	PSI	ISI	SPI	NLC	MLC	NI	MI	NT	MT	CI	NLCP	MLCP	NIP	MIP	RLC	DLC	RI	DI	PI	RLCP	DLCP	RIP	DIP	VOTE
PSI	—	.32	.16	-.03	-.17	.02	-.13	.45	-.43	-.33	-.35	.38	-.20	.17	-.05	-.21	.00	-.08	-.40	-.32	.35	-.19	.15	-.50
ISI		—	.32	-.03	-.10	-.05	-.15	.39	-.41	-.31	-.52	.67	-.27	.25	-.07	-.02	-.09	-.08	-.26	-.49	.53	-.26	.15	-.47
SPI			—	-.12	-.04	.52	.36	.20	-.25	-.39	-.29	.26	.27	.63	-.13	-.04	.46	.43	-.15	-.24	.19	.25	.56	-.28
NLC				—	-.17	.19	.01	-.16	.09	.08	.24	.07	.15	.00	.63	-.13	.17	.02	.04	.18	.08	.17	.02	.14
MLC					—	-.09	.26	-.08	.22	.12	.04	-.44	.04	-.19	-.20	.48	-.07	.13	.13	-.01	-.28	-.07	.13	.18
NI						—	.55	-.05	.00	-.11	.14	.03	.65	.62	.12	-.07	.84	.62	.05	.12	.04	.63	.62	.05
MI							—	-.13	.20	.06	.15	-.23	.53	.37	.00	.18	.54	.72	.15	.13	-.20	.50	.36	.20
NT								—	-.40	-.46	-.51	.36	-.36	.15	-.16	-.10	-.10	-.08	-.35	-.44	.23	-.32	.08	-.64
MT									—	-.47	.40	-.49	.25	-.31	.09	.17	.03	.10	.35	.34	-.36	.03	.10	.64
CI										—	.34	-.34	.19	-.34	.10	.08	-.06	-.01	.56	.30	-.22	.18	.26	.48
NLCP											—	-.34	.42	-.16	.20	-.01	.16	.10	.29	.79	-.17	.38	-.04	.56
MLCP												—	-.22	.34	.03	-.26	.00	-.14	-.30	-.29	.67	-.19	.24	-.50
NIP													—	.41	.10	.03	.62	.53	.28	.38	-.12	.85	.47	.36
MIP														—	.00	-.13	.58	.44	-.17	-.14	.27	.42	.79	-.25
RLC															—	-.22	.12	-.02	.06	.22	.09	.12	-.02	.12
DLC																—	-.05	.18	.12	-.04	-.41	-.05	.18	.18
RI																	—	-.72	.07	.14	.00	.71	.70	.08
DI																		—	.12	.08	-.15	.61	.52	.11
PI																			—	.26	-.25	.27	-.13	.36
RLCP																				—	-.12	.36	-.05	.46
DLCP																					—	-.12	.24	-.36
RIP																						—	.57	.32
DIP																							—	-.17

Grouping labels within the matrix: SELF; CANDIDATE (Cognitive / Affective / Conative); PARTY (Cognitive / Affective / Conative).

Source: Survey Research Center/Center for Political Studies, University of Michigan, and the Interuniversity Consortium for Political Research.

there is a considerable range in the size and direction of the coefficients—some are low, others high, and there are both positive and negative correlations. With respect to the size of the coefficients, all but two are sufficiently large to be statistically significant. Those not significant are respondents' perceptions of Nixon's issue stands ($r = .05$) and of the issue stands of the Republican party ($r = .08$).[21] Of the remaining significant coefficients, only a few are notably high. These include the ratings respondents gave to each candidate on the "feeling thermometer," the respondents' proximities to each candidate's perceived ideological position, the respondents' partisan and ideological self-images, and the respondents' proximities to what they perceive as the Republican party's ideological position. Thus, our measures of political self-, candidate, and party images do bear a marked correlation with voters' candidate choices. However, aside from the fact that respondents' thermometer ratings of the candidates have the highest relationship with the vote (that is, the better McGovern's rating, the more likely the vote for him; and the better Nixon's rating, the less likely a vote for McGovern), we have learned little as yet about the relative effects of our three major sets of influences.

Table 8.3 also indicates that many of the twenty-three image variables are intercorrelated. In some instances these correlations are very high. For example, there is the expectedly high correlation between the respondents' perceptions of the candidates' issue stands and those of the candidate's party (rNI, RI $= .84$ and rMI, DI $= .72$). This carries over into high correlations between respondent-candidate proximities and respondent-party proximities (rNIP, RIP $= .85$ and rMIP, DIP $= .79$).

We have arranged the matrix in Table 8.3 by our three sets of political self-, candidate, and party image variables, and within the last two sets by cognitive, affective, and conative aspects. The boxed clusters contain the correlations between measures of a given image set or aspect. In some instances the size and direction of the coefficients within clusters confirm expectations, in others there are surprises. It is no surprise, for instance, that the Candidate Index (on which the higher the score, the more favorable it is to McGovern) has a moderately positive correlation with McGovern's thermometer ratings and a negative correlation with Nixon's ratings. This suggests partial confirmation for our view that both the thermometer ratings and the Candidate Index measure an affective aspect of candidate images. Moreover, given the identification in various voting studies of the function of partisan self-images in influencing perceptions of political objects, it is reasonable to find sizable correlations between the partisan measure of voters' self-images and measures of candidate affect. For instance, note that the more Republican a respondent, the fewer his likes and the lower his thermometer rating of McGovern, while the warmer his feeling toward Nixon. In the same vein, partisan self-images correlate as expected with the Party Index—that is, the more Republican the self-image, the greater the affect toward the GOP.

Within and between boxed sets of image variables as displayed in Table 8.3, however, are unexpected coefficients. For example, note that between perceptions of Nixon and McGovern stands on issues, NI to MI is .55. Similarly there is a marked relationship between respondents' perceptions of the Republican and Democratic stands on issues (.72). And, whereas voters' self-locations on issues (SPI) correlate to what they believe to be Nixon's stands (NI) to the tune of .52, when they perceive that Nixon is opposed to federal action on issues they tend to place increasing distance between themselves and the Republican contender (rNI, NIP = .65).

We suspect that the unanticipated relationships arise in two ways. First, although a variety of studies such as those cited in the previous section indicate an important role for issues in elections (through the increasing tendency of voters to discern differences between candidates and parties on those issues), on several of the nine issues for which we have measures in 1972, our correlations indicate little perceived differences. On the issue of withdrawal from Vietnam, for example, voters' beliefs of Nixon's position correlated −.15 with their view of McGovern's stand; their perception of the two parties' stands was −.02. While these coefficients are negative, they indicate a negligible difference in voters' perceptions of issue stands. Similarly, voters perceived only a slight difference in the Nixon and McGovern stands on bussing ($r = -.13$). On the question of legalization of marijuana the correlation of perceived Nixon and McGovern positions measured .18, and for the two parties there was a correlation of .33; on the issue of rights of persons accused of crimes, the correlations were .24 for the candidates and .36 for the parties.

Second, however, the comparison of measures through correlation probably underestimates the degree of difference voters perceived between the candidates and/or between the parties on issues. Again taking Vietnam as an example, on the seven-point scale running from withdrawal to military victory as alternatives, the mean rating given Nixon's perceived stand was 4.6 compared to 1.7 for McGovern (the mean for Republicans was 4.5 and for Democrats, 2.7). On bussing voters believed Nixon more opposed (mean of 5.0) than McGovern (mean of 3.4), and on legalization of marijuana Nixon's mean rating was 5.5 (opposed) compared to McGovern's 3.6 (for legalization). Bear in mind that we have used correlations covering *composite* indexes of all nine issues; this we suspect obscures perceived differences even more. In sum, our measures are simply not as sensitive in tapping the cognitive and conative aspects of candidate and party images as they are in detecting the affective content of those images.[22]

Thus, the correlation matrix in Table 8.3 tells us certain things but leaves unanswered our most important questions. The coefficients indicate that almost all of our image measures correlate with the vote but do not tell us which contribute the most to it. Further, our twenty-three image measures are intercorrelated to various degrees. In some cases the measures within any one set of image variables (political self-, candidate, or party images) or set of variables representing image aspects (cognitive, affective, or conative) are more highly correlated than are measures across sets, but this is not always the case. This raises the question of whether or not our three categories of image variables and of image aspects are as distinct as we originally classified them. Are they, in short, the principal components of the voting decision?

Components of the Voting Decision

To deal with this question we turn once again to techniques of factor analysis. Recall that factor analysis enables us to reduce a large number of measures, such as our twenty-three image variables and the vote, into a smaller number of principal components, each component consisting of variables arrayed along a common dimension. We can compare the results of such a factor analysis with the components of the voting decision that we originally expected, as a check on our preconceptions. To do this we have conducted two factor analyses, one limited to the twenty-three image variables and the other adding the vote as a twenty-fourth variable.

Table 8.4 displays the results of our factor analysis of all variables, excluding the vote.[23] The four-factor solution vividly refutes our original three-component model of voters' self-, candidate, and party images. Factor I consists of nine variables, each pertaining to *issues*. Respondents' self-locations on issues, their perceptions of both candidate and party issue stands, and their proximities to candidates and parties on issues—all load on this factor, and it thus exhausts all measures that in any way deal with issues. The second

Table 8.4 Factor Loadings of Measures of Image Variables, Excluding Voting Behavior

VARIABLES	FACTOR LOADINGS				COMMUNALITIES
	I	II	III	IV	
RI	88	02	04	−02	78
NI	87	02	04	−02	77
DI	83	02	04	−02	69
DIP	79	−30	01	−05	72
RIP	74	43	07	02	75
MIP	72	−37	00	−06	65
MI	71	24	06	00	57
NIP	64	48	07	03	77
SPI	59	−43	−01	−06	53
MT	−02	63	04	07	40
CI	−02	62	04	07	39
NLCP	00	62	32	−08	42
MLCP	03	−61	14	−17	49
NT	02	−60	−04	−06	36
RLCP	00	58	30	−08	44
PI	−02	56	04	06	32
PSI	02	−55	−04	−06	30
ISI	02	−55	−04	−06	30
NLC	04	13	67	−17	49
RLC	04	12	64	−16	45
MLC	−03	22	−21	43	27
DLC	−03	22	−22	42	27

Note: Decimals have been omitted from factor loadings.

Source: Survey Research Center/Center for Political Studies, University of Michigan, and the Interuniversity Consortium for Political Research.

factor has ten variables and is bipolar, with five measures at the positive end and five at the negative. It is the most difficult factor to interpret, for two of the items measure voters' self-images (partisan and ideological), three measure the affect aspect of candidate images (the two thermometers and the Candidate Index), the Party Index relates to affect toward parties, and the remaining variables are the respondents' proximities to the candidates' and parties' perceived ideological positions. If anything, this factor is a *candidate-party* factor. The remaining two factors offer no such interpretive problems. Factor III is a *conservative* factor consisting of respondents' perceptions of the ideological positions of Nixon and the Republicans; the fourth factor is a *liberal* factor made up of perceived ideologies of McGovern and the Democrats. To round out our picture of the principal components of the 1972 vote decision, we present a second factor solution in Table 8.5, this time with the vote added to all other variables.[24] With the exception of the vote, which enters as the highest loading variable on the second factor, the same four factors emerge as in the earlier analysis.

The results of these factor analyses warrant two assertions. First, as summarized in Table 8.2, our preconceived categories of the components of electoral decisions are not distinct dimensions in the factor analyses. Nor, we must add, are the electoral components derived from earlier voting studies—especially party, candidate, and issue orientation—clear-cut, either. To be sure, issue orientations stand out loud and clear. However, party and candidate orientations meld, and conservative and liberal orientations join the list as identified dimensions of the voting decision. Second, the vote appears on the candidate-party factor. Combining most strongly with it are three affective measures—the ratings of McGovern and Nixon on the candidate thermometers and the Candidate Index. This suggests a strong relationship between affective orientations toward the candidates and the vote in 1972, perhaps even that these affects were the overriding components of the electoral decision. Further, to unravel some of the intricacies in the complex relationship between these electoral components and voting, we turn to the third and final stage of our analysis, multivariate regression.

The Effect of Candidate Image

Multiple regression is a statistical procedure that permits us to ferret out the relative impact of each of several influences having direct effects on the vote. In this case, such influences consist of the twenty-three variables already discussed in our correlation and factor analyses. We treat these as independent variables and enter them into a regression formula designed to explain the variation in a single, dependent variable—that is, the vote in 1972.

Table 8.6 presents the results of our regression analysis for 1972. Although we have reported regression analyses earlier (in Chapter 4), let us briefly review the major types of information contained in Table 8.6. The column labelled "Variables" lists each of our twenty-three independent variables in descending order of how important each is to explaining how respondents voted in 1972. The "Simple R" column states the correlation

Table 8.5   Factor Loadings of Measures of Image Variables and Voting Behavior

VARIABLES	FACTOR LOADINGS				COMMUNALITIES
	I	II	III	IV	
RI	88	06	01	02	78
NI	87	06	01	02	76
DI	83	06	01	02	69
DIP	78	−26	−04	00	67
RIP	75	47	07	06	80
MI	72	28	04	04	60
MIP	70	−33	−04	−01	60
NIP	65	49	08	09	84
SPI	56	−39	−06	−02	48
VOTE	00	74	11	06	57
MT	−01	65	08	06	43
CI	−01	63	08	06	41
NT	01	−61	−08	−06	39
NLCP	−01	61	36	−09	51
MLCP	−01	−60	13	−16	41
RLCP	−01	57	34	−08	45
PI	01	57	07	05	33
DLCP	01	−55	22	−19	39
PSI	01	−54	−07	−05	30
SPI	01	−52	−06	−05	28
NLC	00	08	68	−14	49
RLC	00	08	65	−13	45
MLC	−01	23	−21	43	29
DLC	−01	23	−21	42	28

Note: Decimals have been omitted from factor loadings.

Source: Survey Research Center/Center for Political Studies, University of Michigan, and the Interuniversity Consortium for Political Research.

coefficient of each variable taken alone, with the vote. The column headed "Multiple R" contains multiple correlation coefficients. Each coefficient represents the total strength of all measures listed to that point when related to the vote. (For example, the correlation of respondents' thermometer ratings of both Nixon and McGovern to the vote is .77; the correlation of those two measures plus respondents' proximities to Nixon's perceived ideological position is .79). "R^2" is a coefficient of determination, a statistic that indicates how much of the variance in the vote can be explained by all of the independent variables listed to that point. The coefficient of determination is the square of the multiple correlation coefficient expressed as a percentage (the thermometer rating of the candidates, for instance, is .77 squared, or 59 percent). The column labelled "Change" indicates how much additional variance in the vote is explained by a given variable when it enters the regression formula. Thus, whereas the Nixon thermometer ratings explain 41 percent in the vote's variance, the addition of McGovern's ratings explains an

Table 8.6   Regression Analysis of Image Variables Explaining 1972 Presidential Vote

VARIABLES	SIMPLE R	MULTIPLE R	R²	CHANGE	BETA
NT	−.64	.642	41.3%	.413	−.29
MT	.64	.768	58.9	.176	.30
NLCP	.56	.786	61.8	.029	.15
MLCP	−.50	.794	63.1	.013	−.10
PSI	−.50	.799	63.8	.007	−.11
CI	.48	.801	64.1	.003	.06
NIP	.36	.802	64.4	.003	.09
DIP	−.17	.804	64.6	.002	−.06
DLC	.18	.805	64.8	.002	.05
PI	.36	.806	64.9	.001	−.05
SPI	−.28	.807	65.0	.001	−.05
NLC	.14	.807	65.1	.001	.03
RLC	.12	.807	65.2	.001	−.02
MLC	.17	.807	65.2	***	−.03
NI	.05	.807	65.2	***	.02
ISI	−.47	.808	65.2	***	.02
MI	.20	.808	65.2	***	.03
DI	.11	.808	65.2	***	−.03
RIP	.32	.808	65.2	***	.02
RLCP	.46	.808	65.2	***	−.01
RI	.09	.808	65.2	***	.01
MIP	−.25	.808	65.2	***	−.01
DLCP	−.36	.808	65.2	***	.00

***The amount of additional variance in the vote explained by these variables is too minute to report.

Source: Survey Research Center/Center for Political Studies, University of Michigan, and the Interuniversity Consortium for Political Research.

additional 18 percent. Finally, the "Beta" column contains a coefficient that permits us to examine the relative importance of each listed variable on the vote.

The regression reported here is a stepwise regression, that is, one that enables us to say which variable is most important in explaining the 1972 vote, and then, how much additional variance in the vote can be explained through successively adding other independent variables in order of importance. The most important variable is that with the highest simple correlation to the vote (recall from Table 8.3 that this was the Nixon thermometer rating, −.64). The stepwise procedure then selects the next variable with the highest partial correlation with the vote, that is, its correlation with the vote once its relationship with all other variables has been taken into account. Each variable is tested to determine how significant it is for explaining the vote, and the procedure continues until no significant variables remain.

The Beta coefficients in Table 8.6 indicate that numerous variables had some independent effect on the vote, but the relative importance of some (such as the respondent's proximity to the Democratic party's perceived ideological position) was virtually nonexistent. The most important component of the vote was the affective evaluation of the candidates on the feeling thermometers. The affective orientation to Nixon explained 41 percent of the variance in the vote, and affect toward McGovern added another 18 percent; the Beta coefficients of both are strong: $-.29$ for Nixon and .30 for McGovern. (Since the vote was coded as "0" for Nixon and "1" for McGovern, the negative Beta for Nixon indicates the negative correlation between his thermometer ratings and the vote.) Following affective orientations toward the candidates in explaining the vote are two variables we originally conceived of as conative aspects of the candidate images, and which our factor analyses indicated as with the candidate-party factor—that is, respondents' proximities to each of the candidate's perceived ideological predilections.

Finally, the only other variable explaining an additional 1 percent of the vote variance is the voter's partisan self-image (Beta = $-.11$). Thus, in a regression solution in which all independent variables explain a total of slightly less than two-thirds of the variance in the 1972 vote, five variables provide 64 percent. All five of these were on the candidate-party factor derived from our factor analyses, and four of those pertain directly to our preconceived categories of affective and conative aspects of candidate images.

We noted earlier that our composite indexes of the perceived issue stands of the candidates and the parties obscure some of the differences that voters detect between contestants on specific issues. This may account for the fact that issues (highlighted as the first factor in Tables 8.4 and 8.5) do not emerge as relatively important explanations of the 1972 vote. To test for this possibility we ran another regression analysis, this time maintaining the separation of five key issues (thus entering respondents' self-locations, perceived candidate stands, and proximities to candidates on each issue as separate variables in the analysis). Only the voters' proximities to the perceived Republican position on the Vietnam issue (addition of 4 percent variance and Beta = .20) and proximities to McGovern on Vietnam (1 percent variance and Beta = $-.16$) had relatively important effects. Thus, the farther respondents deemed themselves from perceived Republican policy on Vietnam and the closer to McGovern's perceived stand, the more likely they voted for McGovern. Again, however, the thermometer ratings emerged as the two most important variables; in a regression explaining 72 percent of the variance in vote, the feeling ratings explained 59 percent.

Since regression analysis permits us to probe only the direct relationships between selected variables and the vote, we are still uncertain what indirect effects such influences as partisan and ideological self-images have as they condition voters' perceptions of candidates. The moderate correlations (see Table 8.3) between these variables and affective orientations toward the candidates lead us to suspect such an indirect, conditioning effect. In addition, other analyses comparing the direct and indirect effects of partisan self-images on the 1972 vote suggest considerable indirect effects.[25]

In sum, despite the restrictions inherent in our available measures and techniques, we think we have achieved at least partial answers to the questions we posed at the outset of this presentation of findings regarding the 1972 presidential election study. First, we have seen that most of the twenty-three image measures exhibit a direct relationship to the vote. Second, we have uncovered discrepancies between our early conceptions of the components of the vote as being the voter's self-image, candidate images, and party images and the principal components derived from our factor analyses. Instead, the components appear to be issue, candidate-party, and liberal-conservative orientations. Yet our stepwise regression indicates that candidate image concerns on the candidate-party factor, especially in their affective and conative aspects, are the most important explanations of the 1972 vote. Unanswered are such questions as the relative magnitude of indirect effects from other influences such as party identification on candidate images and voting behavior.

Certainly, in interpreting our findings we must keep in mind that the presidential election of 1972 had special characteristics—a greatly divided majority party, an incumbent running to "Re-elect the President," a landslide victory, and so on. These unique features of 1972 make it impossible to generalize our findings to other elections. We focus on 1972 because it was the first election in which a wide variety of image measures made it possible to assess the relative effects of the kinds of images we deemed important. However, analyses of previous elections employing a more limited set of image measures indicate that our findings may indeed pertain to more than 1972. Using regression analyses for each presidential election in the 1952-1972 period, Kirkpatrick, Lyons, and Fitzgerald attempted to weight the relative impact of candidate and party images on the vote. Restricting themselves to measuring candidate and party images by relying solely on the standard SRC survey question about likes and dislikes of the candidates and parties (the type of data we have employed in constructing our Candidate Index and Party Index), they note "a consistent linear decrease over the twenty-year period in the relative influence of total party images on the vote" and a "uniform linear increase in the role of candidate images." Candidate images have nearly doubled in their explanatory power, but the effect of party images has been cut in half over the six elections. Kirkpatrick, Lyons, and Fitzgerald found this trend to hold generally for the major partisan subgroups, whether defined by strength or direction of partisan self-images.[26]

Similarly, Hinckley, Hofstetter, and Kissel in their study of voting behavior in 1968 also demonstrate the importance of candidate images. Relying on data generated by the SRC's national sample and thirty-one statewide samples compiled by the Comparative State Election Project, they explored what image traits influenced voting in presidential, senatorial, and gubernatorial contests. Their critical measure was the series of open-end questions tapping voters' likes and dislikes about the two major party candidates for president, U.S. senator, and governor. For each candidate they identified four sets of traits referring to the candidate's (1) political party, (2) experience or incumbency, (3) personal qualifications, and (4) issue positions. Using these traits in a regression analysis as independent variables against the dependent variable of the vote, they found that candidate image explained 57 percent of variance in presidential voting, 59 percent in senatorial contests, and 69 percent in gubernatorial races. Personal qualifications were the most important. And, they concluded that the role of partisan self-images as compared to candidate images becomes more important in influencing the vote if voters have little information about candidates (such as poorly articulated images) or the voter's image of candidates is indeterminant, that is, balanced between positive and negative perceptions.[27] These studies by Kirkpatrick, Lyons, and Fitzgerald and by Hinckley, Hofstetter, and Kessel—taken within the context of others we have noted—suggest that our findings regarding the influence of candidate images in 1972, regardless of the special qualities of that election, extend to other instances of voting behavior.

Current Conclusions and Future Research

The images of political candidates are the sets of perceptions people have of them—the subjective knowledge, feelings, and predispositions toward candidates held in the minds of prospective voters. These images are the joint products of the kinds of attributes candidates present to voters and such conditioning pictures as the voters already have in their heads. The relationship between the candidate and voter, then, is a special case of the message-image transactions previously described by Kenneth Boulding[28] (see Chapter 1). It is a relationship that, as we reviewed in Chapter 2, has been studied from various perspectives, sometimes emphasizing what the candidate brings to it and at other times focusing on what the voters contribute. At this point we shall abstract the most general of our conclusions that emerge from the series of investigations of candidate images reported for the first time in this volume and call attention to a few key areas for further research.

(1) Candidate images are multifaceted subjective phenomena; a candidate typically has various images, not a single one, and each image differs in its cognitive, affective, and conative content, its key dimensions, and the combination of attributes that define it. (2) Despite the complexity of any given image, however, there are regularities in the ways voters assign traits, or attributes, to candidates, so much so that research reveals a relatively small number of image traits that voters perceive in candidates. Of these, voters seem especially fond of appraising candidates from the standpoint of perceived strength, integrity, and empathy (Chapter 3). (3) Regularities obtain also in the ways voters order or combine image traits into image types in viewing a candidate. We saw in Chapter 3 that it is not unusual for as few as three to five such orderings to exist within much larger numbers of voters. Yet in spite of the regularities in image traits and image types, much of the variation in voters' perceptions remains unexplained.

Previous research points to a few major influences that shape a candidate's image. A stimulus-determined view notes the attributes, style, and techniques the candidate brings to the campaign. A perceiver-determined view focuses on selected voters' predispositions—partisanship, ideology, sociodemographic characteristics, social influences, and conceptions of self and ideal. Corollary to each view is a theory. First, the *image theory* says that the "image emanating from the candidate accounts for the public's perception"; second, the *perceptual balance theory* says that voters avoid stress in their predispositions by seeing in a preferred candidate what they wish to see, "even if it is unrelated to objective reality."[29] Our research (see especially Chapter 4) leads us to conclude that neither theory constitutes an adequate explanation in the absence of the other. Political images are not the products either of what candidates "project" or voters "hold." Rather they reflect relationships in which leaders not only project selected attributes but must also imagine how followers perceive them; followers not only perceive leaders but also imagine how leaders perceive them. (4) Therefore, the differing imagining, or imaginations, yield different images of political candidates, with quite distinct attributes and/or predispositional influences determining each.

(5) The images of political candidates change. They are not, as much early research suggested, fixed in all voters' minds before, or early in, the campaign. We have seen that changes take place in candidate images during and between election campaigns, although sometimes the change is slight or confined to only a few perceived traits (Chapter 6). The relationship between candidate images and campaign changes is probably a function of the candidate's campaign style in transaction with the voters' preferences at any given moment for certain types of appeals, campaign messages, and media—the various elements of campaign style (Chapter 4). And, we know that image changes result from the candidate's victory or defeat, especially as partisans faced with certain defeat of their preferred candidate grope for ways to adjust both to that prospect and thereafter to the accomplished fact (Chapter 7).

(6) Finally (using 1972 as our example), we have seen that one aspect of candidate images—that of *affect,* or feelings of liking and disliking—was the single most important explanation for voting behavior in the presidential contest. Such a conclusion is consonant with other studies cited in this chapter and also with the results of analytical efforts to develop sophisticated ways to predict how people vote in elections. The work of Kelley and Mirer is a notable case in point. Relying on the same type of data we have used to develop the candidate and party indexes we used in this chapter and to measure images changes in Chapter 6 (that is, responses to open-end questions probing voters' likes and dislikes of the various candidates and political parties), Kelley and Mirer developed a general rule for predicting voting behavior. Simply put, it is:

> The voter canvasses his likes and dislikes of the leading candidates and major parties involved in the election. Weighing each like and dislike equally, he votes for the candidate toward whom he has the greatest number of favorable attitudes, if there is such a candidate. If no candidate has such an advantage, the voter votes consistently with his party affiliation, if he has one. If his attitudes do not incline him toward one candidate more than toward

another, and if he does not identify with one of the major parties, the voter reaches a null decision.[30]

Restated in the language we have been using, the voter formulates images of the candidates and parties and votes for the candidate with the more favorable image; if there is none, his partisan self-image takes over. Voters with neither candidate nor partisan self-images reach null decisions; that is, there is no basis to say how they will vote. The "voter's decision rule" of Kelley and Mirer enabled them to "predict" (actually postdict, or predict after the fact) decisions for more than 98 percent of respondents in the presidential election surveys of 1952-1964. They predicted correctly, on the average, the votes of 88 percent. Most of their predictions were based on attitudes toward candidates and parties—in other words, candidate and party images—and only a few were based on respondents' partisan self-images. The evidence is that voters do formulate images of candidates, at least presidential candidates, and that these images play the key role in their voting behavior.

As noted above, we have been interested in the dimensions of candidate images as well as the effects of those images on voting. We can now summarize what we think those dimensions are, as based on our analysis in Chapters 3 and 5 through 8. First, we have found it useful (as discussed in Chapter 3) to divide a candidate's image into two principal role dimensions—a political role and a stylistic role. Each of these roles consists of a cluster of attributes, which we have dealt with as traits. Originally we considered the political role of the candidate as consisting of two traits, leadership and partisanship. On the basis of our analysis in Chapters 5 and 8 we think it necessary to add a third, a cluster of traits associated with voters' perceptions of where a candidate stands on issues and whether they agree with that stand. (7) Thus, we summarize the political role dimension of the candidate's image as made up of leadership, partisanship, and issue traits. On the stylistic side, we retain our view that voters perceive the personality of the candidate, that is, his personal attributes and his style as a dramatic performer. We have seen personality emphasized as an important role trait in Chapters 5, 6, 7, and 8, and our discussion of the media aspects of campaign style (in Chapter 5) highlights the role qualities of the candidates' dramatic performance. (8) Let us summarize stylistic role, then, as the candidate's personality and performance traits.

We cannot say definitely which of these dimensions and traits voters deem most important in constructing an overall image of a candidate; nor can we say which they respond to most in making their voting decisions. (9) We suspect, however, that personality, a decidedly affective aspect of a candidate's image, dominates voters' responses to candidates during campaigns, and the political qualities return to greater influence, as we noted in Chapter 7, after the election. We hope that future research will probe more deeply into the two dimensions of candidate images and ferret out other traits, as well as coming to some conclusions regarding the relative influence of perceptions of the candidate as a leader, partisan, articulator of issues, personality, and performer.

Obviously in this volume we have only been able to scratch the surface of what is apparently a very thick outer layer of hard questions about candidates and their images. Certainly we make no claim to having penetrated to the core of understanding all the reasons for how and why people subjectively respond to political candidates as they do. Some of the areas that require considerably more attention than we have, as yet, had the opportunity or resources to explore are described below.

First, how can we refine the concept of image, especially that of a candidate's image, to make it more subject to inquiry? We have relied primarily on Boulding's early conceptualizing. However, we realize that in doing so, images come to include so much that they really explain nothing. We think it unfortunate that two decades ago, when Boulding called for a science of "eiconics" (surely, partly in jest) to study and explain the message-image relationship, so few took him up on it. Heightened, even anxious, popular concerns with images abound—self-images, company images, institutional images, brand images, product images, national images, presidential images, as well as candidate images—but in a context of relatively little scholarly interest in the concept.

Second, how determinant are images? Boulding theorized that, "Behavior depends upon the image." We have seen that there is a close relationship between candidate images and voting behavior and have teased out of our available data the suspicion that voting may depend on the image. But what alternative forces have direct effects on voting? And, of more interest to us, what forces operate indirectly on voting through images? In our concern primarily with psychological forces we have scarcely touched on the vast array of socioeconomic, generational, communication (mass, organizational, and interpersonal), institutional, and socializing forces that shape political images over long periods.

Third, how important are candidate images in different kinds of elections? Most voting studies wherein we have examined candidate images have dealt with statewide, congressional, or presidential contests in which the major political parties play an important organizational and dispositional role. But what of the importance of candidate image in those races where partisanship is, in effect, controlled (such as party primaries) or removed altogether (as in nonpartisan elections)?

Fourth, what of the practical applications of learning about candidate images? Certainly, candidates and their various managers and consultants intuitively feel that the candidate's image is important to victory. But if this is so, what can the candidate *do* about his image? Much campaign advising respecting images is on an obvious level. Few candidates need to be told to be honest, sincere, and trustworthy. What they need to know is what attributes voters assign to them, which are accurate, which are fixed, and which can be changed (or publicized, if not assigned), and how. Despite a spate of books pouring forth from professional campaign consultants, an applied science (or ethics) of eiconics respecting candidate imagery scarcely exists.

Finally, how can we study the images of political candidates? From our own efforts and those of others to investigate the phenomenon of candidate-voter imagery, we are greatly impressed by the inadequacy of the dominant approaches to measuring candidate images. Although the use of open-end questions and rating scales in large-sample surveys provides valuable insights into how people think, feel, and respond in relation to candidates, we need other kinds of data-gathering techniques to explore important questions. Our reported efforts to adapt *Q*-methodology to the task have been useful, but we realize that we and others must go farther. For one thing, other procedures for measuring the salience of candidates to voters (especially the Sherifs' own categories technique)[31] need to be adapted to the study of political images. For another, we need to develop techniques for discovering how people construct political reality in relation to the persons and things they regard as important in their daily lives. (For example, an application of the techniques derived from psychologist George Kelley's Personal Construct Theory appears promising in this respect.)[32] Finally, we think it time to take Peter Rossi's advice seriously, a suggestion he made more than fifteen years ago in criticizing studies of voting. Although representative sampling of the population through probability designs provides insights, we need more of what Rossi called "factorial" sampling, "in which the objective is to obtain specified types of individuals in given populations without regard to their actual population under study," in order to provide a sufficient number of them to permit in-depth study. Factorial samples would help us pinpoint the relationship of image types and traits to conditioning variables (when extended to survey designs), to examine image change (when extended to panel designs), and to explore how candidate-voter imagery responds to manipulation (when extended to experimental designs). But, as Rossi indicates, "As a corollary, the suggestion implies that the researcher will have to restrict the scope of his interests" in each study.[33]

What we urged at the close of Chapter 1 warrants repeating. In this volume we have tried to begin understanding the images of political candidates by accumulating a few relatively microcropic studies. In the process we have identified the types of conceptual, technical, and focal problems that pertain to understanding candidates and their images. We hope that our efforts will prompt other researchers "to restrict" (and thus expand) the scope of their interests to *direct* and *explicit* studies of images in politics. Finally, we think Boulding's call for a science of eiconics sufficiently serious to justify a more widely shared concern among students of politics.

Notes

1. Angus Campbell et al., *The Voter Decides* (New York: Harper & Row, 1954).

2. ———, *The American Voter* (New York: John Wiley, 1960). The incorporation of the normal-vote concept into the SRC description of voting appears in Angus Campbell et al., *Elections and the Political Order* (New York: John Wiley, 1966), chap. 2. The treatment of party identification as partisan self-image appears in David Butler and Donald Stokes, *Political Change in Britain* (New York: St. Martin's Press, 1969), pp. 37–43.

3. Arthur H. Miller et al., "A Majority Party in Disarray: Policy Polarization in the 1972 Election" (paper delivered at the Annual Meeting of the American Political Science Association, New Orleans, Louisiana, September 4–8, 1973).

4. Peter B. Natchez, "Images of Voting: The Social Psychologists," *Public Policy* 18 (Summer 1970): 575.

5. Peter B. Natchez and Irvin C. Bupp, "Candidates, Issues, and Voters," *Public Policy* 17 (Summer 1968): 409–37.

6. David M. Kovenock et al., "Status, Party, Ideology, Issues, and Candidate Choice: A Preliminary, Theory-Relevant Analysis of the 1968 American Presidential Election" (paper delivered at the Eighth World Congress of the International Political Science Association, Munich, Germany, September 5, 1970).

7. Miller et al., "A Majority Party in Disarray."

8. Natchez and Bupp, "Candidates, Issues, and Voters," pp. 446–47.

9. Richard W. Boyd, "Presidential Elections: An Explanation of Voting Defection," *American Political Science Review* 63 (June 1969): 498–514.

10. Roy E. Miller and John S. Jackson, III, "Split-Ticket Voting, An Empirical Approach Delivered from Psychological Field Theory" (paper delivered at the Annual Meeting of the Southern Political Science Association, Atlanta, Georgia, November 1–3, 1973).

11. Donald E. Stokes, "Some Dynamic Elements of Contests for the Presidency," *American Political Science Review* 60 (March 1966): 22.

12. We think that there is sufficient evidence to indicate that the nine issues selected were sufficiently salient to voters in 1972 to warrant their inclusion. See, for example, Miller et al., "A Majority Party in Disarray." Yet we realize that had we employed other issues, differences in findings could result. However, we selected these nine issues because they were the ones available on the same interview forms used to elicit respondents' likes and dislikes about the candidates and parties. Not all interview forms included these latter matters but did include queries on other policy issues. Since we were interested in employing data from questions concerning likes and dislikes, we were forced to limit ourselves to the nine policy issues as designated.

13. We collapsed the actual thermometer readings into a seven-point scale as follows: 1 = 0–15, 2 = 16–28, 3 = 29–43, 4 = 44–55, 5 = 56–69, 6 = 70–84, and 7 = 85–100.

14. The index was calculated by taking the number of likes and dislikes (up to three each were coded) expressed by each respondent about each candidate; pro-Republican plus anti-Democratic responses were subtracted from pro-Democratic plus anti-Republican responses. The range of scores (−6 to +6) was then converted to a 1–13 scale.

15. M. Brewster Smith, "The Personal Setting of Public Opinions: A Study of Attitudes toward Russia," *Public Opinion Quarterly* 11 (Winter 1947): 507–23. Also see Thomas M. Ostrum, "The Emergence of Attitude Theory: 1920–1950," in Anthony G. Greenwald et al, eds., *Psychological Foundations of Attitudes* (New York: Academic Press, 1968), pp. 16–18.

16. John Harding et al., "Prejudice and Ethnic Relations," in Gardner Lindzey and Elliot Aronson, *The Handbook of Social Psychology* (Reading, Mass.: Addison-Wesley, 1969), pp. 4–13; see also Angus Campbell, *White Attitudes toward Black People* (Ann Arbor, Mich.: Institute for Social Research, 1971), chap. 1.

17. Other studies have employed the concept of proximity and derived such measures. See Kovenock et al., "Status, Party, Ideology, Issues, and Candidate Choice"; Miller and Jackson, "Split-Ticket Voting"; and Miller et al., "A Majority Party in Disarray."

18. See John H. Kessel, "Comment: The Issues in Issue Voting," *American Political Science Review* 66 (June 1972): 460–61.

19. See Eugene J. Webb et al., *Unobstrusive Measures: Nonreactive Research in the Social Sciences* (Chicago: Rand McNally, 1968), p. 184.

20. The reader will recall the use of correlations elsewhere in this text. The correlations reported in Table 8.3 are Pearsonian coefficients (r's) and may range from +1.0 to −1.0. In reporting the coefficients between variables, we simply use the abbreviated labels appearing in Table 8.2. Hence, the correlation between partisan self-image (PSI) and the Party Index (PI) is reported as rPSI, PI = −.40, indicating that the more Republican the respondent, the less favorable he is toward the Democratic party.

21. Correlation coefficients of .08 or above are significant at the .001 level, thus indicating they could have occurred by chance in only very rare instances.

22. Miller et al., "A Majority Party in Disarray," report greater differences in their correlations of respondents' perceptions of candidate stands in relation to respondents' self-locations on issues. This is partly due to an adaptation by Miller et al. of the seven-point issue scales whereby they assigned to each respondent a score of −1 if the respondent was closer to Nixon, 0 if equidistant from both candidates, and +1 if closer to McGovern. This procedure tends to obscure differences in mean ratings less than basing correlations on the original seven-point scale, but raises analytical problems associated with collapsing categories. In any event, their procedure does not differ in results substantially from the one we employ.

23. Although five factors were extracted and rotated to simple structure by the varimax method, no variables loaded highest on the fifth factor. Hence, Table 8.4 reports only factor loadings for the four key factors, but for comparison commonalities refer to the five factors. The four factors explain 49.6 percent of the variance, and the five factors explain 50.0 percent.

24. As in the case of the factor solution with vote not included, five factors were extracted and rotated but only four emerged with highest loadings for any variable. The four factors explain 50.2 percent of the variance, the five factors, 50.7.

25. See Miller et al., "A Majority Party in Disarray."

26. Samuel A. Kirkpatrick, William Lyons, and Michael R. Fitzgerald, "Candidate and Party Images in the American Electorate: A Longitudinal Analysis" (paper presented at the Annual Meeting of the Southwestern Political Science Association, Dallas, Texas, March 28–30, 1974), pp. 26–27. A revised version of this paper appears in Samuel A. Kirkpatrick, William Lyons, and Michael R. Fitzgerald, "Candidates, Parties and Issues in the American Electorate: Two Decades of Change," *American Politics Quarterly* 3 (July 1975): 247–83. In the same volume appear the results of a regression analysis that also offers evidence of the increasing importance of candidate images in voting: see Eugene Declercq, Thomas L. Hurley, and Norman R. Luttbeg, "Voting in American Presidential Elections," *American Politics Quarterly* 3 (July 1975): 222–46.

27. Barbara Hinckley et al., "Information and the Vote: A Comparative Election Study," *American Politics Quarterly* 2 (April 1974): 131–58.

28. Kenneth E. Boulding, *The Image* (Ann Arbor, Mich.: University of Michigan Press, 1956).

29. Roberta S. Sigel, "Effect of Partisanship on the Perception of Political Candidates," *Public Opinion Quarterly* 28 (Fall 1964): 484.

30. Stanley Kelley, Jr., and Thad W. Mirer, "The Simple Act of Voting," *American Political Science Review* 68 (June 1974): 574.

31. See Carolyn W. Sherif et al., *Attitude and Attitude Change* (Philadelphia: W. B. Saunders, 1965), chap. 4.

32. Fay Fransella and D. Bannister, "A Validation of Repertory Grid Technique as a Measure of Political Construing." *Acta Psychologica* 26 (1967): 97–106.

33. Peter H. Rossi, "Four Landmarks in Voting Research," in Eugene Burdick and Arthur J. Brodbeck, eds., *American Voting Behavior* (Glencoe, Ill.: The Free Press, 1959), pp. 45–46. See also David Willer, *Scientific Sociology* (Englewood Cliffs, N.J.: Prentice-Hall, 1967), especially chap. 6, "Conditional Universals and Scope Sampling."

Appendix A

The Q-Sort as a Procedure in Candidate Image Research

One of the principal difficulties apparent to anyone interested in studying candidate images is the problem inherent in achieving an operational definition of "image." This was clear to the authors after reviewing the literature on the subject prior to undertaking the inquiries discussed in this volume. The procedures commonly used typically had two major limitations. In the first place, all were essentially normative measures in that, to the extent they were quantitative (that is, provided scores measuring perceived traits of candidates), their interpretation involved noting deviations from the mean level of scores of whatever population group a selected respondent represented.[1] Second, with the exception of the semantic differential, the candidate attributes measured were often selected in a haphazard manner, which at best represented campaign messages current at the moment and at worst, the unannounced, biased predilections of the researcher.[2] It seemed preferable to use a procedure that allowed respondents to represent their views in a self-referent manner that, at the same time, ranged over a wide variety of primary characteristics of candidates (within clearly stated limits). More than this, we desired a procedure that would measure "whole" images of candidates, not just bits and pieces—such as individual traits—of those projected images. The Q-sort is just such a procedure.

The Q-sort derives from the contributions of William Stephenson to factor-analytic theory. Stephenson recognized that in factoring data matrices derived from the scores of persons on educational tests, either the test items or the persons responding to those items might be clustered on the basis of similar distributions. Moreover, these clusters derived from factor-analytic

213

techniques are not merely reciprocals of one another but represent complementary perspectives of the same general phenomenon.[3] R-factor analysis locates the major dimensions of variation among attributes, resulting in a *typology of traits* from among the universe of possible traits from which the chosen attributes derive. Likewise, Q-factor analysis locates the major dimensions of variation among the persons having those attributes, resulting in a *typology of persons* from among the universe of possible people from which the sampled persons derive. Just as the traits vary across the persons in the extent to which they are possessed by those persons, the individuals themselves vary in the relative importance they assign to certain traits in their personal rank orderings of perceived attributes.

Ipsative measurement, or assigning scores reflecting the deviation of particular performances about the mean of each individual's level of performance, is one aspect of the Q-perspective in data analysis. But Stephenson, not satisfied with just ipsative measurement that could be obtained through statistical artifice (that is, standardization), devised the Q-sort as a tool that also allowed respondents to describe in a truly self-referent way the events involving them. In addition to being an ipsative, self-referent tool of data gathering, Stephenson's Q-sort procedure reflects other methodological interests that concerned him as well. While the procedure provides a respondent considerable latitude to display his uniqueness through idiosyncratic arraying of the statements comprising the Q-sort, it can nevertheless be used for developing and testing theoretical explanations. Stephenson's concern for theory via the Q-sort is apparent in his careful attention to the potential for experimental design in selecting statements for the Q-sort.[4]

Stephenson had still another concern in devising the Q-sort, which was to promote economy of both time and resources in behavioral research. He accomplished this by associating the procedure with the small-sample doctrine emerging from the statistical work of R. A. Fisher.[5] On the one hand, the small-sample doctrine suited Stephenson's demand for economy in scientific research; on the other, it suited well the complex, intensive character of the Q-sort. In sum, then, Stephenson's Q-methodology is an approach to behavioral science research that emphasizes theoretical explanation through experimental design, intensive data gathering built on small-sample theory, and an inductive typal analysis based on self-referent observations from respondents themselves. The Q-sort procedure potentially incorporates all these features.[6]

The Q-sort is particularly suited to the purpose of measuring candidate images. Respondents can be requested to rank a relatively large number of attribute statements according to their appropriateness in describing a given political figure. The resulting array of statements for each respondent represents, through his own operations, that person's perception of the political figure as projected to him through whatever media and whatever filtering attitudes he may hold. For each person, then, the resulting scores for statements are arrayed around his idiosyncratically determined center to reflect his own conception of the object and not some a priori norm provided by the investigator. (Strictly speaking, the Q-sorter assigns the statements to scores, whereas other procedures typically require the subject to assign scores to statements.)

In *Q*-methodology, respondents' sorts can be systematically compared, usually through factor-analytic procedures, to determine whether there are psychological patterns underlying similar descriptions of the same objects. If no pattern emerges, images of the object in question are probably so idiosyncratic, or subjective, as to preclude psychological investigation of images except on an individual-by-individual basis. (Such a finding would be of general interest, unlikely as its occurrence is, since it would strongly indicate that campaign messages had failed to constrain images in the ways intended by candidate projections.) If, on the other hand, all *Q*-sort arrays are essentially similar, then only a single factor will emerge. Such consensus is unlikely for a specific political actor, but images of an ideal public official tend toward such a consensual public image. If there are only some persons whose *Q*-sorts are alike, then a basic image type emerges. There may be several such basic types of images of a given political candidate. The varying ways in which these diverse types array the statements can be analyzed to understand why some subjects perceive a candidate in essentially similar fashion, but different from the perceptions of others. Such differences are actual differences among the subjects rather than variations from some statistical norm of an a priori social group.[7] In some instances, of course, people constituting a factor may largely derive from some particular group—for example, women or doctors or Republicans—but such a finding would only enhance the analysis.

To give the reader a better understanding of the applicability and the limitations of the *Q*-sort in candidate image research, a step-by-step discussion of the procedure follows. Our very brief discussion in this appendix should in no way be taken as an adequate substitute for a much more thorough and comprehensive introduction to *Q*-methodology. We offer only the essentials of *Q*-methodology here as a means of facilitating the reader's understanding of our discussion in previous chapters. We urge readers interested in learning more about the procedure to consult the various works cited in the notes to this appendix.

The construction of the *Q*-sort is a critical first step, in which the investigator must make the important decision of whether to incorporate an experimental design as a guide in selecting statements or to follow a less structured approach. Once the *Q*-sort is constructed, the researcher must determine how and to whom it is to be administered. The critical questions here are those of distribution of statements by respondents—that is, free versus forced sorting—and of sampling persons so as to maximize potential sources of relevant experimental variation. Finally, the researcher is confronted with the problem of analyzing a very rich body of data. The initial purpose of the investigation will probably point to either item (statement) analysis or more global procedures such as factor analysis, but the data matrix remains susceptible to both analytical approaches. The appendix closes with a discussion of the role and limitations of the *Q*-sort in image research.

Constructing the *Q*-Sort

Attribute statements comprising a *Q*-sort are a sample selected to represent a potential population of attributes of the object under investigation, just as in survey research a sample of respondents (whether selected through a random procedure or not) represents in some fashion a population of persons. Whether sampling attributes or persons, the investigator must decide whether to structure the sample for specified theoretical purposes or to forego structuring, presumably to obtain as representative a sample of the general population as possible. Quota sampling and stratified random sampling in survey research are common techniques to obtain structured samples of persons, whereas a completely randomized procedure of sampling is an unstructured technique. In principle, sampling attribute statements is no different.[8]

An unstructured *Q*-sort is a set of attribute statements characterized by relative homogeneity, in that they are all directed toward a common object but gathered without any specific regard to any basic dimensions or variables that may underlie them. Presumably the set of statements adequately represents the population of statements from which it is drawn. This type of *Q*-sort appears most frequently in published research. The unstructured *Q*-sort is of particular heuristic value where no theoretical or analytical framework exists, but where propositional statements abound.[9] The investigator who uses the unstructured *Q*-sort has a special obligation, however, to make every effort to cast a wide net in selecting statements that sufficiently cover the domain of inquiry.

The structured *Q*-sort, while not so widely used, has some especially desirable properties. Principles of Fisherian experimental design can be incorporated into the selection of statements so as to partition the single domain of referent statements in accordance with a theory or hypothesis. Thus, the *Q*-sort is not only a heuristic device for exploratory investigation but where it can be structured, it also becomes a valuable tool for theory or hypothesis testing. And since the structured *Q*-sort incorporates factorial design, analysis of variance—a powerful analytical tool—can be brought to bear on the resulting data.

The question may arise, however, whether the investigator is imposing his predilections on the subjects by employing the constraints of a structured *Q*-sort. The answer is in the affirmative insofar as the researcher seeks to represent a theory by way of the statements selected for respondents' sorting. Yet the subjects remain free to interpret and array the statements as they see fit. Indeed, the investigator cannot be certain that all, or even any, subjects will interpret any given statement in the same way that led him to include the statement as representative of a particular theoretical notion (or "treatment level," in the parlance of experimental design). This suggests a further value of the *Q*-sort: its unobtrusiveness. Even with the structured *Q*-sort, the subject does not need to be made aware of the exact focus of the inquiry or the particular treatment levels reflected in the partitioning of statements. And usually the subject will not recognize the focus of the inquiry. Thus, the structured *Q*-sort loses none of the features of the unstructured *Q*-sort and gains capabilities for theory testing and the use of more powerful analytical tools.[10]

If the decision is to use a structured *Q*-sort, a one-way design or any of several multifactorial designs may be chosen; in the latter case, a two-way design is most common as more complex designs are likely to inflate the number of statements required beyond a practicable level.[11] Our study of images of McGovern and Nixon among college students during the 1972 presidential campaign is an example of the one-way design in which statements reflect the partitioning of candidate attributes according to the role orientations of candidates (see Appendix D). The earlier studies of candidate images in the 1970 senatorial campaign and of Democratic presidential contenders' images in 1971, on the other hand, follow a two-way design in which all statements are partitioned to reflect both role orientation and semantic qualification (see Appendixes B and C).

Whether or not the *Q*-sort is structured, a relatively large number of statements, say 50 to 120, is necessary. This requirement follows from the concern in *Q*-methodology with the need for adequate representation of the universe of possible attributes that the object may project and the subject may perceive. Moreover, where a factorial design is incorporated, adequate replication of treatment levels within partitions is essential. This requirement for replication is an important limiting element in the complexity of multifactorial designs. For example, in the 1971 study of Democratic contenders, the role effect was partitioned into four treatment levels—leader, politician, person, speaker—and the semantic variable into three—evaluative, potency, activity. Twelve (4 × 3) statements were required, then, merely to reproduce the design once. A decision was made to use a balanced set of six replications, increasing the number of statements required to seventy-two. A larger number of replications would have increased confidence. However, a tradeoff was necessary as respondents were asked to sort the same set of statements for each of six objects. The burden placed on respondents was heavy, and any further replications might have been excessively demanding. Hence, a large number of statements is desirable, but other research goals and the cooperation and comfort of research subjects must be weighed by the investigator before deciding the matter.

Such a large number of statements, particularly when a complex factorial design is involved, may tax the investigator's imagination. Fortunately, however, there are a variety of sources to look at when collecting statements for a *Q*-sort. Existing literature and psychological tests are often replete with statements that may be extracted for use. Perhaps most valuable of all are open-end responses to questions eliciting comments on the topic of concern from people similar to those likely to be used in the subsequent *Q*-sorting. These open-end responses are likely to be rich in just the kinds of self-referent statements by which people ordinarily describe objects.

In the final selection of statements, whatever their original source, a number of considerations must enter. For example, in a structured design, statements may be either balanced in number—that is, an equal number of replications for treatment levels—or unbalanced. Balanced sets are preferable in most circumstances. Unbalanced designs require an averaging procedure that weakens the application of statistical analysis of variance and may affect the interpretation of the relative importance given to the partitioning elements in the design. Other considerations noted by practitioners and critics include the necessity of avoiding antonyms, using statements having the same average level of desirability over the entire population of persons to be sampled, and using statements that at the same time exhibit substantial variance in placement by sorters.[12] But these oft-noted "rules" probably ought not to be taken as inviolate; research purposes and contingencies may make it necessary to discard them. More importantly, the investigator should recognize that what is antonymal or desirable to one person may not be to another. For example, in the Symington-Danforth study, persons who ranked "liberal" items strongly and positively in describing their ideal senator did not negatively displace "conservative" items to the same degree, suggesting that "liberal" and "conservative" are not antonyms for these people. Persons preferring "conservative" attributes, however, did displace "liberal" items negatively. And, as to expected placement variance, the research problem should dictate. In all the Q-sort candidate image studies, there was relatively low variance in describing ideal public officials but high variance in describing actual candidates, a finding of substantive importance (see Chapter 3).

Constructing a Q-sort is a task not to be undertaken casually. Certainly, any set of statements reflecting a single domain of concern can be typed one to a card—the usual mode of presentation to respondents—and administered as a Q-sort to whatever persons desired. But as with any interviewing procedure the quality of the final product, the data, is dependent on the skill and judgment of the creator in devising the research instrument.

Administering the Q-Sort

Once the researcher decides on the sample of statements (the Q-sample), two major research decisions must be confronted in administering the sort. Obviously, the question arises: *To whom* is the sort to be administered (that is, the P-sample)? But a second question must also be answered: *How* are the subjects to rank the statements?

Deciding whom the Q-sort is to be administered to is a matter of sampling. However, statistical theories of sampling are of little avail, since the Q-sort is so complex as to preclude its use ordinarily in cross-sectional surveys where the representativeness of person samples is a basic concern. Instead, small-sample doctrine must be looked to for guidance. Under this doctrine, substantive theoretical concerns are basic, pointing to a concern for locating sources of variance. Since the primary purpose of the Q-sort is to determine basic psychological types of persons by way of their self-referent descriptions of a common domain, the best advice for sampling persons is to look where maximum variance may be expected. Thus, in the Symington-Danforth campaign, a partisan contest, Republicans, Democrats, and Independents in equal numbers were administered the sort, not necessarily because these three groupings of people were really expected to emerge as three distinct psychological types, but because the research on voting behavior indicates a stronger likelihood of differences among these groupings than is likely to occur for any other social categories.[13]

Once such potential sources of variation among groups of people are determined, an equal number of people from each group should usually be administered the sort. This structure of an equal number from each group should be adhered to whenever possible to maximize the likelihood of finding variation among the groups. At the same time, no illusion should exist that the person samples are *truly representative* of the population groups from which they derive.

As with determining person samples, the instructions for the sortings by respondents should be based on the particular needs of the study at hand. The one critical aspect of these instructions is whether respondents will be allowed to rank the statements freely in whatever distribution suits them, or if all will be required to follow the same distribution of rankings.[14] Statistically, a forced distribution is preferable for its convenience in data manipulation. Moreover, the free (unforced) sort appears to obscure correspondences among sorts, usually provides fewer discriminations, and exhibits no greater degree of reliability.

There may be some questions about which procedure provides more information. For example, the forced distribution results in a common mean (elevation) and standard deviation (scatter) across all respondents. And, the free sort is likely to result in superficial discriminations such that clearly obvious determinants emerge and camouflage other distinctions of importance.[15] There are differences between the two procedures, then, but the differences statistically and psychologically tend to favor the forced distribution. If elevation and scatter are indeed important to a study, the researcher might be well advised to look to some procedure other than *Q*-sort. There is no ground for dogmatism, of course, but the weight of both evidence and practice clearly favors the forced sort.

How, then, is the forced distribution achieved? The forced distribution specifies the number of ranks with a predetermined number of statements to be placed in each rank. A respondent then sorts the statements along the continuum established by the forced distribution under some instruction, such as ranking them from "most like" to "least like" the object to be described. The investigator has previously assigned scores to the ranks. A typical distribution of frequencies of statements across ranks and the assigned scores with a *Q*-sort of seventy-two items might be:

	(most like)							(least like)			
Score	+5	+4	+3	+2	+1	0	−1	−2	−3	−4	−5
Frequency	3	6	6	8	8	10	8	8	6	6	3

This distribution is platykurtic—that is, less centered on the mean than nor-mal—and this seems to be the preferred pattern. However, the number of ranks is more important than the distribution of frequencies among the ranks. This number should be large enough to require a sufficient number of discrimina-tions for meaningful analysis of psychological differences, but not so many as to overtax respondents; nine to eleven ranks are typical and produce generally satisfactory results without raising resistance from respondents to an unaccept-able level.

After the investigator has decided on the person sample and the sorting instructions, the Q-sort is ready for actual field administration. The usual prac-tice is to have a single subject sort the statements, typed one to a card. How-ever, multiple decks of cards may be prepared for administration in group settings, and alternative formats for preparing statements, such as printing them on adhesive labels and allowing the subjects to place the labels directly on a preprinted chart showing the required distribution, further enhance group administration. A rich body of data can then be acquired in a reasonably short period of time, whether through individual or group administrations of the Q-sort. Still more decisions have to be made by the investigator, however, in confronting a large number of bits of information, even if from a relatively small number of people.

Analyzing the Data

The basic choice confronting the analyst of Q-sort data is whether to make item-by-item comparisons or to make global comparisons—that is, to analyze the similarity across or between entire Q-sorts. Item-by-item comparisons are more likely to be of interest in clinical research, especially where the same Q-sort is administered to a single individual on two or more occasions. Still, item-by-item analysis may be of more general value, even if global com-parisons are of more interest and more commonly used in areas other than clinical application.

Item-by-item comparisons for groups included in the person sample can be accomplished in a number of ways, but all such procedures involve the computations of means and standard deviations for each group on each item of interest. For example, in the Symington-Danforth campaign study, the means ($\overline{X}$), standard deviations (s), and coefficients of variability ($s/\overline{X}$) were computed for every statement across each of three groups (Republicans, Democrats, and Independents); the t statistic was also used as a measure of significant differences among means. The user of such procedures should bear in mind that in most instances, such comparisons within person samples are of dubious value in forming generalizations about social groups since the representativeness of respondent groups is unknown. Still, the information is useful as a source of research hypotheses and as a check on the assumptions leading to the selection of the person sample: the groups should vary most on key statements specifically reflecting the differences among the groups. As an example, Republicans and Democrats did not differ greatly on most items in the 1970 *Q*-sort study, but most of those few differences involved partisan or ideological statements.

Several procedures can also be used to make global comparisons, depending primarily on what kinds of conclusions the investigator wants to draw. If a structured *Q*-sort is the data-gathering instrument, there would presumably be an interest in the effect that the variable(s) reflected in the selection of statements has on that array. Analysis of variance (ANOVA) tests for such effects and may be applied to the sorts of individuals and to the average (whether mean, modal, or median) sorts of group samples, although this latter type of analysis should be treated circumspectly. The appropriate ANOVA design is determined by the factorial design used in structuring the *Q*-sort.

If the primary research interest is in determining similarities among *Q*-sorts, then correlational analysis is the appropriate procedure. Whole *Q*-sorts, thus, are correlated pairwise. A number of possible comparisons may be made: the sorts of a single person under the same condition of instruction on two different occasions,[16] the sorts of a single person under different conditions of instruction on the same occasion,[17] or two persons' sorts under the same condition of instruction on the same occasion. The latter type of comparison is usually made only as a preliminary step to factor-analytic comparisons. But the comparisons that can be made with a single person's sorts should not be passed over lightly. In the first case the *Q*-sort provides a rigorous measure of image change over time, and in the second case it provides a more detailed measure of similarities and differences of two related objects, such as opposing candidates, as perceived by a single individual. These are valuable measurements in the analysis of campaign effects.

Still, factor-analytic comparisons are those most often made in global analyses of *Q*-sort data. Essentially, *Q*-factor analysis requires the correlation of entities across their attributes and the subsequent clustering of persons with similar attributes.[18] *Q*-sort data are especially amenable to this procedure as the attributes are contained in the statements arranged or ranked via the *Q*-sort, and each person is represented by a set of assigned scores across all the statements. The resulting clusters, or factors, represent groupings of persons who are similar because they tend to array the *Q*-statements in basically the same way. Most important, the factor analyst can create a "new," synthetic *Q*-sort for each factor by computing a factor score matrix. This matrix presents an array of scores for all the statements across all the basic factors (image types). Each factor array represents a "modal," or synthetic, *Q*-sort when ranked in the original distribution of the *Q*-sort and tells the analyst how the persons of each type tend to imagine the perceived object.

The interpretation of these factors and their arrays can be pursued in as many as three different ways, depending on the decisions made at earlier stages in the construction and administration of the *Q*-sort. If a structured design is built into the *Q*-sort, each of the factor arrays can be examined as if it were the sort of an individual, making not only item-by-item comparisons across the factor arrays possible but also ANOVA of the theoretical partitions within each array to determine the effects of the theoretical dependencies incorporated in the design. If ANOVA is carried forward on each factor array, then conclusions can be drawn not only as to the power of such theoretical explanations in general but also about particular types of observable variants as well. Such information, in turn, is likely to aid psychological understanding of the bases for typal variation.

If the person sample is structured as well, then the tendency of particular a priori categories of people to cluster together can be examined. For example, the tendencies of Republicans to cluster together, and likewise Democrats, in their perceptions of the ideal public official were noticeably weak in the 1970 campaign study. On the other hand, the a priori categories used in the 1971 study, three groups of Democrats, exhibited much stronger group tendencies to cluster together, with members of the regular Democratic party organization tending to have one image of the ideal president and members of the New Democratic Coalition and ordinary party identifiers tending to share an alternate image. These differences supported earlier research comparing attitudes of persons having different organizational affiliations with the Democratic party.[19]

Finally, in analyzing factor array scores, the contextual arrangement of particular statements embedded within the array should not be overlooked. Although most attention will focus on the two or three extreme ranks on either end of the Q-sort continuum, the image common to a type is operationally defined as the entire array of statements. Comparisons between factor arrays are necessary to locate the widest points of divergence between types, yet subtle but important nuances that distinguish types may be found if intensive examination of the contextual patterning of statements is conducted.

Other statistical procedures can be used in analyzing Q-sort data, as seen at various points throughout this volume, but those described in this appendix are the ones most commonly used. Q-sort data are so rich, however, that the possibilities may at times seem limitless. Limits to the potentialities of the Q-sort are discussed below, but just as the Q-perspective in research implies, the user of the Q-sort should approach his data experimentally, pushing them to those limits.

The Role of the Q-Sort in Image Research

The Q-sort, most simply, is a modified ranking procedure in which a respondent arrays a set of stimuli, usually written statements, according to some condition of instruction. Yet, since the respondent assigns these stimuli to "scores," operant definitions of the research topic are supplied by him, instead of the investigator's providing a priori operational definitions. The Q-sort is thus an inductive, intensive approach to behavioral research that comes closer to the "real" images of people than is true of most procedures.[20] More than this, the Q-sort allows an openness, indeed, invites heuristic, in the freedom given to respondents to reveal their images of phenomena. As such, and given its relative complexity, the *Q*-sort points to what should be looked for, rather than being used as simply another head-counting device in cross-sectional studies.

While its heuristic, probing character is most often emphasized, the *Q*-sort is a valuable tool for theory and hypothesis testing. On the one hand it is highly adaptable to experimental design, and on the other its economy in requirements for time and other scarce resources points to repetitive testing, a sine qua non of scientific research. When these features are combined with the richness and flexibility in statistical manipulation of the resulting data, the affinity of the *Q*-sort to theoretical research is clearly apparent.

At the same time, the *Q*-sort typically provokes a cooperative response from research subjects. Not only are many of the barriers that interviewers frequently confront broken down but the data are less likely to be contaminated due to defensive reactions from subjects who regard many interview questions as unduly obtrusive. The *Q*-sort may easily be designed to mask the primary focus of inquiry. And subjects most often become so involved with the immediate stimuli and the sorting task itself as to preclude searches for underlying psychological variables ("hidden meanings").

Finally, the *Q*-sort is a valuable tool in image research not only for its potentialities in obtaining substantive results but also as a preliminary instrument in constructing other testing devices. These other devices include variants of the *Q*-sort more amenable to cross-sectional research,[21] as well as more traditional formats.[22]

With all these strengths, why has the *Q*-sort not been adopted more widely in the study of candidate images? While a number of harsh criticisms have been made of the procedure,[23] it has one major drawback that is of great importance, given the proclivities of behavioral researchers today. As has been stressed repeatedly here, the *Q*-sort is not very amenable to large-sample, cross-sectional research purposes. The requirement for large samples of persons runs counter to the complexities, and perhaps more importantly, to the thrust of the *Q*-sort procedure. The answer, however, follows from that old ritualistic plea arguing for methodological openness among behavioral scientists, an openness that keeps us from being the prisoners of our research instruments. The solution, then, is to use *Q*-sort where exploratory investigation of underlying psychological factors is the central concern, conserving our scarce resources for cross-sectional research when we have significant questions framed in the soundest formats. Moreover, as we tried to demonstrate in Chapter 5 and others have demonstrated in related research,[24] it is possible to adapt certain features of the *Q*-sort to the concerns of designs employing large-sized, representative samples in survey research.

Notes

1. See Donald M. Broverman, "Normative and Ipsative Measurement in Psychology," *Psychological Review* 69 (July 1962): 95–305.
2. Open-end questions are, of course, clearly self-reflective impressions of respondents, but they are not conducive to much in the way of quantitative

analysis except in constructing a general index value of support for a candidate, and are further limited by their bias in favor of more articulate respondents.

3. The reciprocity controversy continued for about two decades, with Sir Cyril Burt as the chief proponent of the view that inverted (Q) factor analysis is only the reciprocal of R-factor analysis; the battle lines are nowhere so clearly and responsibly drawn as in Cyril Burt and William Stephenson, "Alternative Views on Correlations between Persons," *Psychometrika* 4 (December 1939): 269–81. Convincing demonstrations in favor of the complementarity of the two approaches are to be found in George Stern et al., *Methods in Personality Assessment* (Glencoe, Ill.: The Free Press, 1956), pp. 235–38; and Steven R. Brown, "A Fundamental Incommensurability between Objectivity and Subjectivity," *Science, Psychology, and Communication: Essays Honoring William Stephenson,* Steven R. Brown and Donald J. Brenner, eds. (New York: Teachers College Press, 1972), pp. 57–94.

4. See William Stephenson, *The Study of Behavior: Q-Technique and Its Methodology* (Chicago: University of Chicago Press, 1953), pp. 62–85, 101–13.

5. On the small-sample doctrine, see R. A. Fisher, *Statistical Methods for Research Workers* (London: Oliver and Boyd, 1930); William Stephenson, *The Play Theory of Mass Communication* (Chicago: University of Chicago Press, 1967), pp. 17–22; David Bakan, *On Method: Toward a Reconstruction of Psychological Investigation* (San Francisco: Josey-Bass, 1968), pp. 1–29; and David O. Arnold, "Dimensional Sampling: An Approach for Studying a Small Number of Cases," *American Sociologist* 5 (May 1970): 147–50.

6. Unfortunately, no technical manual or handbook exemplifying the procedures of the Q-sort has been published. With all of its concern and argumentation regarding larger methodological questions, Stephenson's *The Study of Behavior* remains the most succinct formulation and description of the technical aspects of the Q-sort. Brief treatments are found in two research methods textbooks: Fred N. Kerlinger, *Foundations of Behavioral Research,* 2nd ed. (New York: Holt, Rinehart and Winston, 1973), pp. 582–600; and William D. Brooks, "Q-Sort Technique," *Methods of Research in Communication,* Philip Emmert and William D. Brooks, eds. (Boston: Houghton-Mifflin, 1970), pp. 165–80. We have also profited from having access to an unpublished manuscript by Steven R. Brown, entitled "Small-Sample Behavioral Research: Procedures for Employing Q-Technique in Political Science"; see also his "Bibliography on Q-Technique and Its Methodology," *Perceptual and Motor Skills* 26 (April 1968): 587–613.

7. Cf. William Stephenson, "Methodology of 'Imagery' Measurement: MR Redefined" (paper presented at the Sixth Annual Public Utilities Seminar, American Marketing Association, St. Louis, Missouri, February 18, 1960).

8. On the methodological need for attribute sampling, see William Stephenson, "A Statistical Approach to Typology: The Study of Trait-Universes," *Journal of Clinical Psychology* 6 (January 1950): 26–38; and Steven R. Brown and Thomas D. Ungs, "Representativeness and the Study of Political Behavior: An Application of Q-Technique to Reactions to the Kent State Incident," *Social Science Quarterly* 51 (December 1970): 514–26.

9. See Sylvan H. Cohen and Richard W. Taylor, "An Experimental Study of the Public Mind: An Application of Q-Methodology to a Local Referendum Situation," *Experimental Study of Politics* 1 (July 1971): 82–117, as an example of the use of an unstructured Q-sort in political research.

10. In addition to the work cited in note 4 above, see Steven R. Brown, "On

the Use of Variance Designs in *Q*-Methodology," *Psychological Record* 20 (Spring 1970): 179–89; and his "Experimental Design and the Structuring of Theory," *Experimental Study of Politics* 1 (February 1971): 1–41.

11. See, for example, Roger E. Kirk, *Experimental Design: Procedures for the Behavioral Sciences* (Belmont, Calif.: Brooks/Cole, 1968), for a review of design possibilities that might be applied in the construction of *Q*-sorts.

12. Cf. Brooks, "*Q*-Sort Technique," p. 169. These are questions deserving of methodological research, as most technical studies of the *Q*-sort and its construction focus on reliability of a set of statements as a whole.

13. Indeed, while partisan differences did emerge as sources underlying image types, such tendencies were weak. See Dan Nimmo and Robert L. Savage, "Political Images and Political Perceptions," *Experimental Study of Politics* 1 (July 1971): 12–31; and "Image Typologies in a Senatorial Campaign: A Comparison of Forced vs. Free Distribution Data" (paper presented at the Annual Meeting of the American Political Science Association, Chicago, September 10, 1971).

14. On the forced-free distinction in *Q*- sort research generally, see the review in Brooks, "*Q*-Sort Technique," pp. 169–73. More recent comparisons are Nimmo and Savage, "Image Typologies in a Senatorial Campaign"; and Steven R. Brown, "The Forced-Free Distinction in *Q*-Technique," *Journal of Educational Measurement* 8 (Winter 1971): 283–87.

15. See Nimmo and Savage, "Image Typologies in a Senatorial Campaign."

16. Cf., for example, Dan Nimmo et al., "Effects of Victory or Defeat upon the Images of Political Candidates," *Experimental Study of Politics* 3 (February 1974): 7–10.

17. Cf., for example, Nimmo and Savage, "Political Images and Political Perceptions," pp. 17–18.

18. On factor-analytic design, see William Stephenson, "The Foundations of Psychometry: Four Factor Systems," *Psychometrika* 1 (September 1936): 195–209; Franklin C. Shontz, *Research Methods in Personality* (New York: Appleton-Century-Crofts, 1965), pp. 107–28; and R. J. Rummel, *Applied Factor Analysis* (Evanston, Ill.: Northwestern University Press, 1970), pp. 192–202.

19. See Dan Nimmo and Robert L. Savage, "The Amateur Democrat Revisited," *Polity* 5 (Winter 1972): 268–76.

20. However, for other possibilities, see Francesca M. Cancian, "New Methods for Describing What People Think," *Sociological Inquiry* 41 (Winter 1971): 85–93.

21. See, for example, Everett F. Cataldo et al., "Card Sorting as a Technique for Survey Interviewing," *Public Opinion Quarterly* 34 (Summer 1970): 202–15; Fred E. Fiedler, *Leader Attitudes and Group Effectiveness* (Urbana, Ill.: University of Illinois Press, 1958); and David E. Jackson and Charles E. Bidwell, "A Modification of *Q*-Technique," *Educational and Psychological Measurement* 19 (Summer 1959): 221–32.

22. See, for example, Fred N. Kerlinger, "*Q*-Methodology in Behavioral Research," *Science, Psychology, and Communication*, pp. 3–38.

23. These criticisms focus largely on highly technical matters of statistical theory; see Kerlinger, *Foundations of Behavioral Research*, pp. 595–97, for a balanced assessment of these criticisms.

24. See David M. Jackson and Charles E. Bidwell, "A Modification of *Q*-Technique"; and Leonard V. Gordon, "*Q*-Typing: An Exploration in Personality Measurement," *Journal of Social Psychology* 78 (1969): 121–36.

Appendix B

Procedures for a Q-Sort Study of Candidate Images: The 1970 U.S. Senatorial Campaign in Missouri

We initiated our first Q-sort study of candidate images with several research concerns, both substantive and methodological. These included our desire to explore the dimensions underlying candidate images, the sources of those images, and the substantive consequences of using free and forced distributions in Q-sorting. Such research interests could have been advanced by studying any contest for public office, but the 1970 campaign in Missouri for a U.S. Senate seat was particularly suited to our concerns. First, as researchers we were located in the state at that time and could take advantage of that fact. Beyond that, however, it appeared very early that the campaign was to be an "image campaign" involving "image makers," extensive political advertising, and a classic confrontation between a widely recognized incumbent of the state's hitherto dominant Democratic party and a much lesser-known challenger relying on paid consultants, pollsters, and media experts.[1] Previous studies of candidate images, moreover, had typically been limited to examining presidential contests; an off-year race for the U.S. Senate would allow us to examine political perceptions in an election of less general salience to voters.

The Research Setting

The 1970 contest for the U.S. Senate seat from Missouri pitted the Democratic incumbent, Stuart Symington, and his Republican challenger, John C. Danforth, state attorney general. Both men were members of wealthy, prominent families, but as candidates they differed in several key respects. Symington

227

was completing his third term in the Senate; Danforth, in contrast, had held public office less than two years, being the first Republican elected to a state-wide office in Missouri since 1946. Symington was well known in Missouri and national politics; Danforth, even after his impressive 1968 victory, possessed considerably less recognition. Symington was, at age sixty-nine, the elder statesman of active Missouri politics; Danforth was only thirty-four.

These differences were clearly reflected in the opposing campaigns of the two candidates. The Symington campaign effort was traditional, mixing innumerable personal appearances, speeches at fairs and rallies, and moderate use of mass-media advertising. Moreover, his campaign messages minimized issues for which he had received national publicity, such as stands on defense spending and the war in Vietnam, and were addressed instead to more localized issues, especially the state's agricultural problems. Danforth, in contrast, employed mass advertising extensively, particularly through the medium of television. His media campaign featured saturation showings of a half-hour television documentary, "The Uphill Run," and numerous spot commercials; these shorter spots frequently were brief clips edited from the documentary. Thus, definite themes projecting a consistent image emerged from Danforth's advertising campaign. These themes included Danforth's youth, independence, and opposition to the Democratic "Establishment" in Missouri politics. As Missouri was earmarked as a target state by national Republicans, Danforth received considerable support from both within and without the state. Ultimately, Symington won reelection, but only with 51.1 percent of the vote, his closest margin in Missouri electoral politics.

The Research Design: Constructing the Q-Sort

The primary focus in this study centered on the stimulus-versus-perceiver controversy reviewed in Chapter 4—that is, are candidate images simply accommodated from campaign stimuli or are they assimilated on the basis of some prior attitude of the perceiver? A subsidiary but critical question guiding the research turned on the problem of adequately representing the key facets of candidate images. Both research problems, in conjunction with previous studies by others and a pilot study conducted by the authors in the summer of 1970, pointed toward a two-way factorial design that would potentially capture both stimulus- and perceiver-determined effects.

The images projected by candidates are presented to voters from alternative social roles played by candidates in their various activities, whether directly related to the campaign or not. Previous research and our pilot study suggested that underlying these projected roles were two principal dimensions: political role and stylistic role. Yet each of these dimensions is itself multifaceted. The political role includes the candidate's efforts to project himself as a leader, an individual who can successfully carry out the duties of a public office; but the political role includes also any partisan activities that mark the candidate as a representative of a political party or other political interests. The stylistic role, likewise, includes at least two aspects: the manner in which the candidate communicates with voters to impress them with his capabilities (through the mass media or in personal appearances) and the qualities he exhibits as a singular individual, a person. Thus the projected image (stimulus determined) of the candidate may appear across four levels—leadership role, partisan role, communicator role, and personal qualities.

Similarly, the voter's image of the candidate (perceiver determined) is multidimensional. Previous image research suggested the special importance of semantic (meaning) categories in gauging perceiver reactions to objects-of-perception. While various studies using the semantic differential had uncovered alternative semantic modalities, the basic tripartite dimensionality of semantic qualification uncovered by the creators of this instrument had stood the test of time and general experience.[2] Thus, the treatment levels selected for perceiver-determined effects were evaluation, potency, and activity.

As Table B.1 indicates, the resulting two-way factorial design yields $4 \times 3 = 12$ combinations of candidate image attributes, each combination reflecting both role orientation and semantic qualification. These combinations were used to select two-word statements for the *Q*-sort. In each statement one word ("Leader," "Politician," "Speaker," or "Person") represents one of the candidate's projected roles. The other word is a modifier drawn from studies employing the semantic differential and represents one of the semantic categories of the voter's image—evaluative, potency, or activity. Each of these factorial combinations was replicated by five such two-word statements. These replications and combinations appear in Table B.2.

Table B.1   Factorial Design for *Q*-Sort Study of Candidate Images in 1970 Senatorial Campaign

EFFECTS	TREATMENT LEVELS		*n*
1. Stimulus determined	(a) Leadership role	(b) Partisan role	4
	(c) Communicator role	(d) Personal qualities	
2. Perceiver determined	(e) Evaluation (f) Potency	(g) Activity	3

Additionally, to provide the types of clearly partisan cues characteristic of campaigns, both "Democratic" and "Republican" were included as modifiers for both "Leader" and "Politician." Ideological overtones were tapped by employing "Liberal" and "Conservative" as modifiers of both "Leader" and "Politician" also. The five replications of each of the twelve factorial combinations plus the eight additional statements reflecting partisan and ideological modifiers resulted in an unbalanced sample of sixty-eight *Q*-statements. These statements are factorially categorized in Table B.1.

The Research Design: Administering the *Q*-Sort

Since the purposes of this study were to explore the nature of candidate images, note the relationship of stimuli and perceivers in political imagery,

Table B.2   Q-Sample of Candidate Image Attributes in 1970 Senatorial Campaign

(ae)	(af)	(ag)
Able Leader	Inexperienced Leader	Active Leader
Trained Leader	Tenacious Leader	Hard-working Leader
Pessimistic Leader	Flexible Leader	Indecisive Leader
Sophisticated Leader	Tough Leader	Rash Leader
Deceitful Leader	Feeble Leader	Resolute Leader
(be)	**(bf)**	**(bg)**
Dissenting Politician	Strong Politician	Dead Politician
Unselfish Politician	Aggressive Politician	Temperate Politician
Reputable Politician	Hard Politician	New Politician
Clean Politician	Brave Politician	Loyal Politician
Two-faced Politician	Productive Politician	Aimless Politician
(ce)	**(cf)**	**(cg)**
Insincere Speaker	Uninspiring Speaker	Energetic Speaker
Awkward Speaker	Persuasive Speaker	Emotional Speaker
Skillful Speaker	Colorless Speaker	Warm Speaker
Lucid Speaker	Serious Speaker	Hesitant Speaker
Informative Speaker	Organized Speaker	Tense Speaker
(de)	**(df)**	**(dg)**
Disagreeable Person	Practical Person	Calm Person
Conceited Person	Confident Person	Mature Person
Noble Person	Small Person	Stingy Person
Fair Person	Insensitive Person	Rebellious Person
Kind Person	Wealthy Person	Quiet Person

PARTISAN AND IDEOLOGICAL ATTRIBUTES

Democratic Leader
Republican Leader
Conservative Leader
Liberal Leader
Democratic Politician
Republican Politician
Conservative Politician
Liberal Politician

and examine the impact of partisanship on images, a large-sized, representative sample of Missourians was unnecessary. Instead a "theoretical sample" of small size and balanced to assure inclusion of relevant variables was deemed more desirable (as well as more practical). Partisanship was clearly the single best criterion variable in a study of candidate images in a partisan contest; hence the small sample ($n = 36$) was selected to represent equally Republicans, Democrats, and Independents. The sample of persons reflected different levels of political involvement (eleven respondents were activists with varying organizational responsibilities), occupations (businessmen, housewives, teachers, etc.), ages (ranging from twenty-one to fifty), and both sexes (fifteen males, twenty-one females). While the person sample cannot be credited as representative of any larger population, no particular bias seemed to affect the results, with the exception, of course, of the balancing of partisans. The

Q-sorts were administered during the period between mid-September and election day in November 1970.

Each respondent was asked to sort the statements for each of three different political figures—the ideal U.S. senator, Stuart Symington, and John Danforth—under conditions of both forced and free distributions for a total of six sorts. The ideal senator sorts served as criterion sorts against which the candidate sorts might be compared, in addition to providing data for an initial exploration of social typing in American politics.[3]

Under the forced condition respondents ranked the statements from "most desirable" to "least desirable" according to the following distributions:

Score	(most desirable)						(least desirable)				
	+5	+4	+3	+2	+1	0	−1	−2	−3	−4	−5
Frequency	3	5	5	8	8	10	8	8	5	5	3

For purposes of methodological comparison the respondents were also required to use the same statements for the same three political figures under the instruction to "Sort the cards into as many piles as you feel are necessary to characterize the favorable or unfavorable qualities of _____." Respondents were thus permitted to use as many categories as they wished and to place as many statements in each category as they deemed appropriate. This procedure actually gave respondents more discretion than that typically allowed in free-sort studies, since respondents were free to select the number of categories as well as the number of items per category. The range in number of categories actually used varied from a low of two to a high of thirteen. The order of the six sorts was systematically rotated across the thirty-six respondents as a check for bias resulting from the order of presentation; no systematic effect for the order of sorts was found.

The free sorts required normalization based on the assumption of equal intervals in assigning the original raw scores, prior to any statistical analyses. When compared with the free sorts, the forced sorts also required normalization. Ultimately, then, we acquired a large amount of data from thirty-six respondents performing six sorts of sixty-eight items each. These are the data—submitted to correlational, factor-analytic, and variance analysis techniques—that we have reported in summary fashion in Chapters 3 and 4 of this text, and in more detail elsewhere.[4]

Notes

1. As examples of the journalistic concern for "image" politics, see "Electronic Politics: The Image Game," *Time* 96 (September 21, 1970): 43–48; and "The Selling of the Candidates 1970," *Newsweek* 76 (October 19, 1970), 34–43.

2. See Charles E. Osgood et al., *The Measurement of Meaning* (Urbana, Ill.: University of Illinois Press, 1957); and James G. Snider and Charles E. Osgood, eds., *Semantic Differential Technique: A Sourcebook* (Chicago: Aldine-Atherton, 1969).

3. See Orrin E. Klapp, *Heroes, Villains, and Fools: The Changing American Character* (Englewood Cliffs, N.J.: Prentice-Hall, 1962).

4. The data were subjected to intensive analysis through most of the procedures discussed in Appendix A, as well as others. The interested reader should consult Dan Nimmo and Robert L. Savage, "Political Images and Political Perceptions," *Experimental Study of Politics* 1 (July 1971): 1–36; and "Image Typologies in a Senatorial Campaign: A Comparison of Forced vs. Free Distribution Data" (paper presented at the Annual Meeting of the American Political Science Association, Chicago, September 10, 1971), for more detailed discussion of these results.

Appendix C

Description of the Q-Sort Study
of Images of 1971 Democratic
Presidential Contenders

The study of candidate images has generally been tied to electoral situations involving voters confronted with alternative candidates. Those candidates usually come to the campaign with a previous history and some degree of public recognition. Although students of voting behavior have been aware that they must take account of this precampaign background of particular candidates and elections,[1] studies in which they do so are rare.[2] Moreover, most research has examined partisan campaigns so that the interactive effect of partisanship and candidate preference tended to obscure the nature and role of candidate images in shaping electoral choice. The jockeying among Democratic aspirants to the presidency in the summer of 1971 afforded an excellent opportunity for a Q-sort study, which would help fill these gaps in the literature of candidate image research.

The events of the time included a large array of contenders holding varying positions on issues and organizational affiliations, all vying for nomination as the Democratic party's presidential candidate in 1972. This wide array of potential nominees emerged among Democrats of different levels of involvement and bases of partisan commitment. By focusing on the images among their own partisans of the leading Democratic presidential contenders, it was possible to examine candidate images while controlling for partisanship. In this inquiry we shifted our research purposes from those that guided the 1970 study of the Missouri senatorial campaign. Satisfied that candidate images are transactional phenomena linking the projected images of candidates with voters' subjective constructs, we dropped stimulus-determined and perceiver-determined dimensions from our design while retaining the actual

operational variables.[3] Moreover, the earlier study led us to a definite preference for the forced distribution of Q-sorts. These results of the earlier study
thus contributed to subtle, but important, modifications in the design and
execution of this second Q-sort study of candidate imagery.

The Research Setting

In the chaos of Democratic party politics of 1968, a "traditional" schism
within the party took national organizational shape through the creation of
the New Democratic Coalition (NDC). This new organization was based on
local chapters in a fashion corresponding to that of the regular Democratic
party structure. Members of the NDC exhibited much the same characteristics that James Q. Wilson attributed to a new breed of political activists,
whom he labelled "amateur Democrats."[4] These "amateurs" defined politics
of intrinsic importance, in that it provides a means for expressing principles
related to a conception of the larger public interest. Those politicians whom
Wilson designated "professionals," in contrast, define politics in more personal
terms, focusing on the extrinsic rewards such as power, income, or simply the
fun of the political game. These "professionals" constitute the core of the
regular party organization.

In 1971 the NDC was still a viable organization in many local communities. Given their differing political outlooks, it seemed safe to assume that
NDC members and party regulars differed also in their orientation toward
prominent national leaders in the Democratic party. A number of contenders
had indeed emerged to seek the Democratic nomination for the presidency,
including John Lindsay, George Wallace, Fred Harris, Wilbur Mills, Henry
Jackson, George McGovern, Edward Kennedy, Hubert Humphrey, and Edmund Muskie. The last four of these appeared to be the frontrunners and,
more importantly, seemed to be differently aligned with respect to the major
organizational factions within the party. McGovern and Kennedy had been
more closely tied to the political style urged by the NDC, whereas both
Humphrey and Muskie carried the banner of the regular Democratic party in
1968. Thus, these four potential presidential candidates were selected as
focal objects of a Q-sort study of political imagery among the members of a
single political party, well before the ensuing primary campaigns.

The Research Design: Constructing the Q-Sort

At the time of this second Q-sort study, there was no basis for dissatisfaction
with the operational design used previously in the construction of the Q-sort.
Hence, we retained the basic factorial design incorporating four levels of role
orientation and three levels of semantic qualifications (see Table B.1). We
again used two-word statements representing all the $4 \times 3 = 12$ factorial combinations. The major difference in the Q-sample used in this study was our
decision to use a balanced sample of statements across the combinations.
Each of the twelve combinations we replicated six times, resulting in a Q-
sample of seventy-two statements (see Table C.1).

Table C.1   *Q*-Sample of Candidate Image Attributes in 1971 Democratic Contenders

(ae)	(af)	(ag)
Trained Leader	Tenacious Leader	Active Leader
Sophisticated Leader	Flexible Leader	Hard-working Leader
Liberal Leader	Tough Leader	Resolute Leader
Deceitful Leader	Inexperienced Leader	Indecisive Leader
Incompetent Leader	Feeble Leader	Rash Leader
Conservative Leader	Severe Leader	Timid Leader

(be)	(bf)	(bg)
Clean Politician	Strong Politician	New Politician
Unselfish Politician	Brave Politician	Temperate Politician
Democratic Politician	Productive Politician	Loyal Politician
Dissenting Politician	Lax Politician	Dead Politician
Shady Politician	Cowardly Politician	Aimless Politician
Republican Politician	Belligerent Politician	Obstructive Politician

(ce)	(cf)	(cg)
Skillful Speaker	Persuasive Speaker	Energetic Speaker
Lucid Speaker	Serious Speaker	Warm Speaker
Informative Speaker	Organized Speaker	Dynamic Speaker
Insincere Speaker	Uninspiring Speaker	Hesitant Speaker
Awkward Speaker	Colorless Speaker	Emotional Speaker
Unintelligible Speaker	Boring Speaker	Tense Speaker

(de)	(df)	(dg)
Noble Person	Practical Person	Calm Person
Fair Person	Confident Person	Mature Person
Kind Person	Masterful Person	Stable Person
Disagreeable Person	Small Person	Stingy Person
Conceited Person	Insensitive Person	Rebellious Person
Ugly Person	Shallow Person	Oily Person

The Research Design: Administering the *Q*-Sort

As in the earlier study, the research purposes were exploratory in nature—to assess further the dimensions underlying candidate images, to examine image typologies under a variety of political circumstances, and to relate social psychological variables to the structuring of image types. Hence, a small, balanced sample of respondents selected for possessing characteristics relevant to the general research problem was preferable to a large, representative sample of persons. Given those purposes and the research setting at hand, we deemed it appropriate to interview only Democrats; accordingly, we selected a balanced sample ($n = 36$) of three categories of party members.

First, we chose twelve officers and members of a local chapter of the New Democratic Coalition in a medium-sized county in the Midwest. Of the dozen NDC subjects, seven were members of the organization's county executive committee and another was a former NDC state chairman. Second, the P-sample contained twelve members of the county's Democratic party organization who were not NDC members. This subsample included four precinct committee men and women, two members and one former chairman of the county Democratic committee, a county official, a city council member, a district court judge, and a former state legislator. Finally, the P-sample included twelve people who identified themselves as Democrats but who were not active in organized party politics. All subsamples included persons of various occupations and both males and females. The mean age for the entire sample was forty years, but as should be expected, the party regulars deviated from this sample mean considerably—for the regulars, the mean age was fifty; for NDC members, thirty-five; and for identifiers, thirty-four.

The seventy-two statements of the Q-sample, typed one to a card and randomly ordered prior to each sorting, were given to each of the thirty-six respondents with the instruction to sort the statements from "most characteristic" to "least characteristic" under each of six separate conditions: (1) "Yourself as You Actually Are," (2) "The President as He Ideally Should Be," (3) "Hubert Humphrey," (4) "Edmund Muskie," (5) "Edward Kennedy," and (6) "George McGovern." The first two conditions were presented in the above order, but the sequence of the remaining four sorts was systematically rotated throughout the entire P-sample; no order effects were evident. Respondents were required to place statements in a forced distribution scored as follows:

Score	(most characteristic)						(least characteristic)				
	+5	+4	+3	+2	+1	0	−1	−2	−3	−4	−5
Frequency	3	6	6	8	8	10	8	8	6	6	3

These interviews were conducted in June and July of 1971.

Analyzing the Data

Although the data were subsequently analyzed in the same thorough manner as the data in the 1970 study had been, the decision to balance the Q-statement subsamples, in conjunction with the hypotheses proffered by Wilson regarding the distinctive traits of amateur and professional politicians, pointed especially to a comparison of Q- and R-factor analytic results.[5] Q-factor analysis uncovers basic dimensions among the persons (or more strictly, the Q-sorts as whole arrays), which are reasonably interpreted as image types. R-factor analysis, on the other hand, lays bare the dimensions among the statements comprising the Q-sorts; these R-factors represent image traits, that is, related clusters of attributes of a given image object. These comparisons provided a substantively complementary examination of the differences among Democratic partisans in their perceptual responses to the activities of their leaders.[6]

Notes

1. Cf. Peter H. Rossi, "Four Landmarks in Voting Research," *American Voting Behavior,* Eugene Burdick and Arthur J. Brodbeck, eds. (Glencoe, Ill.: The Free Press, 1959), pp. 5–54.

2. One such study, although a secondary analysis of data gathered for other purposes, is Herbert H. Hyman and Paul B. Sheatsley, "The Political Appeal of President Eisenhower," *Public Opinion Quarterly* 19 (Winter 1955–56): 26–39.

3. The transactional character of both role orientation and semantic qualification was, of course, recognized at the outset of the earlier study, but the results of that study clearly underlined this fact. Not only do voters represent candidates dimensionally across semantic modalities, but they differentially stress candidate role orientations as well. And candidates may wittingly or unwittingly stress semantic modalities differentially in their campaign messages; for example, Symington, the incumbent, stressed his potency in reciting his accomplishments; and Danforth, the youthful challenger, stressed activity in his "uphill run."

4. *The Amateur Democrat* (Chicago: University of Chicago Press, 1962).

5. This complementary use of *Q*- and *R*-factor analysis had been used earlier, but the unbalanced *Q*-sample and the lack of theoretically anticipated differences among Republicans and Democrats in that study made such comparisons of less intrinsic value than was the case in the 1971 study.

6. See Dan Nimmo and Robert L. Savage, "The Amateur Democrat Revisited," *Polity* 5 (Winter 1972): 268–76. For a similar outlook on the complementary use of *Q*- and *R*-factor analysis, cf. Harry E. Heller, "Defining Target Markets by Their Attitude Profiles," *Attitude Research on the Rocks,* Lee Adler and Irving Crespi, eds. (Chicago: American Marketing Association, 1968), pp. 47–57.

Appendix D

Description of the Q-Sort Study of Candidate Images in the 1972 Presidential Election

In many respects the 1972 election for the American presidency generated less public enthusiasm than had any other recent presidential election. Yet there was a considerable interest among one set of potential voters: college students, especially those voting for the first time. That a large portion of these new voters supported George McGovern, the Democratic candidate, seemed apparent. What was less certain were the perceptions college students had of McGovern and his opponent—the incumbent, Richard Nixon—and how these students resolved the perplexities of a campaign in which many of them had assisted their choice, McGovern, in obtaining his party's nomination only to be confronted so shortly with his almost certain defeat in the general election.

Two general research problems were suggested by this set of circumstances. In the first place, Do new voters of the 1970s differ from experienced voters in the perceptual categories they bring to political campaigns, and do they array political stimuli so differently as to have very different types of images of candidates? Second, What is the impact of a major political event, specifically, the victory of one candidate over the other, on candidate images?

The Research Setting

There were sharp contrasts in the prospects for victory of the respective candidates in the 1972 presidential election, approximating the same conditions that existed in the 1964 election to a considerable extent. As in that earlier election, the 1972 campaign featured an incumbent president employing the many resources and advantages that his position afforded, confronting a challenger with much less public recognition—and that recognition tainted by the lack of solidarity in support of him by his own partisans. Yet important differences between the two campaigns existed. Lyndon Johnson had embarked on the campaign trail less than a year after his inauguration and was near the height of his public popularity at the outset of his campaign. Richard Nixon in 1972, on the other hand, had been in office for four years, registering both successes and losses; none of his successes, however, had led to a widespread popular appeal on a personal basis (see Chapter 6). Nixon retained much of the aura that led to the celebrated poster with his picture and the caption "Would you buy a used car from this man?" George McGovern, like Barry Goldwater, found himself hailed as an extremist, a spokesman for a minority interest within his own party. But unlike Goldwater, McGovern was plagued by a series of campaign debacles that all reflected badly on him regardless of his own culpability. The most notable of these were the events surrounding the withdrawal of Thomas Eagleton as McGovern's vice-presidential running mate.

Campaign strategies and tactics of the candidates echoed these contrasting circumstances. Nixon followed the traditional strategy of incumbent presidents by remaining aloof from partisan strife and left campaign rhetorics to lieutenants such as Vice-President Spiro Agnew. McGovern, in contrast, had the twin problems of insufficient financial resources for a sustained media campaign and a lack of enthusiastic support from many state and local Democratic party organizations. These weaknesses forced McGovern and his staff toward even greater reliance on the tactics used in the primary elections, local canvassing and voter registration drives by volunteer supporters, many of whom were college youths. Such tactics very likely served to reinforce a prevalent description of McGovern as a spokesman for a minority—and disaffected—interest.

Nixon and McGovern differed considerably in their issue positions as well. Particularly divergent, at least at the outset of the campaign, were their views on the questions of American involvement in the Vietnamese war. McGovern's "dovish" position on the issue seemed to consolidate further his popularity with college students. The 1972 campaign, then, was a particularly appropriate setting for examining the political perceptions of college students, although just how appropriate it would turn out to be was not quite so clear at the initial stages of this Q-sort investigation of candidate images.

The Research Design: Constructing the Q-Sort

While not dissatisfied with the Q-sample used in the 1971 study of Democratic presidential contenders, a considerable departure seemed in order. In the first place there was a desire to break away from the rigidity of the two-word statements. The simplicity of these brief statements assuredly promoted the interpretability of results, but a richness and openness that could be obtained from more complex statements was thereby sacrificed. Second, as the semantic qualification variable added relatively little to the statistical explanation of variance within most image types, and by way of compensation for moving to longer, more complex statements, we decided to simplify the factorial design that had guided previous efforts in the construction of *Q*-sorts. Thus, we adopted a one-way design using candidate role orientation as the main effect and incorporating four treatment levels—public official, party leader, person, and dramatic performer (or communicator). The overall *Q*-sample was smaller than those employed in earlier studies, having thirteen replications of each of the four levels (4 × 13 = 52 statements).

The decision to use complete sentences as *Q*-statements was not so easy to implement, however. A larger body of such statements from which to extract a sample that adequately represented the design simply did not exist. Thus, on two college campuses in the spring of 1972, hundreds of students were asked to write brief essays on the qualities they expected or observed in the "ideal president," Richard Nixon, and in their preferred Democratic party contender for the nomination. These questionnaires required the students to describe these qualities for the three political figures separately as public officials, party leaders, persons, and communicators. From the many hundreds of discrete statements obtained from these questionnaires, more than two hundred distinguishable statements were generated. The final *Q*-sample of fifty-two statements (see Table D.1) was selected from this briefer list.

Table D.1   *Q*-Sample of Candidate Image Attributes in 1972 Presidential Campaign

PUBLIC OFFICIAL

1. He is a good administrator.
2. He can unite people in support of his policies.
3. He is not fearful of criticism.
4. His words, actions, and manner always reflect the dignity and honor of the office.
5. He takes a firm stand on pertinent issues but does not disregard the views of others.
6. He is concerned with the public as a whole, not a collection of minority and majority groups.
7. He is a statesman and a leader who explains to the people as much as possible the reasons behind his actions or proposals.
8. He arrives at decisions through careful consideration and analysis of all available information.
9. He sticks to his decisions once they are made.
10. He is a middle-of-the-roader.
11. He attempts to bring people together in common goals.
12. While qualified in terms of education, he has common sense.
13. He makes only those promises he has the ability to keep.

PARTY LEADER

14. He articulates what his party stands for and always tries to show how each action or proposal is moving toward that goal.
15. He is capable of maintaining party unity on major issues.
16. He carries the image and platform of his party to the people.
17. He represents the major policy stands of his party but he is flexible as the situation and public mood change.
18. He has a record of good, honest service for his party.

Table D.1 (continued)

19. He is grateful for his party's support but not controlled by their demands.
20. He is a leader and open-minded when it comes to his party's varying membership and interests.
21. He is able to get things done and this means charming and motivating people.
22. He does not mirror the policies of any one party.
23. He should be elected as a result of his party allegiance because talk is cheap and all candidates promise great things.
24. He makes deals without compromising his principles.
25. He listens to other advisors' opinions first and then feels free to do what he thinks is best for everyone.
26. He does not sling mud at his fellow party members or members of the opposing party whose ideas do not coincide with his own.

PERSON

27. He has the highest degree of honesty, integrity, and intelligence.
28. When he is wrong, he admits it.
29. He is calm, analytical, and cautious, yet bold and decisive in carrying out his plans.
30. He earnestly wants to be liked and respected.
31. He has a faith in God and is not afraid to express it.
32. He is a good family man.
33. He is of high moral character.
34. He is imaginative, experimental, and hip.
35. He is natural and sincere and does not appear to be trying to impress people.
36. He has a sense of humor.
37. He is ambitious.
38. He is a person capable of deep emotion and warmth.
39. The central quality which gives depth and substance to all the others is his quality of caring.

COMMUNICATOR

40. He is cool, calm, and collected in front of an audience.
41. He does not read his speeches; he delivers them!
42. He exhibits warmth and personal appeal on television.
43. His personal magnetism and physical attractiveness are positive assets.
44. He is expressive but not overdramatic.
45. He expresses himself intelligently and clearly so that the educated and uneducated alike understand what is said.
46. His perseverance, firmness, coolness, and aggressiveness clearly project a take-charge image.
47. He is sure of what he is saying and is ready for anything.
48. He is himself; he is genuine in television appearances since people today sense insincerity.
49. He is able to hold his audience's interest.
50. His voice, speech patterns, expressions, and cool appearance are more important than the mere words of his speech.
51. He is proof that Madison Avenue advertising techniques make television appearances more effective.
52. He appeals to reason rather than people's emotions and prejudices.

The Research Design: Administering the *Q*-Sort

To satisfy our research purposes we once again selected a small P-sample (but larger than those used in the earlier *Q*-sort studies). In drawing that sample we faced two problems not previously encountered, however. First, the complexity of the research problem, requiring a P-sample of college students compared to new-versus-experienced voters, could easily have led to a sample too large for conventional *Q*-analysis. Moreover, at least some of the respondents would have to be reinterviewed after the election to obtain the necessary longitudinal data. We therefore proceeded as follows: The basic data for college students we obtained from a P-sample of one hundred college students eligible to vote for the first time. Ten students were selected from each of ten college campuses[1] and interviewed during a two-week period just prior to the November 1972 election. Six students from each of three campuses (located in a midwestern state, a border state, and a southern state) were reinterviewed following the election. From each of the same communities in which these three campuses were located, six experienced voters were also interviewed before and after the election. Thus, the one-time P-sample ($n = 100$) of college students voting for the first time partially incorporated a panel design to provide the new voters for a second, balanced P-sample ($n = 36$) of new and experienced voters.

In each interview session subjects were required to sort the fifty-two statements under three separate conditions of instruction, ranking the statements from "most characteristic" to "least characteristic" to describe their conceptions of (1) the ideal president, (2) Richard Nixon, and (3) George McGovern. Subjects sorted the statements along a forced-choice, nine-point continuum scored as follows:

	(most characteristic)						(least characteristic)		
Score	+4	+3	+2	+1	0	−1	−2	−3	−4
Frequency	4	5	6	7	8	7	6	5	4

In addition to the three *Q*-sorts, each subject completed a brief questionnaire indicating age, sex, race, party self-identification, ideological self-identification, and candidate preference.

Analyzing the Data

The resulting data had to be analyzed in a number of different ways, since there were both the one-shot sample of one hundred students and the overlapping panel subjects. The $n = 100$ data were subjected to both *Q*- and *R*-factor analyses. While these results were of intrinsic interest (see Chapter 3), the *Q*-factor loadings were also used as quantitative measures of candidate images to permit regression analyses, with images as dependent variables to be statistically explained by the variables contained in the brief questionnaire (see Chapter 4).

The panel data, bringing together new and experienced voters, required separate *Q*-factor analyses for the pre- and postelection sorts. As the differences between new and old voters were relatively slight, the data for the two subsamples were merged. Comparisons of pre- and postelection images were based on these two, merged sets of data (see Chapter 7).[2]

Notes

1. The ten colleges and universities from which the student sample was drawn were located in virtually all areas of the nation. Still, the sample should not be considered as representative of any larger population.

2. The results of the panel study are also reported in Dan Nimmo et al., "Effects of Victory or Defeat upon the Images of Political Candidates," *Experimental Study of Politics* 3 (February 1974): 1–30.

Index

Index 247

Fisher's z, 182 n
Fitzgerald, Michael R., 204, 205, 211 n
Flanigan, William H., 40 n, 105 n
Fox, James W., 12 n
Frame of reference, 90–92
Fransella, Fay, 212 n
Free-choice techniques, 28–30
Freedman, Jonathan, 15, 40 n

Gaitskell, Hugh, 54
Gallo, Philip S., 42 n, 43 n, 77–78 n,
    105 n, 162–63, 168, 170, 182 n
Gallup Poll, 136
Getter, Russell, 168–69, 181, 182 n
Glock, Charles Y., 42 n
Goffman, Erving, 82, 103 n
Goldwater, Barry, 18, 23, 25, 38, 50–52,
    77 n, 94–95, 105 n, 133–240
    passim
Goldwater Coalition, The (John Kessel),
    128
Gordon, George N., 126 n
Gordon, Leonard V., 29–30, 42 n, 63–64,
    79 n, 225 n
Governor, ideal, 93, 189
Graber, Doris A., 12 n, 28, 42 n, 48, 59,
    76 n, 108 n, 129, 159 n
Greenberg, Bradley S., 41 n
Greenstein, Fred I., 79 n
Greenwald, Anthony G., 210
Grimond, Jo, 19, 164
Gruenbaum, Werner F., 75 n
Guggenheim, Charles, 111

Haas, Michael, 12 n
Hagerty, Jim, 49
Halo effect, 54. See also Stereotypy, image
Harding, John, 193, 210 n
Harris, Fred, 234
Harris, Louis, 160 n
Harris Poll, 136
Hart, Phillip, 163
Heaton, Eugene E., Jr., 77 n
Heider, Fritz, 182 n
Heller, Harry E., 237 n
Hidden Persuaders, The (Vance Packard), 4
Hiebert, Ray, 77 n
Hill, Winfred F., 103 n
Hinckley, Barbara, 77 n, 129, 159 n,
    204–5, 211 n
Hirsch, Robert O., 78 n, 105 n
Hodgson, Godfrey, 126 n
Hofstetter, C. Richard, 204–5
Horowitz, Mardi Jon, 11 n
Horton, Donald, 103 n
Hovland, Carl I., 41 n, 42 n, 96–97, 107 n
Hughes, Colin A., 12 n, 43 n, 76 n, 77 n,
    103 n
Humphrey, Hubert, 18, 23, 50, 52, 56–57,
    64, 65, 68, 95, 135–236 passim
Hursh, Gerald D., 42 n

Hyman, Herbert, 12 n, 76 n, 105 n, 237 n

Ideal public official, 47, 63–69, 74–75, 81,
    90, 93–94, 189, 215, 218, 222.
    See also Governor, ideal; Presi-
    dent, ideal; Senator, ideal
Ideal types, 112
Ideological anchoring; ideological self-
    images, 81, 90–93, 96, 180,
    192, 200, 218
Illinois, voter studies in, 93, 190
Image
    affective aspects of, 9, 21, 30, 32,
        91–92, 94–95, 99, 129, 163–64,
        188, 192–93, 198, 200, 203–4,
        206, 207
    cognitive aspects of, 9, 21, 30, 91–92,
        99, 129, 164, 192–93, 198
    conative aspects of, 9, 30, 91, 192–93,
        198, 203–4
    corporate, 4–5
    as creations of the mass media, 6–7,
        98–100, 143–55, 158
    defined, 7–9
    of Democratic candidates, 52, 133–34,
        152, 155, 158
    party, 36–38, 190–91, 194, 204
    of Republican candidates, 52, 133–34,
        152, 155, 158
    roles, as components of, 31, 46–59, 62,
        92, 113, 117, 119–20, 123–24,
        128, 133, 135–38, 152, 154,
        164–65, 175–77, 179–81,
        207–8, 228–29, 237 n, 241–42
    semantic dimensions, as components of,
        30–31, 47, 59–63, 85–87, 162–
        64, 229, 237 n, 240–41
    thesis, 31–32, 81–90, 206
    traits, 47–48, 53–66, 69–75, 205
    types, 45, 48, 63–75, 205
Image, The (Kenneth Boulding), 8, 143
Image: Or What Happened to the American
    Dream, The (Daniel Boorstin), 5
Imagery and reality, 3–4, 9
Imago, 5
Imitari, 5
Influence, personal, 20
Interpersonal Values, Survey of, 29–30,
    63–64
Ipsative measurement, 214
Issue
    candidates, 32
    orientations, 192, 199, 200, 207
    voting, 37–38, 120, 189–90

Jackson, David E., 225 n
Jackson, Henry, 234
Jackson, John S., 111, 44 n, 93–94, 106 n,
    190, 210 n
Johada, Gustav, 108 n
Janowitz, Morris, 106 n